Unreal Engine 5 Shaders and Effects Cookbook

Over 50 recipes to help you create materials and utilize
advanced shading techniques

Brais Brenlla Ramos

BIRMINGHAM—MUMBAI

Unreal Engine 5 Shaders and Effects Cookbook

Group Product Manager: Rohit Rajkumar
Publishing Product Manager: Nitin Nainani
Senior Editor: Hayden Edwards
Technical Editor: Simran Udasi
Copy Editor: Safis Editing
Project Coordinator: Aishwarya Mohan
Proofreader: Safis Editing
Indexer: Manju Arasan
Production Designer: Ponraj Dhandapani
Marketing Coordinator: Nivedita Pandey

First edition published: May 2019
Second edition published: May 2023

Production reference: 2050924

Published by Packt Publishing Ltd.
Livery Place
35 Livery Street
Birmingham
B3 2PB, UK.

ISBN 978-1-83763-308-1
www.packtpub.com

To my mother, Mari Carmen Ramos, and my father, Manuel Brenlla; for their love and support throughout the years. To my partner, Celia, for brightening up my days. To my cousin, Fran, for instilling a love for video games in me.

- Brais Brenlla

Foreword

Unreal Engine has evolved over 25 years to become the leading game engine for AAA studios and powers some of the most exciting projects. As the Architecture Industry Manager at Epic Games, I've had the opportunity to see how the engine has spread beyond games and now empowers a new generation of content creators in the Architecture, Engineering, and Construction industry, as well as in other domains.

Brais Brenlla is a great example of one of these new creators: working for AccuCities, a company specializing in capturing the built environment, Brais is someone who has leveraged the power of Unreal Engine to construct one of the best examples of applying the power of real-time technology to the problem of urban planning. With over 60 square kilometers of London buildings loaded into Unreal Engine, he's figured out how to make it all run smoothly on a small laptop. Optimizing for such a large, city-scale project while still allowing for smaller-scale objects requires a deep understanding of Unreal Engine to make things work efficiently and with fidelity.

With that experience under his belt, you'll find this book contains a wealth of knowledge to help the reader tackle many of the materials and effects required in a real-world environment. Starting from the basics, you'll soon start to work with some of the techniques that are used daily in any real-time production, followed by more advanced concepts that tackle some of the more interesting rendering features that the engine has to offer – all whilst following an accessible learning curve.

You're in good hands with Brais at your side.

Kenneth Pimentel

Architecture Industry Manager, Epic Games

Contributors

About the author

Brais Brenlla Ramos is an architect, 3D artist, and Unreal Engine developer, but above all else, he is a passionate follower of all things 3D. His interest in that field likely started when he first laid his eyes on a video game advert, with this "love at first sight" moment leading him to explore architectural design and visualization, 3D modeling, game development, and more.

He is currently focused on developing digital city twins in Unreal Engine, where he gets to sharpen his UI design, visual effects, and tool development skills every day. He also loves sharing all this knowledge with others, something that has turned him into a little bit of an author in recent times!

I want to thank everyone that has played a role in supporting me throughout the writing of this book: my partner Celia, who suffered from my sometimes lack of free time; my parents, who taught me the value of hard work; the team behind the scenes at Packt, who have made this possible; and my editor Hayden, who has caught more than one mistake along the way. Thank you all for being there.

About the reviewer

Deepak Jadhav is an XR developer, with a master's in game programming and project management, based in Germany. He has been involved in developing augmented reality, virtual reality, and mixed reality enterprise applications for various industries, such as healthcare, pharmaceuticals, and manufacturing. He has also developed digital twin apps for the manufacturing and healthcare industries.

Before that, Deepak was a game developer who developed games for multiple platforms, such as mobiles, PCs, and consoles. He has a strong background in C#, C++, and JavaScript, as well as years of experience in using game engines such as Unity and Unreal Engine for XR and game development.

Deepak has experience in reviewing many technical books on Unity and Unreal Engine.

Table of Contents

2

Customizing Opaque Materials and Using Textures   43

3

Making Translucent Objects   81

4

Playing with Nanite, Lumen, and Other UE5 Goodies   137

5

Working with Advanced Material Techniques   173

6

Optimizing Materials for Mobile Platforms   221

9

Adding Post-Processing Effects   329

Preface

This *Unreal Engine 5 Shaders and Effects* cookbook aims to be your companion on your quest to master the world of materials in Unreal Engine. Throughout its pages, you'll be able to discover many of the techniques that enable the powerful graphics on display in many of the latest interactive experiences, from the cutting-edge graphics present in some of the most renowned triple-A video games made today to some of the innovative effects making their way into the movie and visualization industries.

The importance that Unreal has gained in recent years across several industries can't be overstated, and that prominence is a direct result of both the powerful visuals it offers and the ease of use across many different fields. Taking advantage of that situation, we'll take a look at some of the most used techniques that will make you, the reader, capable of handling any material-related challenge you face within Unreal.

By the end of this book, you'll become competent in a wide variety of techniques that will allow you to create many different materials, which will put you at an advantage when working with any type of real-time rendering project.

Who this book is for

Anyone interested in Unreal Engine, rendering, real-time graphics, or simply passionate about creating stunning visuals will be more than capable of following the contents of this book. This will be possible thanks to a gentle learning curve that takes you from the basics of the rendering pipeline to some of the most advanced techniques in that area. This journey provides a wealth of information that will see you become proficient in this topic, no matter your background knowledge!

What this book covers

In *Chapter 1, Understanding Physically Based Rendering*, we will look at the basics of working with materials (and lights!) in Unreal Engine.

In *Chapter 2, Customizing Opaque Materials and Using Textures*, we'll learn how to create some of the most common materials available in Unreal.

In *Chapter 3, Making Translucent Objects*, we'll discover how to work with translucent shaders, a very important type that allows us to create substances such as glass or water.

In *Chapter 4, Playing with Nanite, Lumen, and Other UE5 Goodies*, we'll explore some of the new rendering features that made it into the latest version of the engine.

In *Chapter 5*, *Working with Advanced Material Techniques*, we'll take a look at some of the more advanced features available in the Material Editor.

In *Chapter 6*, *Mobile Shaders and Optimization*, we'll understand how to optimize our materials when targeting mobile devices.

In *Chapter 7*, *Exploring Some More Useful Nodes*, we'll see how to take advantage of some nifty nodes that defy categorization.

In *Chapter 8*, *Going Beyond Traditional Materials*, we'll work with some types of shaders that challenge the logic of where materials should be applied!

In *Chapter 9*, *Adding Post-Processing Effects*, we'll tap into Unreal's post-processing pipeline to create shaders that simultaneously affect the whole scene.

To get the most out of this book

You will need a version of Unreal Engine, preferably the latest one. At the time of writing, this is Unreal Engine 5.1.1, even though 5.2.0 is already on the horizon. The topics covered in this book should be applicable to all Unreal Engine 5.X versions.

Software/hardware covered in the book	Operating system requirements
Unreal Engine 5	Windows 10 or 11
Optional – ray tracing capable graphics card	
Optional – image editing software (Photoshop, GIMP)	

All of the topics can be tackled using only the default Unreal Engine installation. Access to a hardware-enabled ray tracing graphics card can be useful in a specific recipe where we discuss hardware and software ray tracing. Access to image editing software is not needed, but we recommend it in case you want to modify or author your own textures.

Download the project files

You can download the project files for this book from the following link: `https://packt.link/20u7B`

If there's an update to the project files, they will be updated through that link as well.

We also have other project and code bundles from our rich catalog of books and videos available at `https://github.com/PacktPublishing/`. Check them out!

Download the color images

We also provide a PDF file that has color images of the screenshots and diagrams used in this book. You can download it here: `https://packt.link/ia7i3`.

Conventions used

Bold: Indicates a new term, an important word, or words that you see onscreen. For instance, words in menus or dialog boxes appear in **bold**. Here is an example: "Select **System info** from the **Administration** panel."

> **Tips or important notes**
> Appear like this.

Get in touch

Feedback from our readers is always welcome.

General feedback: If you have questions about any aspect of this book, email us at `customercare@packtpub.com` and mention the book title in the subject of your message.

Errata: Although we have taken every care to ensure the accuracy of our content, mistakes do happen. If you have found a mistake in this book, we would be grateful if you would report this to us. Please visit `www.packtpub.com/support/errata` and fill in the form.

Piracy: If you come across any illegal copies of our works in any form on the internet, we would be grateful if you would provide us with the location address or website name. Please contact us at `copyright@packt.com` with a link to the material.

If you are interested in becoming an author: If there is a topic that you have expertise in and you are interested in either writing or contributing to a book, please visit `authors.packtpub.com`.

Share Your Thoughts

Once you've read *Unreal Engine 5 Shaders and Effects Cookbook*, we'd love to hear your thoughts! Scan the QR code below to go straight to the Amazon review page for this book and share your feedback.

https://packt.link/r/1-837-63308-8

Your review is important to us and the tech community and will help us make sure we're delivering excellent quality content.

Download a free PDF copy of this book

Thanks for purchasing this book!

Do you like to read on the go but are unable to carry your print books everywhere?

Is your eBook purchase not compatible with the device of your choice?

Don't worry, now with every Packt book you get a DRM-free PDF version of that book at no cost.

Read anywhere, any place, on any device. Search, copy, and paste code from your favorite technical books directly into your application.

The perks don't stop there, you can get exclusive access to discounts, newsletters, and great free content in your inbox daily

Follow these simple steps to get the benefits:

1. Scan the QR code or visit the link below

https://packt.link/free-ebook/9781837633081

2. Submit your proof of purchase
3. That's it! We'll send your free PDF and other benefits to your email directly

1
Understanding Physically Based Rendering

Welcome to *Unreal Engine 5 Shaders and Effects Cookbook*!

In this first chapter, we'll begin by studying the PBR workflow. **PBR** is an acronym that stands for **physically based rendering** – an approach to rendering a scene that takes into account how light behaves when it encounters 3D objects. This is at the core of the rendering pipeline, and the focus of the recipes you'll encounter in the next few pages. In them, we'll work with the building blocks of the PBR workflow – lights and materials – while also studying their impact on performance – things that we need to be aware of if we want to succeed in the rendering arena. With that in mind, this is what we are going to do:

- Setting up a studio scene
- Working inside the Material Editor
- Creating our first physically based material
- Visualizing a simple glass
- Using IBL and Lumen to light our scenes
- Using static lighting in our projects
- Checking the cost of our materials

Here's a little teaser of what we'll be doing:

Figure 1.1 – A look at some of the things we'll work on in this chapter

Technical requirements

To complete this chapter, you'll need to get a hold of Unreal Engine 5, the main star of this book!

This can be done by following these simple steps:

1. Download the Epic Games Launcher from the engine's website, https://www.unrealengine.com/en-US/download, and follow the installation procedure indicated there.

2. Once installed, get the latest version of the engine. You can do so by navigating to the **Unreal Engine** section of the launcher, and then to the **Library** tab. There, you'll be able to see a + icon, which lets us download whichever version of Unreal we want.

3. Launch the newly downloaded version of Unreal by clicking on the homonymous button:

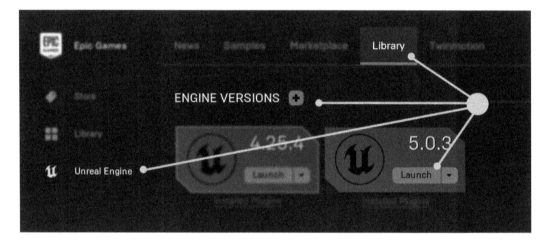

Figure 1.2 – Location of the buttons you'll need to click to complete the previous set of actions

Apart from that, we have provided all the assets that you'll see me use throughout the different recipes in this book. You can download all of the ones used in this chapter at https://packt.link/20u7B.

And don't worry – I'll make sure to remind you in the relevant *Getting ready* sections of the recipes if there is an already created file that you can use to follow along with.

Setting up a studio scene

The first objective that we are going to tackle in this book is creating a basic scene, one that we'll be able to reuse as a background level for some of the materials we'll be creating later. This initial step will allow us to go over the basics of the engine, as well as get familiar with some of the tools and panels that we'll revisit multiple times throughout the next few pages. Let's jump right in!

Getting ready

First, we need to download Unreal Engine 5 before we can start creating the basic studio scene. I've started writing this book with version 5.0.3, but don't hesitate to use the latest iteration available to you whenever you read this. Instructions on how to download and install it can be found on the previous page!

How to do it...

Having launched the engine as the last of the previous steps, let's look at how to set up a studio scene right from the beginning:

1. Go to the **Games** category in the Unreal Project Browser and select the **Blank** option.

2. Leave the standard options applied, except for **Starter Content** – untick that checkbox. We'll be taking a further look at the options mentioned in this section both later in this recipe and in the *How it works...* section.

3. Select a location for your project, give it a name, and click on **Create**:

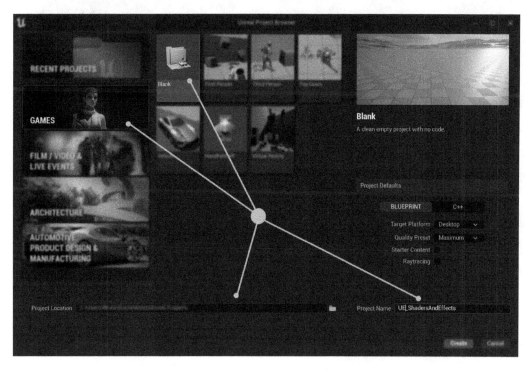

Figure 1.3 – The Unreal Project Browser

4. Once the editor loads, click on the **File | New Level…** option, and select the **Basic** map. This is the quickest option to create a new, empty level without any preexisting objects. Adjust the save location to your liking, and feel free to check out the *See also* section to learn about more ways you can organize your project.

5. Once done, we are now ready to start spicing things up. Erase everything from the World Outliner (which you can find in the top-right corner of the screen if you're using the default layout) – we are not going to be using any of that for our studio scene.

    The next thing that we are going to do is add the assets of the Starter Content to our project. This is something that could have been done when we created the project, but I intentionally left this for us to do now so that we know how to achieve this.

6. Navigate to the Content Drawer by hitting the *Ctrl* and *Space* keys at the same time.

7. Right-click on the main **Content** folder and select the **Add/Import Content** option. Then, select **Add Feature or Content Pack….**

8. In the newly opened panel, open the **Content** tab and choose the **Starter Content** option:

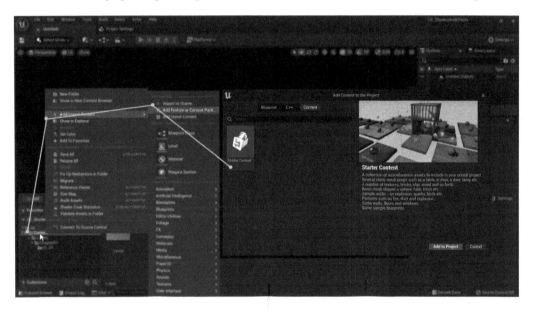

Figure 1.4 – Sequential steps to import the Starter Content assets into any project

That will cause the selected content to be downloaded, giving us the same assets we first discarded at the project creation stage. We'll now be using them to populate our temporarily empty level.

9. With that done, we can see that the Starter Content includes a **Blueprint**, which can be quite useful for setting up the lighting in our scene. You can look for this inside the **Content Browser | Starter Content | Blueprints** folder, under the name **BP_ Light Studio**.

> **Important note**
>
> The asset called **BP_LightStudio** is a Blueprint that Epic Games has already created for us. It includes several lighting settings that can make our lives easier – instead of having to adjust multiple lights, it automates all of that work for us so that we can focus on how we want our scene to look. Making a simple studio scene will be something very easy to achieve in this way.

10. Drag the Blueprint into the scene.

11. Select the **BP_LightStudio** Blueprint from the World Outliner so that we can tweak several of its settings, allowing us to make the scene look the way we want.

    In the **Details** panel, the first of the parameters we can look at is the **HDRi** tab. **HDRi** is short for **High Dynamic Range imaging**, which is a type of texture that stores the lighting information from the place at which the photo was taken. Using that data as a type of light in 3D scenes is a very powerful technique, which makes our environments look more natural and real.

12. Tick the **Use HDRi** checkbox to be able to use that type of lighting and set the **Alexs_Apt_2k** texture as the value for the **HDRi Cubemap** parameter. Doing that will allow us to light the scene through that asset, which is exactly what we are after. You can see this in the following figure:

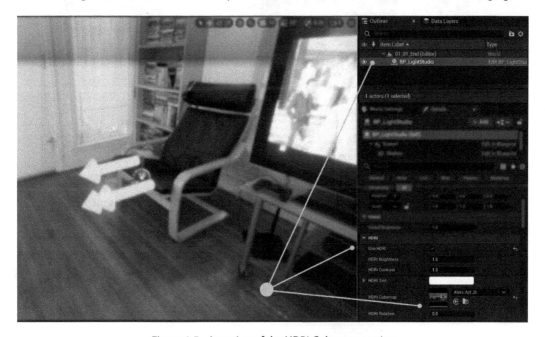

Figure 1.5 – Location of the HDRi Cubemap setting

> **Tip**
>
> HDRi images are very useful for 3D artists, though they can be tricky to create on your own. There are many websites from which you can buy them, but I like the following one, which gives you free access to some very useful samples: `https://polyhaven.com/hdris`.

13. You can now untick the **Use Light Sun and the Use Atmosphere** options found under the **Sun** and **Atmosphere** sections too. As we mentioned earlier, using an HDRi sometimes renders the use of other lights to be optional.

14. Once you've done that, create a basic plane so that we have a surface where we can lay out any objects that we might want to display later. You can do this by accessing **Quickly add to the project | Shapes | Plane**, as seen in the following figure:

Figure 1.6 – Creating a plane by accessing the appropriate option in the Shapes menu

15. Now that we have a plane, let's assign an interesting material to it. With the plane selected, scroll down to the **Materials** section of the **Details** panel and change its default value to **M_Wood_Pine** (this material is part of the Starter Content, so make sure you have it installed).

We should now be looking at something similar to the following:

Figure 1.7 – The final look of our first scene!

With that out of the way, we can say that we've finished creating our basic studio scene. Having done that will enable us to use this level for visualization purposes, kind of like having a white canvas on which to paint. We will use this to place other models and materials as we create them, to correctly visualize our assets. Fun times are ahead!

How it works...

There are at least two different objectives that we can complete if we follow the previous set of instructions – creating our first studio scene is the first one, while the second is simply getting familiar with the engine.

With regards to the first objective, the first step that we took was to create the Unreal Engine project on which we'll be working throughout this book. Then, we added the assets present in the Starter Content package that Epic Games supplies, as it contains useful 3D models and materials that we can use later in other recipes. The most important bit we did was probably the lighting setup, though. This is because having a light source is vital for visualizing the different models that we create or add to the scene. Lighting is something that we'll explore more later, but the method we've chosen in this recipe is a very cool technique that you can use in your projects. We are using an asset that Unreal calls a Blueprint, something that allows you to use the Engine's visual scripting language to create different functionalities within the game engine without using C++ code. This is extremely useful as you can program different behaviors across multiple types of actors to use to your advantage – turning a light on and off, opening a door, creating triggers to fire certain events, and so on. Again, we'll explore these more as we go along, but at the moment, we are just using an already available Blueprint to specify the lighting effects we want to have in our scene. This is a good example of what Blueprints can do, as it they allow us to set up multiple different components without having to specify each one individually – such as the HDRi image, the Sun's position, and others that you can see if you look at the **Details** panel.

The second objective – getting familiar with the engine – is something that will continue to happen over time. Despite that, something that could also speed that up further is reviewing the steps we follow in each recipe, just like we've done before. Not only will we learn things faster, but knowing why we do things the way we do them will help us cement the knowledge we acquire (so expect to see a *How it works…* section after each recipe we tackle!).

Before we move on, I also want to revisit the initial project settings that we chose in *step 2* when we created the project. We chose the **Blank** template out of the games section, and we created a Blueprint-based project with no Starter Content and no ray tracing – all while leaving the target platform as the desktop and the quality preset set to the maximum. The reason we did that was to simply retain as much control over our project as possible, as templates only differ between themselves in terms of the specific plugins that they have activated or the type of settings that are enabled by default. These are all things that can be changed at a later stage, and we will also be able to review some of them before we finish this book – especially the ray tracing one! Having said that, it's important to lose any fear we might have after seeing the sheer amount of initial available choices, so don't worry thinking that we might be missing out on some cool advanced feature simply because we went with the simplest option.

See also

In this recipe, we created and saved a new level within Unreal's Content Browser. If you care about organization (and you should if your project becomes something bigger than a quick testing exercise!), know that there are many different styles that you can follow to keep things tidy. One is what I like to refer to is Allar's style guide, which you can find out more about here: `https://github.com/Allar/ue5-style-guide/tree/main`.

Working inside the Material Editor

Let's get started working inside the Material Editor! This is where the magic explained in this book happens, as well as the place where we'll spend most of our time, so we'd better get well acquainted with it, then! As with everything inside Unreal, you'll be able to see that this space for creating materials is a very flexible one – full of customizable panels, rearrangeable windows, and expandable areas. And the best thing is you can place them wherever you want!

Because of its modular nature, some of the initial questions that we need to tackle are the following: how do we start creating materials? And where do we look for the most used parameters? Having different panels means having to look for different functionalities in each of them, so we'll need to know how we can find our way around the editor. We won't stop there, though – this new workspace is packed with plenty of useful little tools that will make our jobs as material creators that much easier, and knowing where they live is one of the first mandatory steps we need to take.

So, without further ado, let's use the project we set up in the previous recipe as our starting point, and let's create our first material!

Getting ready

There's not much that we need to do at this point – all thanks to having previously created the basic blank project in the previous recipe. That's the reason we created it in the first place – so that we can start working with materials straight away. Setting up the studio scene is all we need to do at this point.

Despite this, don't feel pressured to use the level we created in the first recipe. Any other one will do, so long as there are some lights in it to help you visualize the world. That's the advantage of the PBR workflow – whatever we create while following its principles will work across different lighting scenarios.

How to do it...

Let's quickly tackle this recipe by creating our first material in Unreal Engine 5. This won't take us long, but it will give us a good excuse to look at the Material Editor – a panel that we'll revisit plenty of times throughout this book and one that you will explore in further detail in the *How it works...* section of this recipe.

With that said, let's learn how to create our first material:

1. Navigate to the Content Browser and right-click in an empty area. Create a new material by selecting the **Material** option from the expandable menu that should have appeared upon right-clicking on the screen. Alternatively, you can create a new material by clicking on the + **Add** button located at the top-left corner of the Content Browser.

2. Give the material a simple name, such as **M_BasicPlastic**, as that's what we'll be creating.

3. Double-click on the newly created asset to bring up the Material Editor, the place where we'll be able to alter the appearance of our materials.

4. Right-click on the main graph, preferably to the left of the **Main Material** node, and start typing Constant. Notice how the auto-complete system starts showing several options as you start typing: **Constant**, **Constant2Vector**, **Constant3Vector**, and so on. Select the last one, **Constant3Vector**.

5. You will see that a new node has now appeared. Connect its output pin to the **Base Color** property of the **Material** node.

6. With the **Constant** node selected, take a look at the **Details** panel. You'll see that we can tweak the **Base Color** property of the material. Since we want to move away from its current blackish appearance, click on the black rectangle to the right of where it says **Constant** and use the color wheel to change its current value. Let´s choose a different color just to demonstrate the effect, as shown in the following figure:

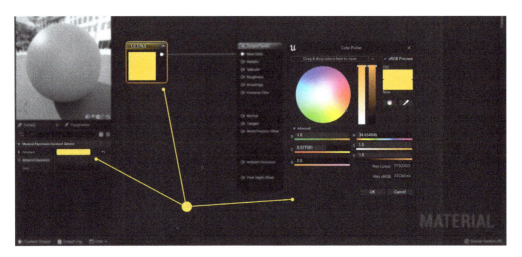

Figure 1.8 – The look of the material graph so far and the color selection process

---

Important note

You can read more about the **Constant** nodes and the **Base Color** property in both the *How it works…* and *See also* sections of this recipe!

As you can see, we have managed to modify the color of the preview sphere. We should also adjust the sharpness of the reflections given by the material, given how we want to go for a plasticky look. To do so, we need to modify the **Roughness** parameter using a different constant. Instead of right-clicking and typing, let's choose it from the palette menu instead.

7. Navigate to the **Palette** panel and scroll down until you reach the **Constant** category. We want to select the first option in there, simply named **Constant**. Alternatively, you can type its name in the search box at the top of the panel instead of scrolling.

8. A new and smaller node should have appeared. Unlike the previous one, we don't have the option to select a color – we need to type in a value. Let's go with something low for now, something like **0.2** – a value that will produce clear reflections on the material. Connect the output of the new node to the **Roughness** input pin of the main material.

You might notice how the appearance of the material has changed after applying the previous **Roughness** value; it looks like the reflections from the environment are much sharper than before. This is happening thanks to the previously created **Constant** node, which, using a value closer to 0 (or black), made the reflections a lot clearer. Whiter values (those closer to 1) decrease the sharpness of the reflections or, in other words, make the surface appear much rougher.

Having done so, we can now apply this material to a model inside our scene. Let's go back to the main level and add a new shape onto which we can apply our new asset.

9. Open a level that you can use to visualize models (such as the one we created in the previous recipe).

10. Create a cube and place it on top of the existing plane, if you're working on the level we created in the previous recipe, or simply somewhere you can see it if you're working with a custom map. You might need to adjust its scale if you're working with our scene, as the default size of the cube will be too big.

11. With the cube selected, head over to the **Materials** section of its **Details** panel and click on the first drop-down menu available. Look for the newly created material and assign it to our cube.

The scene should now look something like this:

Figure 1.9 – The way our scene should look by the end of this recipe

And there it is – we have applied our material to a simple model, being displayed on the scene we created previously. Even though this has served as a small introduction to a much bigger world, we've now gone over most of the panels and tools that we'll be using in the Material Editor.

How it works...

We've just created our first material in Unreal Engine 5 – the first of many more! But what better opportunity than now to take a further look at the Material Editor itself? Let's dive right in and see the different sections that we'll have to pay attention to throughout the rest of our journey.

> Tip
> Remember that you can open the Material Editor by double-clicking on any material that you want to modify.

The Material Editor

Given how we are going to be studying the Material Editor, let's make sure that we are looking at roughly the same screen. We can do this by resetting the appearance of the Material Editor itself, which can be done through **Window | Load Layout | Default Editor Layout**. Remember that resetting things to their default state doesn't necessarily mean that our screens will look the same, mainly because settings such as the screen resolution or its aspect ratio can hide panels or make them imperceptibly small. Feel free to move things around until you reach a layout that works for you!

Now that we've made sure that we are all looking at the same panels, let's turn our attention to the Material Editor itself and the different sections that are a part of it. By default, we should all be looking at something very similar to the following screenshot:

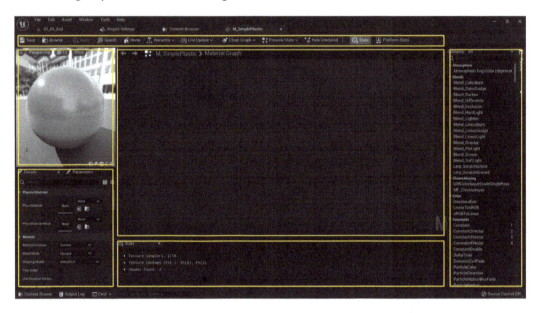

Figure 1.10 – The default look of the Material Editor and several of its subcomponents

Let's break down the previous interface:

- The first part of the Material Editor is the **Toolbar** area. Some of the most essential controls can be found here, such as saving the material on which you are working or finding it in the Content Browser.

- The second panel is the **Viewport** area, where we'll be able to see what our material looks like. You can rotate the view, zoom in or out, and change the lighting setup of that window.

- The **Details** panel is a very useful one, as it is here where we can start to define the properties of the materials that we want to create. Its contents vary depending on what is selected in the main graph editor.

- The **Stats** panel will give you information regarding the resources used by the current material. You can also open the **Search** panel next to this one by clicking on that option in the **Toolbar** area.

- **Material Node Palette** is a library that contains all the nodes and functions available in the material graph. It's collapsed by default, but you can access it by clicking on the text with your mouse.

- The **Main Editor Graph** area is where the magic happens, and where you'll be working most of the time. It is in this area where we arrange the different components that make up a material.

Constants

We've used this recipe to learn about the Material Editor and created our first material along the way. Knowing what each section does within the editor will help a lot in the immediate future, as what we've just done is but a prelude to our real target – creating a PBM. Now, we are in a much better position to tackle that goal, which we will look at in the next recipe!

Before moving on, though, let's review the nodes that we have used to create this simple material. From an artist's point of view, the names that the engine has given to something such as a color value or a grayscale one can seem a bit confusing. It might be difficult to establish a connection between the name of the **Constant3Vector** node and our idea of a color. But there is a reason for all of this.

The idea behind the naming convention is that these nodes can be used beyond the color values we have just assigned them. At the end of the day, a simple constant can be used in many different scenarios – such as depicting a grayscale value, using it as a brightness multiplier, or as a parameter inside a material function. Don't worry if you haven't seen these other uses yet; we will – the point is, the names that these nodes were given tell us that there are more uses beyond the ones we've seen. Going a bit further than that, we should know that they are also called constants because their values don't change: they are the ones that they receive when we create them!

With that in mind, it might be better to think of those elements we've used in more mathematical terms. For instance, you can think of color as a **Red, Green, and Blue (RGB)** value, which is what we defined using that previous **Constant3Vector** node. If you want to use an RGB value alongside an alpha one, why not use **Constant4Vector**, which allows for a fourth input? Even though we are at a very early stage, it is always good to familiarize ourselves with the different expressions the engine uses.

See also

You might have noticed that I've named our first material in a very particular way, using the *"M_"* prefix before the actual name of the asset. The reason for that is that I like to stick to clear naming conventions that I can rely on for better readability throughout the project. The one I use follows a very famous one called Allar Style Guide, and you can find more about it at `https://github.com/Allar/ue5-style-guide`.

I also want to point you to the official documentation regarding the importance of the **Base Color** property in Unreal, as well as some of the other common attributes that you'll find when working inside the Material Editor: `https://docs.unrealengine.com/5.0/en-US/physically-based-materials-in-unreal-engine/`. You'll see that certain real-world materials have a specific measured value that should be used to act as the **Base Color** property's intensity. Be sure to check that information out if you are interested in using realistic values in your projects!

Creating our first physically based material

PBR is, at its core, a principle that several graphic engines try to follow. Instead of being a strict set of rules by which every rendering program needs to abide, it is more of an idea – one that dictates that what we see on our screens is the result of a study on how light behaves when it interacts with certain surfaces.

As a direct consequence, the so-called PBR workflow varies from one rendering engine to the next, depending on how the creators of the software have decided to program the system. For us, that means that we are going to be looking at the implementation that Epic Games has chosen for the rendering pipeline in Unreal Engine 5.

However, we are going to do so in our already established recipe process – that is, by creating materials that follow the PBR workflow so that we can see the results. Let's get to it!

Getting ready

We don't need a lot to get started working on this recipe – just a basic scene like the one we created in the previous recipe with some basic lights and models in it. You can simply continue using the previous level and repurpose the model we placed there.

> **Tip**
> We are going to create multiple materials in this section, so duplicating and modifying an already existing asset is going to be faster than creating several ones from scratch. To do this, just select any material that you want to duplicate in the Content Browser and press *Ctrl + D*.

How to do it...

Let's start our journey by creating a new material and looking at the attributes that define it:

1. Right-click anywhere inside the Content Browser and select the **Material** option in the **Create Basic Asset** section. Name it whatever you want – I'll go with **M_PBR_Metal** this time. Double-click on the newly created material to open the Material Editor.

    With that panel open, let's focus our attention on the **Material** section of the **Details** panel. This area contains some very important settings that determine how our material behaves concerning the light that reaches it – from blocking it to allowing it to pass. It also contains some other parameters that affect other properties, such as how the asset is shaded or what type of objects can use it:

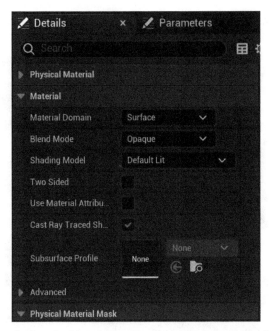

Figure 1.11 – A close-up look at the Material section of the Details panel for the Main Material node

The settings you can see here are the default ones for most materials in Unreal, and they follow the PBR pipeline very closely. The first option, **Material Domain**, is currently set to **Surface**. This tells us that the material we are creating is meant to be used on a 3D model (as opposed to on a light or a UI element). **Blend Mode**, which has a value of **Opaque**, indicates that light won't travel through the objects where this material is applied. Finally, **Shading Model** is set to **Default Lit**, which indicates a particular response to the light rays that the material receives. This configuration is considered the standard one and is useful when you're trying to create materials such as metals, plastics, walls, or any other ones that represent everyday opaque surfaces.

2. With that bit of theory out of the way, create a **Constant3Vector** node anywhere in the graph and plug it into the **Base Color** input pin of our material. We used the **Base Color** attribute in the previous recipe, which defines the overall color of our creation.

3. The next item we have to create is a simple **Constant**. You can do so by holding the *1* key on your keyboard and clicking anywhere within the Material Editor graph. Assign a value of **1** and plug it into the **Metallic** attribute of our material.

Important note

The **Metallic** attribute controls whether the material we create is a metal. That is true if we plug a value of 1 into the slot, and false if we choose 0 or if we leave it unconnected. Values between the previous two should only be used under special circumstances, such as when dealing with corroded or painted metallic surfaces.

4. Next, create another **Constant** and plug it into the **Roughness** slot. This gives it a value of **0.2** instead. The final material graph should look something like this:

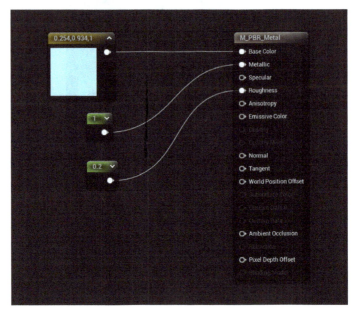

Figure 1.12 – The look of the material graph so far

Important note

The attribute we are controlling through the previous **Constant** defines how rough the surface of a given material should be. Higher values, such as 1, simulate the micro details that make light scatter in all directions – which means we are looking at a matte surface where reflections are not clear. Values closer to 0 result in those imperfections being removed, allowing a clear reflection of the incoming light rays and a much clearer reflected image.

With that, we have looked at some of the most important attributes that are used to define a PBR material by creating a metal, a type of opaque material. Despite that, it will be good to create another quick example that doesn't involve the metallic property – this is because some of the other settings used in the PBR workflow, such as the **Specular** material attribute, are meant to be employed in such cases.

5. Create another material, named **M_PBR_Wood**, and open the Material Editor for that asset.

6. Right-click anywhere inside the main graph for our newly created material and search for the **Texture Sample** option. This will allow us to use an image instead of a simple color as one of the attributes that defines the material.

7. With that new node in our graph, click on it to access its **Details** panel. Click on the drop-down menu next to the **Texture** property located within the **Material Expression Texture Base** section and type Wood. Select the **T_Wood_Floor_Walnut_D** asset and connect the **Texture Sample** node to the **Base Color** material attribute.

With that done, it's time to look at another material attribute – the **Specular** parameter. This is a property that controls how much light is being reflected by the material, in a range of values that go from 0 to 1. The standard figure is 0.5, a value that gets automatically assigned whenever we don't connect anything to that material slot. Even though it doesn't see a lot of use nowadays, it can be of assistance in cases such as this one, where we are trying to replicate a wooden surface – a surface that contains both flat areas that reflect a normal amount of light as well as crevices that trap it. Using a texture to define that circumstance can be helpful to achieve more realistic results, so let's tend to that!

8. Drag a pin from the red channel of the previously created **Texture Sample** node into the **Specular** attribute of the **Main Material** node. This will let us use the black-and-white information stored in the previously selected texture to drive the amount of light reflected by the material. This will work perfectly for our purposes, seeing as the areas of the main wooden boards are clearer than the seams.

You might be wondering why we are using the red channel of the wood texture to drive the specular parameter. The simple answer is that we need a black-and-white texture to drive that setting, and selecting an individual channel out of an RGB texture allows us to do just that. Because seams are going to contain darker pixels than other areas, the result we achieve is still very similar if we use the red channel of the original texture. The following figure shows our source asset and the red channel by its side:

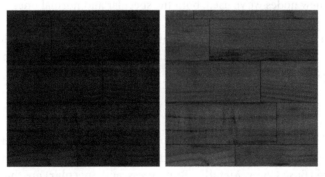

Figure 1.13 – The original RGB wood texture on the left and the red channel to its right

9. Now, copy the **Texture Sample** node twice, since we are going to use more textures for the **Roughness** and the **Normal** material attribute slots.

10. As we did previously, select the **T_ Wood_ Floor_ Walnut_ M** and **T_ Wood_ Floor_ Walnut_ N** assets on each of the new nodes. Connect the first one to the **Roughness** slot and the second one to the **Normal** node.

11. Save the material and click on the **Apply** button.

12. Navigate back to the main level and select the floor plane. In the **Details** panel, scroll down to the **Materials** section and assign the **M_PBR_Wood** material we have just created – feel free to assign the metallic material to the existing box. With that in place, let's see what our scene looks like:

Figure 1.14 – A final look at the scene after applying the different materials

Nice job, right? The new nodes we've used, both the specular and normal ones, contribute to the added details we can see in the previous screenshot. The **Specular** node diminishes the light that is being reflected in the seams between the wood planks, and the **Normal** map modifies the direction in which the light bounces concerning the surface. The combined effect is that our model, a flat plane, looks as if it contains a lot more geometrical detail than it does. Cool stuff!

How it works...

Efficiency and speed are at the heart of any real-time application. These are two factors that have heavily influenced the way that the PBR pipeline works in Unreal. That being the case, the parameters that we have tweaked (**Base Color, Metallic, Roughness**, and **Specular**) are the most important ones when it comes to how Unreal deals with the interaction between light and 3D models. The **Base Color** property gives us the overall appearance of the material, while roughness indicates how sharp or blurry the reflections are. The **Metallic** property enables us to specify whether an object is made from metal, and the **Specular** node lets us influence the amount of light that gets reflected. Finally, normal maps allow us to modify the direction in which the light gets reflected – a useful technique for adding details without using more polygons.

The previous parameters are quite common in real-time renderers, but not every program uses the same ones. For instance, offline suites such as V-Ray use other types of calculations to generate the final output – physically based in their nature but they use other techniques. This shows us that, at the end of the day, the PBR workflow that Epic uses is specific to its engine, and we need to be aware of its possibilities and limitations.

Throughout this recipe, we have managed to look at some of the most important nodes that affect how PBR gets tackled in Unreal Engine 5. The base color, roughness, specularity, ambient occlusion, normals, and the metallic attribute constitute the basics of the PBR workflow.

Having seen all of them, we are now ready to start looking into how to build more complex materials and effects. And even though we still need to understand some of the other areas that affect our pipeline, we can do so with the certainty that the basics are covered.

See also

Now that we have introduced textures and 3D models, we're going to start playing with more complex assets that might feel a bit more complicated than the constants we used in our first material. Creating them requires us to use their specific software packages, and we would need another couple of books to cover the specifics of how to do this.

Having said this, the guys at Epic have recently acquired the amazing company Quixel, which focuses on capturing realistic 3D assets, and made their library free to use for Unreal users – so we can count ourselves lucky! You only need an Epic Games account to enjoy their content, and you can get a hold of those resources through the Quixel Bridge integration, which comes packaged with Unreal. Simply head over to **Window | Quixel Bridge** inside Unreal and explore all the content available there!

On top of that, I also want to leave you with another site that might help you with your work, `https://www.textures.com/`, where you can get useful textures and scans to bring your scenes to the next level.

Visualizing a simple glass

In the previous recipe, we had the opportunity to create a basic material that followed the physically based approach that Unreal Engine uses to render elements on the screen. We saw how we could potentially create endless combinations by simply using nodes and expressions that affected the roughness and the metallic attributes of a material, letting us achieve the look of plastics, concrete, metals, or wood.

Those previous examples can be considered simple ones – for they use the same **Shading Model** to calculate how each element needs to be rendered. Most of the materials that we experience in our daily lives fall into that category, and they can be described using the attributes we studied previously. Despite that, there are always examples that can't be exactly covered with one unique **Shading Model**. The way that light behaves when it touches glass, for example, needs to be redefined in those cases.

The same applies to other elements, such as human skin or foliage, where light distribution across the surface varies from that of a wooden material.

With that in mind, we are going to start looking into these other categories by creating several small examples of materials that deviate from the standard **Shading Model**: one in this recipe, and some others in future chapters. We will start by creating a simple glass, which will work as an introductory level before we can create more complex examples at a later stage.

Getting ready

The sample Unreal project we created previously will serve us fine, but feel free to create a new one if you are starting with this section of this book. It is completely fine to use standard assets, such as the ones included with the engine through the Starter Content, but I've also prepared a few of them that are included alongside this book that you can use if you want to follow me more closely. If so, feel free to open Level 01_04_Start from the download link in the *Technical requirements* section, and buckle up for the ride!

How to do it...

So, to create our simple glass example, follow these steps:

1. Right-click within the Content Browser and create a new material. You can give it an appropriate name (or not!), such as **M_SimpleGlass**, as that's what we'll be creating.

2. Open the Material Editor and select the **Main Material** node.

3. Focus your attention on the **Details** panel. Look under the **Material** section and find the **Blend Mode** option. Change the default **Opaque** value to the **Translucent** one.

4. Now, scroll down to the **Translucency** section. There is a drop-down menu there called **Lighting Mode**, which is currently set to a default value of **Volumetric Non Directional**. Switch to the **Surface Translucency Volume** instead.

> **Important note**
>
> You might have noticed that several input pins in the **Main Material** node became unusable once we changed **Blend Mode** from **Opaque** to **Translucent** and that they became available once again after changing the material's **Shading Model**. These pins define how a material behaves, so it makes sense to have certain ones available depending on what the previous parameters allow our material to do – it wouldn't make sense to have an opacity option in a fully opaque material. Make sure you read the *How it works...* section to learn more about those settings.

5. Create a **Constant4Vector** and connect it to the **Base Color** node of the material.

6. Assign any value you fancy to the **RGB** channels of the previous **Constant** and set the **Alpha** parameter to **0.5**. I've gone for a bluish tone, which will give us the effect of slightly transparent blue glass. Take a look at the values in the following figure:

Figure 1.15 – The values that were chosen for the Constant4Vector variable

7. Plug the output of the previous **Constant4Vector** node into the **Base Color** input pin on the **Main Material** node.

8. Now, it's time to plug the **Alpha** value of our **Constant4Vector** into the **Opacity** slot of our material. So, drag from the output pin of **Constant4Vector** into an empty space in the material graph and release the mouse button. A contextual menu should appear, letting us select a node. Type Component to reveal a node called **ComponentMask**; select it.

9. With the last node selected, take a look at the **Details** panel. You can select which of the four components from the **Constant4Vector** node you want to use. Tick the **Alpha** option as we want to drive the opacity of the material through the value we selected earlier.

10. Connect the output of the previous mask to the **Opacity** pin.

11. Click on the **Apply** button and save the material.

    The preview window may take a moment to update itself but, once it does, we should be looking at a translucent material that lets us see through it.

12. Now that we have our material correctly set up, let's apply it to the model in our scene. If you've opened the level that I've set up for you, `01_04_Start`, you'll find an object there called **SM_Glass**. If you are working on your own project, just create a model to which we can apply this newly created material and see its effects. No matter the option you choose, the scene should look something like the following after you apply the new material:

Figure 1.16 – The final look of our first glass material

Simple but effective! We'll be looking at how to properly set up a more complex translucent material in future chapters, with reflections, refractions, and other interesting effects. But for now, we've taken one of the most important steps in that path – we've started to walk it!

How it works...

Translucent materials are inherently different from opaque ones, both in real life and in Unreal. As a result, the rendering principles that enable us to visualize them are fundamentally different, and that is the reason behind needing to choose a different **Blend Mode** to work with translucent effects.

Having gotten this far, I want to use the opportunity to take a closer look at the **Blend Mode** property, as well as the concept of **Shading Model** – two very important settings located at the top section of the details panel for the **Main Material** node.

First, a **Shading Model** is a combination of mathematical expressions and logic that determines how models are shaded or, in other words, painted with light. The purpose of one **Shading Model** is to describe how light behaves when it encounters a material that uses said shading method. There are multiple options available for this setting, usually as many as needed to describe the different materials we see in our daily lives – for example, the way light scatters on our skin or the way it does the same on a wooden surface constitute two different models. We need to be able to describe that situation in a way that our computer programs can tackle that problem.

With that in mind, you could think that there should be a unique **Shading Model** just to describe translucent materials. However, things get a bit more complicated in real-time renderers, as the calculations that we would have to perform to realistically simulate that model are too expensive in terms of performance. Being always on the lookout for efficiency and speed, the way that Unreal has decided to tackle this issue is by creating a different **Blend Mode**. But what is that?

You can think of **Blend Mode** as the way that the renderer combines the material that we have applied to a model in the foreground over what is happening in the background. We've seen two different types up until now – the opaque and the translucent ones. The opaque **Blend Mode** is the easiest one to comprehend: having an object in front of another will hide the second one. This is what happens with opaque materials in real life – with wood, concrete, bricks, and so on. The translucent mode, however, lets the previously hidden object become partially visible according to the opacity value that we feed into the appropriate slot.

This is a neat way of implementing translucency, but there are some caveats that the system introduces that we must be aware of. One such issue is that this **Blend Mode** doesn't support specularity, meaning that seeing reflections across the surface is a tricky effect that we will have to overcome later. But don't worry – we'll get there!

Using IBL and Lumen to light our scenes

This introductory chapter has covered the foundations of the PBR workflow that Unreal uses for its rendering pipeline. With that as our focus, we have already looked at several of its key components, such as different material parameters and Shading Models. However, we can't dismiss the fact that the PBR workflow takes information from the lights in our scene to display and calculate how everything should look.

So far, we've only focused on the objects and the materials that are being rendered – but that is only part of the equation. One of the other parts is, of course, the lighting information. Lights are crucial to the PBR workflow: they introduce shadows, reflections, and other subtleties that affect the final look of the image. They work alongside the materials that we've previously applied by making sense of some of the properties we set up. Roughness textures and normal maps work in tandem with the lights and the environment itself. All of these things combined are an integral part of the pipeline we are looking at in this introductory chapter.

With that as our background, we'll focus on creating different kinds of lights in this recipe and see how they affect some of the materials that we have created previously. One particular type that we are going to focus on is **Image-Based Lighting**, IBL for short, which is a peculiar kind of light that employs textures to specify how a scene should be lit. These textures are also special, and they are known as HDRIs: 32-bit textures that tell the engine how to light up a scene. We'll learn more about them in the *How it works…* section.

Furthermore, we'll also look at how **Lumen**, Unreal Engine's 5 new dynamic global illumination and reflections system, makes everything look great and simplifies the rendering workflow by quite a bit.

Getting ready

All you'll need to do to complete this recipe is access the scene we created in the first recipe. Please go ahead and open it, or feel free to create a new one that contains the same assets we placed back then – a simple plane and the **BP_LightStudio** Blueprint available through the Starter Content.

How to do it...

We will start this recipe by placing a reflective object in our default scene and looking at how certain aspects of the environment can be seen reflected on its surface. To do this, take the following steps:

1. Open the **Quickly add to the project** panel, select **Place actors | Shapes**, and choose **Sphere**.

2. Assign a scale of **0.1** for all axes and place the sphere on top of the existing plane.

> **Tip**
> Move the sphere over the existing plane and hit the *End* key to have it fall on top of it. This technique can be used with any two models that have collision enabled.

3. Assign a reflective material to it. I'll be using a copy of the metallic shader we created in the recipe, *Creating our first physically based material*, the one we named **M_PBR_Metal**; I've slightly adjusted to make it look more like a chrome ball by simply changing the color to white and setting the roughness value to **0.02** (refer to *steps 1* through *4* of the *Creating our first physically based material* recipe if in doubt). I've gone ahead and called this new copy **M_Chrome**:

Figure 1.17 – Step-by-step guide for adding a Sphere to the scene

We can see the environment properly reflected on the chrome ball thanks to the reflective properties of the material that we've applied. This is happening thanks to the HDRi image we are using. We are now going to replicate this effect without using the Blueprint that was already created for us; instead, we will set up our own lighting.

One of the things that we want to move away from in the setup that we are going to create is always having the environment image visible. This is just so that we can have better control of the level we create without having to rely on pre-made solutions.

4. Delete the **BP_LightStudio** Blueprint. We should now be looking at a completely dark scene.

5. Add a **Sky Light** actor to the scene. To do this, access the **Quickly add to the project** panel, then select **Place actors | Lights**, and choose **Sky Light**.

6. With our new actor selected, navigate to the **Details** panel and look at the options under the **Light** section. The first drop-down menu we can modify is called **Source Type**, which should have a default value of **SLS Captured Scene**; this uses the assets already placed in the level to create lighting information that can be used to light the scene. We want to change that value to the other available option, **SLS Specified Cubemap**.

7. Once that's done, select a **Cubemap** texture from the next drop-down menu – the one simply called **Cubemap** – and choose the same HDR texture we used back in the first recipe, **Alexs_ Apt_ 2k**.

Now that we've gone through the previous set of steps, let's take a moment to look at the scene and see what we have in front of us:

Figure 1.18 – The scene comprised of the reflective Sphere, a plane, and a Sky Light actor

As we can see, we can achieve a very realistic look by simply placing a light in our scene that uses an HDRi image to provide the lighting information. The color, the reflections, and the global illumination all stem from that one source, and we can render everything in real time thanks to the work that Lumen carries out behind the scenes. This differs quite a lot from the previous systems used in earlier versions of the engine, where lighting, global illumination, and reflections all worked as separate elements – and can still do this if we so choose to, as we'll see in the next recipe. Therefore, this new method is great in terms of speeding up the workflow for artists as they can implement and see the changes they make in much less time than before, all while achieving amazing-looking results.

Now, let's learn how to create a different type of light – **Point Light**.

8. Head over to the **Quickly add to the project** panel, go to **Place actors | Lights**, and choose **Point Light**. Place it close to the chrome ball so that we can see some shadows cast from it.

9. In the **Lights** section of the **Details** panel, adjust the **Intensity** and **Light Color** properties. I've chosen a value of **2** candelas for the former and a yellowish color for the latter.

> **Tip**
>
> Changing certain light settings can sometimes reset the units used by your light. If that happens, expand the **Advanced** drop-down option within the **Light** section of your light and select the units you want to use in the last setting, **Intensity Units**.

10. Change the **Mobility** property of the light to **Dynamic**.

11. We can see that the edges of the shadows cast by our **Point Light** show some serrated edges. Increase the **Source Radius** property of the light to something like **2**, which will improve the look of the scene.

Our scene should now look like this:

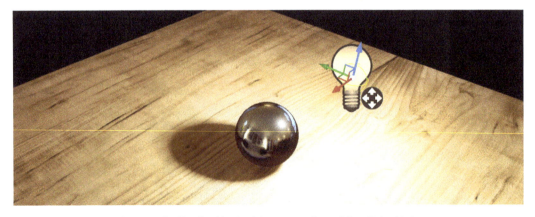

Figure 1.19 – The final look of the scene after adding Point Light

The light we have just added to the scene makes use of the new default shadowing method that was introduced alongside Unreal Engine 5: Virtual Shadow Maps. This is a new technique that aims to deliver high-quality shadows to dynamically lit scenes; you will learn more about it in the next section, *How it works….*

With that out of the way, we have seen how easy it is to work with most of the elements that we need to tackle when rendering a scene: lighting, reflections, and global illumination. Lumen and Virtual Shadow Maps have made our lives way easier than before, as we now only need to focus on the properties of the different elements we create without having to worry about baking important information later. There are many settings and considerations that we need to consider when dealing with these new technologies, but they make our lives easier while making everything look better than before.

How it works…

Seeing as we've introduced quite a few new concepts in this recipe, let's go ahead and focus on each of them.

HDRis and IBL

We've had the opportunity to work with IBL in this recipe, something that we were able to do thanks to the **Sky Light** actor – a particular asset that allows us to use HDR images to light up the scene. The way these textures work is by capturing the lighting state of a particular location and sampling the same environment multiple times under different exposure settings; taking multiple pictures this way allows us to combine them and use the differences in lighting to better understand how the area is lit.

Something that we need to be on the lookout for is that we use the right type of textures. HDRi images need to be in a 32-bit format, such as `.EXR` or `.HDRi`, where each pixel contains multiple layers of information condensed into itself. You might find HDRi images in a non-32-bit format, but they don't contain as much lighting information as the real ones because of the format they use.

Another parameter that we need to take into consideration is the number of f-stops that a given HDRi image is composed of. This number indicates the number of different pictures that were taken to be able to create the final texture. A value of 5 means that the HDRi was created using that same number of interpolated images; the bigger that value, the more information is contained in the texture.

The following figure shows an example of three of the different images that were used to create the Alex's apartment asset:

Figure 1.20 – Example of three different exposure settings for the Alex's apartment asset

Each of them is a shot taken at a specific exposure level, which, when combined, gives us the final .HDRi file. You can take a look at the lighting information contained in each exposure level by looking at the brightness of the areas that emit light.

Lumen

As mentioned previously, Lumen is the new fully dynamic global illumination system that was introduced alongside Unreal Engine 5. This is a rendering feature that has always been very taxing in past game engines – not so much anymore. The way it works is by analyzing the models placed in the scene through a mixture of ray tracing methods that produce high-quality results. The first of these traces happens on the information present on the screen, while secondary analysis happens against either a meshes' distance field or the triangles that comprise the model, depending on whether the hardware running the app is capable of hardware ray tracing or not. These methods, when combined, manage to achieve a system that can target high frame rates on both next-generation consoles and high-end PCs.

Virtual Shadow Maps

Virtual Shadow Maps have come to replace the standard Shadow Maps, and they work so well that Epic has decided to make them the default Shadow Map Method – even though they are technically in a beta stage as of Unreal 5.0.3. Their main advantage is their increase in resolution, which enables them to depict much more accurate shadows that get rid of some of the known inaccuracies of the previous methods while providing multiple benefits, such as soft shadowing and a more simplified approach to lighting the scene. Be sure to check out the links in the next section to take a further look at them.

See also

You can find more information about Lumen and Virtual Shadow Maps through the following links:

- `https://docs.unrealengine.com/5.0/en-US/lumen-global-illumination-and-reflections-in-unreal-engine/`
- `https://docs.unrealengine.com/5.0/en-US/lumen-technical-details-in-unreal-engine/`
- `https://docs.unrealengine.com/5.0/en-US/virtual-shadow-maps-in-unreal-engine/`

Make sure you go over them if you are interested in knowing more about the pros and cons of using these two new techniques!

Using static lighting in our projects

The arrival of Unreal Engine 5 has seen the introduction of very important and groundbreaking technologies into the real-time visualization realm. Nanite, Lumen, and Virtual Shadow Maps… these are all new features that move the industry forward, and they ultimately give us higher-fidelity results when it comes to rendering our scenes.

Perhaps accidentally, we've already looked at one of those new features – Lumen. As the new default dynamic global illumination and reflections system, we've seen its effects when we worked on the previous recipe. We only had to worry about placing a light and adjusting some of its attributes for the scene to look right; we didn't have to create reflection captures, bake lightmaps, or use other techniques that might sound completely alien to you if you are starting your real-time journey with this version of the engine.

Despite that, we are going to take a quick look at the way things used to be done before Lumen. This isn't going to be a mere learning exercise, though – previous lighting techniques are still included in the engine, and for good reason. Even though Lumen can achieve astonishing rendering quality, the truth is that, first and foremost, it is a system meant to be used on high-end PCs and next-generation consoles (Xbox Series and PlayStation 5). Older techniques are still of utmost importance when it comes to producing scenes that can work fast on more humble devices. Furthermore, these previous systems can still be relevant when working toward achieving ultra-realistic results, as they can sometimes be more temporally stable than Lumen, producing better visuals when taken to the extreme.

Having said that, let's look at how to implement those old systems!

Getting ready

This recipe is going to start right at the end of the previous one, simply because we are going to replicate the same look using static lighting instead of relying on Lumen. Feel free to open the `01_06_ Start` level if you want to follow along using the same project as the one we'll be looking at throughout the following pages. If you're using your own assets, try to open a scene that contains some dynamic lights in it and where we can see reflections on some of the objects.

How to do it...

As mentioned previously, we are going to start working on a level that has some lights already in it, as well as reflective objects that show us a bit of their surroundings. With that as our starting point, let's proceed to switch from Lumen to the previous systems for global illumination and reflections:

1. Start by heading over to the **Rendering** section of the **Project Settings** area and navigating to the **Global Illumination** area.

2. Change the **Dynamic Global Illumination Method** option from **Lumen** to **None**.

3. Select the **Sky Light** actor present in the scene and, in the **Details** panel, change its **Mobility** from the current **Stationary** value to **Static**.

   After making the previous change, we should now be looking at a completely black screen. For comparison, this is what the screen looked like before versus now:

Figure 1.21 – Comparison shot of the same scene with and without Lumen

Dramatic, isn't it? Not using Lumen means that our now static **Sky Light** must rely on lightmaps to shade the scene. These textures need to be created, or built, in a process that we'll explore soon. Before that, though, let's take a quick look at the reflections.

4. Head back to the **Project Settings** section we were working on before (**Project Settings | Rendering | Global Illumination**) and revert to using **Lumen** as the **Dynamic Global Illumination Method** option.

5. Head down to the **Reflections** area, right underneath the previous **Global Illumination** one, and change **Reflection Method** from the standard **Lumen** to **None**. Go back to the main viewport once again.

   Back in the viewport, we can see that the scene is once again lit, but this time, we are lacking the reflections present in the chrome ball. This is mainly because the reflection system isn't making use of Lumen, and it requires us to build the reflections much like we also need to cook the lighting. This operation isn't that straightforward, as we are required to place a special type of actor called **Reflection Captures** in the areas where we want to have reflections.

6. So, go back to the **Project Settings** category and set the **Dynamic Global Illumination Method** option back to **None** too.

7. Head over to the main viewport and click on **Build** to expand its drop-down menu (you can find **Build** alongside the typical Windows program options – **File**, **Edit**, **Window**, and so on).

8. Focus on the **Lighting Quality** property and switch from the standard **Preview** setting to **Production**. This is the option that gives the best results when baking lighting information, as well as the most expensive in terms of the required computation power. This shouldn't be an issue for us seeing as we are dealing with a simple scene.

9. Without exiting the **Build** panel, click on the **Build Lighting Only** option. Your computer might take a bit of time to compute, but it should finish rather quickly – giving us similar results to the ones shown here:

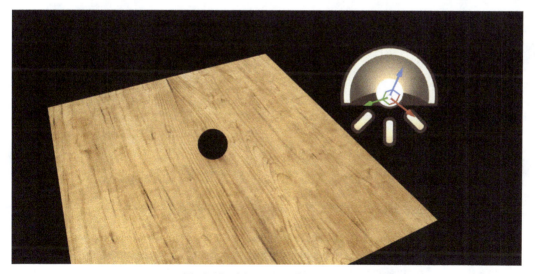

Figure 1.22 – The look of the scene after building the lighting

And there is our missing light! This is the process that we have to follow whenever we want to see the results of new lights added to the scene. The lighting information gets stored in special textures called **lightmaps**, and as such, we can control their quality by adjusting their resolution. We'll look at that in a moment, but let's take care of the reflections first.

10. Place a **Sphere Reflection Capture** actor in the level. You can select it from the **Visual Effects** panel located within the **Quickly add to the project** menu.

11. Place it around the sphere and adjust its **Influence Radius** to a value lower than the default, something like **500**.

> **Tip**
> We used a value of **500** in the previous step, but I wouldn't blame anyone for not knowing the units behind that number – meters? Inches? Miles? Unreal uses centimeters by default, so there is the answer!

12. As we are now closer to seeing something reflected in our non-reflective chrome ball, let's place something interesting around it just so that we can see those objects reflected. I'll place four different planes around the boundaries of the wooden floor, leaving the default material applied to them.

13. Navigate back to the **Build** panel and select the **Build Reflection Captures** option.

We should now be looking at a scene more akin to the following:

Figure 1.23 – The look of the scene after building the reflections

With that done, we've managed to achieve similar results to when we were using Lumen, with the difference that we are baking the lighting information, as well as capturing the reflections in a static and more controlled way. There is one more thing that we can do at this stage, and that is to add a new **Point Light**, just like we did in the previous recipe.

14. Head over to the **Quickly add to the project** panel and select **Point Light** from the **Place actors | Lights** menu. Place it close to the chrome ball, just so that we can see some shadows cast from it.

15. Adjust the **Intensity** and **Light Color** properties of the light. Just like in the previous recipe, I've chosen a value of **2** candelas for the former and a white color for the latter.

16. Change the **Mobility** property of our **Point Light** to **Static**.

17. Proceed to build the lighting, as we did previously, by clicking on the **Build Lighting Only** option located within the **Build** panel and look at the results once that operation finishes. The following figure shows what I'm seeing, given the position of the light I just placed:

Figure 1.24 – The results of the light-baking process

In a similar way to what we previously saw when using Lumen and Virtual Shadow Maps, the initial results don't look that great: the shadows cast on the wooden floor by the previously placed light are quite low-resolution. This is because of a key setting that we need to adjust in the plane – the **lightmap resolution**.

18. Select the floor and look at the **Lighting** section in the **Details** panel. There is a grayed-out property called **Override Light Map Res** that we need to tick and change, as the default value of **64** is too low. Increase the number to **512** and look at the results once again:

Figure 1.25 – The final look of the scene

The shadows should now look much sharper, a direct side effect of increasing the lightmap's resolution. Be sure to adjust this setting in all the static meshes that need it in your project!

How it works...

In this recipe, we looked at some of the previous methods that artists relied on to build their levels, especially before the advent of Lumen and the new rendering techniques that came along with the newest version of the engine and the latest technology in real-time rendering – with ray tracing being one of the most prominent features. Even though these new technologies are cutting-edge and offer several advantages over their predecessors, the now legacy rendering techniques that we've seen in this recipe still see widespread use and will continue to do so in certain types of experiences. These range from those tailored to low-powered devices to ultra-high refresh rate scenarios, without forgetting the demanding virtual and mixed-reality environments.

The reason why we can't use the newest features in those situations is varied, but most of them revolve around performance. Lumen and ray tracing introduce an extra price to pay for the frames that we want to render that can be simply too expensive to use on older devices, or too limiting on even the newest machines when aiming for high frame rates. These constraints will most likely go away at some point, especially thanks to the recent push that we are seeing in the fields of AI image reconstruction techniques, but you can bet that they will continue among us for at least some time.

With that being the state of things, static lighting is there to alleviate calculating what the scene should look like. This removes the need to figure out lighting every frame, making it easier for the graphics units of our devices to correctly render the scene. The same can be said for the reflection captures that we've used. The light-baking process we performed relies on building an extra set of textures (not unlike the **Base Color**, **Roughness**, or **Metallic** properties, which we covered when we built our first materials) that tell the engine how to light the scene. We don't plug these textures ourselves, with the engine being the one that generates and applies that data. We can't forget that, because they are indeed textures; their size contributes to the memory used by our level and therefore can have an impact on performance.

When it comes to reflections, the system works in a slightly different way compared to the method used for baking lighting. As the level can be viewed from many different angles, the reflections system needs to take that into account when baking that data. That being the case, the engine prompts us to place different reflection captures around the areas where we want to display that effect to know where it should focus its efforts when building the reflections. The process Epic recommends is to start with a large **Reflection Capture** actor and continue placing smaller ones in the areas where we want to increase the quality of the reflections.

Staying on topic, the shape of the capture actor plays an important role when it comes to what part of the map gets captured. Not only that but choosing between the **Box** and **Sphere** types affects the shape that is used for reprojecting the area that we capture and the part of the level that receives reflections from the **Cubemap** property. Differentiating between the two, the spherical type provides us with a solution that lacks discontinuities in its reflection data, at the expense of not being perfectly suited to any one object given its perfect shape (except for our little chrome ball!). On the other hand, the box shape tends to be less useful, as it only works correctly in places that are rectangular, such as hallways or box-shaped rooms.

See also

Here's some more information regarding the reflection capture methods:

- `https://docs.unrealengine.com/5.0/en-US/reflections-captures-in-unreal-engine/`
- `https://docs.unrealengine.com/4.27/en-US/BuildingWorlds/LightingAndShadows/ReflectionEnvironment/`

Also, make sure you check out the information regarding the GPU Lightmass Global Illumination method, a faster way to build lighting in your scenes if you have ray tracing hardware on your computer: `https://docs.unrealengine.com/4.26/en-US/RenderingAndGraphics/GPULightmass/`.

Checking the cost of our materials

So far, this introductory chapter has gone over the basics of the rendering pipeline – we've seen how to create a PBM, understood what the different Shading Models were, and saw how light played a key role in the overall look of the final image. However, we can't move on to different topics just yet without learning a bit more about the impact that our materials have on the machines that are displaying them.

The first thing that we need to be aware of is that some materials or effects are more expensive in terms of their rendering cost than others. Chances are you have already experienced that in the past – think, for example, about frames per second in video games. The number of frames per second that are fed into our displays by the graphics cards directly influences how the game plays and feels. Many elements affect performance, but one crucial factor in that equation is the complexity of the materials we create.

With that in mind, we need to be able to control how performant the scene we are creating is. Unreal puts several tools at our disposal that allow us to see how expensive certain effects and materials are, letting us focus on the areas that might not be working as expected. We are going to be looking at those tools in the next few pages, as well as the different visualization options that are available to us whenever we want to check what might be causing issues in our maps.

Getting ready

All we need to do to start working on this recipe is to have a scene that contains multiple different assets that we can compare. It would be useful if the content within the level you work on is varied – one that contains multiple models, materials, and effects. I've prepared such a scene that you can use as a reference in case you want to work on the same project I'll be tweaking: if so, simply open the `01_07_Start` level and we'll take it from there.

Having said that, feel free to bring your own meshes and materials, as we are going to be checking them out from a technical point of view. All we care about at this point is having multiple elements that we can look at to see the difference in performance between them.

How to do it...

This recipe is going to be a bit of a passive one, where we'll be looking at different panels and studying the information that they provide concerning the complexity of the scene that we have in front of us. That being the case, let's take a quick look at the level in which we'll be working just to get familiar with it. It can be seen in the following figure:

Figure 1.26 – Initial look at the objects present in the scene

As you can see, we have many objects placed in this scene, all with different materials assigned to them. We've created some of those before, such as the chrome ball and the glass, while others are there simply for demonstration purposes. We are only going to be looking at several performance indicators at this stage, so we don't have to worry about the assets themselves at this point.

With that out of the way, let's start taking an in-depth look at the cost of the scene in a more technical light:

1. Select the wooden Christmas tree and look at its **Details** panel, and scroll down to the **Materials** section. You should see a shader being used there called **M_ChristmasTree**. Double-click on it to open the Material Editor.

2. Review the information displayed in the **Stats** panel for the current material. You can access it by clicking **Windows | Stats**. This area gives us information regarding different parameters used by this asset, such as the number of textures being used and other important information that will give us an initial idea of how costly the material is.

Comparing the information that you find here against that of the other materials can serve as a good initial approximation of the material's cost. Let's look at the following figure, which shows us the difference between the material being applied to the toy Christmas tree and the brick wall:

Figure 1.27 – Comparison shot of the different stats provided by the Stats panel for two different materials

The previous methodology can give us an idea of the complexity of a given material, all while looking at some of the numbers behind what we can see on the screen. There is also another, more visually interesting way to look at this.

3. Head over to the main viewport and click on **View Modes | Optimization View Modes | Shader Complexity**. The screen should now look something like this:

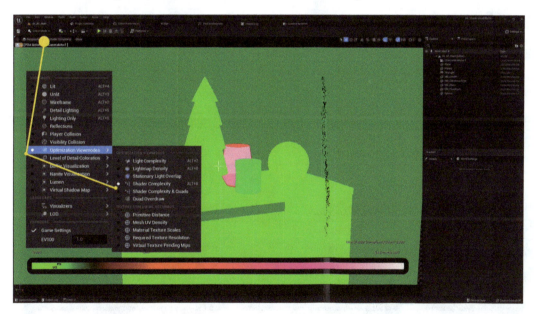

Figure 1.28 – Optimization view modes

The assets that have been placed in the level have now been colored in such a way that they give us an indication of how costly they are to render, all according to the graph that should have also appeared at the bottom of the screen. One of the most expensive assets is the glass as it uses **Translucent Blend Mode** – one that is particularly costly to render.

> **Tip**
>
> While these are by no means definitive ways to measure the complexity of the scene, they do give a good approximation as to those places that we can check, should we need to make any cutbacks.

The complexity of our materials isn't the only indicator of how expensive the scene is to render from the perspective of the disciplines covered in this book. Let's look at some of the other things that we need to be aware of.

4. Go back to **View Modes** and click on **Light Complexity** within the **Optimization View Modes** section (you can also press *Alt + 7* to access this view). You should be looking at a black screen.

5. Revert to the **Lit** view and place a couple of **Point Light** actors around the scene, preferably above the objects already there, and adjust the intensity of the lights to your liking. Remember that you can add these assets just like we've added other models before, by selecting them from **Quickly add to project** | **Lights** | **Point Light**.

6. Click on the **Light Complexity** view mode once again and review the state of the level:

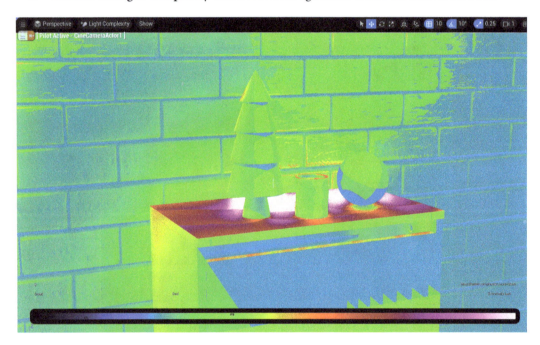

Figure 1.29 – The complexity of the lighting in our scene according to the Light Complexity view mode

The scene should have looked almost completely dark when we first selected the **Light Complexity** optimization view mode, but we should now be looking at a much more interesting picture. The reason why we saw everything shaded in deep blue at first is that the HDRi lighting we were using wasn't too costly to render. Adding a couple of extra lights made our hardware work harder to render the same scene, increasing the complexity of the final picture.

On the topic of lights, there is an extra view mode that you might want to explore – **Lightmap Density**. This is useful whenever we work with static lighting and baking lightmaps as it gives us a good idea of the relative sizes that we have assigned to the textures that will be created once we start baking. Ensuring that the lightmap resolution is even across the entirety of the level is key when trying to achieve great results.

Important note

Remember that you can adjust the size of each asset's lightmap by selecting it from the level and inspecting the **Lighting** section of its **Details** panel. There is an option there called **Override Light Map resolution** that lets you do just that.

---

7. Click on **View Modes | Optimization Viewmodes | Lightmap density** and look at the scene. You'll see that the models are now rendered using different colors and using a grid-like texture that indicates the resolution of the lightmap that they have assigned.

8. Select all of the models and assign a different lightmap resolution to each of them, just so that the color and the spacing of the grid texture look homogeneous across the level. You can do this by going to each model's **Details** panel and searching for the **Override Light Map resolution** property located within the **Lighting** section. The result will look like this:

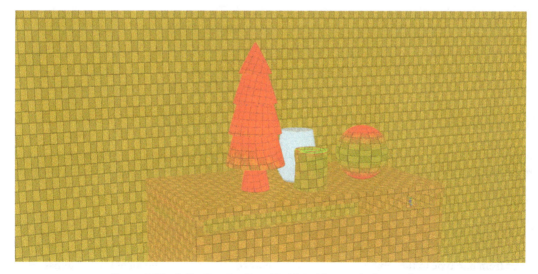

Figure 1.30 – Adjusting the Override Light Map resolution settings

9. Remember that even though we've changed the lightmap resolution, we will only be able to appreciate its effects if we work with static lighting, as we did in the previous recipe.

Now that we've looked at those panels, there are a couple of other ones that I want to at least make you aware of at this point. The new technologies bundled with Unreal Engine 5 (Lumen and **Nanite**) also include visualization modes just so that we can debug if they are working as expected. We will be looking at them next.

10. Drag the **SM_ToyTank** asset into the scene and place it close to the other models. Feel free to delete any if needed. This model has **Nanite** enabled, so we can use it for debugging purposes.

11. Select the **Overdraw** option within the **Nanite Visualization** category inside the **View Mode** panel and focus your attention on the only Nanite-enabled mesh in the scene – the toy tank.

    Even though there are several viewing options for the Nanite system, this is the one that tells us whether our scenes suffer from overdraw. This problem appears when we have several Nanite meshes overlapping with each other – if one isn't far enough from the other, the engine won't be able to cull the triangles of any of them, resulting in us rendering more triangles than we need to. This can cause performance problems that we can avoid using careful placement.

    Having looked at Nanite, let's head over to the Lumen visualization category.

12. Choose the **Surface Cache** option from the **Lumen** section of the **View Mode** panel and focus on the areas rendered in pink.

    Lumen parameterizes the scene we are trying to render to speed up the lighting calculations that it performs. This process is done by creating something called **Lumen Mesh Cards**, of which there are 12 by default. Sometimes, that number isn't enough, especially when dealing with complex interiors and irregularly shaped models. The present view mode allows us to see what they look like and the areas that don't have surface cache coverage, displayed in pink.

Now, we know which areas to check when looking for performance pitfalls. There are more things that we could consider, but being aware of these and acting in consequence already puts us in a good position when it comes to creating performant and optimized levels. Make sure you consider them!

How it works...

Before we move on, I'd like to quickly talk about Shading Models and Nanite – two elements that impact the cost of rendering a scene.

Shading Models

As we've seen in previous recipes, materials are not homogeneous entities. The mathematics and functions used to describe the different shading and Blend Modes carry a weight with them that varies from one type to the next. We don't need to know the specifics behind all of them, but having an overall idea is key to running a well-oiled application.

We've seen some examples in the previous pages, which include an opaque material and a translucent one – examples that we've worked on in the past. However, we need to keep in mind that there are more types that we can (and will) encounter in the future. Unreal includes many different Shading

Models, and we can order the most common ones in the following way according to how expensive they are to render:

Unlit > Default lit > Preintegrated skin > Subsurface > Clear coat > Subsurface profile

Even though there are more of them, the previous ones are some of the most common ones you'll have to deal with. On top of that, we can see that there is a pattern in their complexity concerning how they interact with light: the shaders that don't get affected by it are the simplest ones, while the subsurface profile (which scatters light right underneath the skin) is the most complex of the bunch. Of course, the actual cost of a material depends on how complex its graphs end up being, but that previous order applies to base materials with no extra nodes within them. Also, there are options within each type that can make them more or less difficult to render: having a material being two-sided or using a particular type of translucency effect can increase the cost to the GPU.

Beyond that, Epic has created some performance guidelines for artists that highlight where we should be focusing our attention to keep our applications running well. You can find them here: `https://docs.unrealengine.com/4.27/en-US/TestingAndOptimization/`.

Nanite

Something that we have started to see in this recipe is the new geometry system introduced alongside Unreal Engine 5, Nanite. This is one of the biggest changes that the fifth iteration of the engine has introduced in its pipeline, and it is a real game-changer when it comes to displaying models on the screen. Developers are no longer constrained to meeting specific polycount budgets as they can now create much denser models without worrying whether the engine will be able to handle them. This allows us to display much more detailed assets than before while also simplifying the development process of an app or a game.

As with everything, there are also caveats to this system – for example, it doesn't currently support skeletal meshes (characters or other types of deformable models). Be sure to check out the documentation provided in the next section if you want to be on top of this system.

See also

You can find more information about the different topics that we've covered in this recipe here:

- `https://docs.unrealengine.com/4.26/en-US/BuildingWorlds/LevelEditor/Viewports/ViewModes/`

- `https://docs.unrealengine.com/5.0/en-US/lumen-technical-details-in-unreal-engine/`

- `https://docs.unrealengine.com/5.0/en-US/nanite-virtualized-geometry-in-unreal-engine/`

2

Customizing Opaque Materials and Using Textures

Starting gently—as we always do—we'll begin working on a simple scene where we'll be able to learn how to set up a proper material graph for a small prop. On top of that, we are also going to discover how to create materials that can be applied to large-scale models, implement clever techniques that allow us to balance graphics and performance, study material effects driven by the position of the camera or semi-procedural creation techniques—all of that while using standard Unreal Engine assets and techniques. I'm sure that studying these effects will give us the confidence we need to work within Unreal.

All in all, here is a full list of topics that we are about to cover:

- Using masks within a material
- Instancing a material
- Texturing a small prop
- Adding Fresnel and Detail Texturing nodes
- Creating semi-procedural materials
- Blending textures based on our distance from them

And here's a little teaser of what we'll be doing:

Figure 2.1 – A look at some of the materials we'll create in this chapter

Technical requirements

Just as in the previous chapter (and all of the following ones!), here is a link you can use to download the Unreal Engine project that accompanies this book: `https://packt.link/20u7B`

In there, you'll find all of the scenes and assets used to make the content you are about to see become a reality. As always, you can also opt to work with your own creations instead—and if that's the case, know that there's not a lot you are going to need. As a reference, we've used a few assets to recreate all the effects described in the next recipes: the 3D models of a toy tank, a cloth laid out on top of a table, and a plane, as well as some custom-made textures that work alongside those meshes. Feel free to create your own unique models if you want to test yourself even further!

Using masks within a material

Our first goal in this chapter is going to be the replication of a complex material graph. Shaders in Unreal usually depict more than one real-life substance, so it's key for us to have a way to control which areas of a model are rendered using the different effects programmed in a given material. This is where masks come in: textures that we can use to separate (or mask!) the effects that we want to apply to a specific area of the models with which we are working. We'll do that using a small wooden toy tank, a prop that could very well be one of the assets that a 3D artist would have to adjust as part of their day-to-day responsibilities.

Getting ready

We've included a small scene that you can use as a starting point for this recipe—it's called `02_01_Start`, and it includes all of the basic assets that you will need to follow along. As you'll be able to see for yourself, there's a small toy tank that has two **UV** channels located at the center of the level. UVs are the link between the 2D textures that we want to apply to a model and the 3D geometry itself, as they are a map that correlates the vertex positions in 2D and 3D space. The first of those UVs indicates how textures should be applied to 3D models, whereas the second one is used to store the lightmap data the engine calculates when using static lighting.

If you decide to use your own 3D models, please ensure that they have well-laid-out UVs, as we are going to be working with them to mask certain areas of the objects.

> **Tip**
> We are going to be working with small objects in this recipe, and that could prove tricky when we zoom close to them—the camera might cut through the geometry, something known as **clipping**. You can change the behavior of the camera by opening the **Project Settings** panel and looking inside the **Engine | General Settings | Settings** section, where you'll find a setting called **Near Clip Plane,** which can be adjusted to sort out this issue. Using a value of **1** will alleviate this issue, but remember to restart the editor in order for the changes to be implemented.

How to do it...

The goal of this recipe is going to be quite a straightforward one: to apply different real-world materials to certain parts of the model with which we are going to be working using only a single shader. We'll start that journey by creating a new material that we can apply to the main toy tank model that you can see at the center of the screen. This is all provided you are working with the scene I mentioned in the *Getting ready* section; if not, think of the next steps as the ones you'll also need to implement to texture one of your own assets. So:

1. Create a new material, which you can name **M_ToyTank**, and apply it to the toy model.

2. Open the **Static Mesh Editor** for the toy tank model, by selecting the asset from the level and double-clicking on the **Static Mesh** thumbnail present in the **Details** panel.

3. Visualize the UVs of the material by clicking on the **UV | UV Channel 0** option within the **Static Mesh Editor**, as seen in the next screenshot:

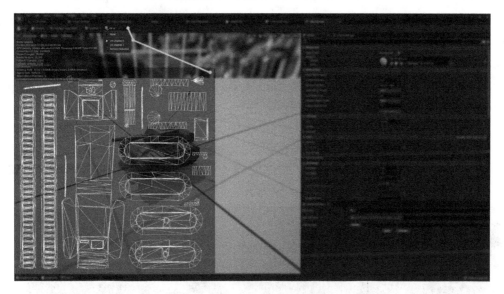

Figure 2.2 – Location of the UV visualizer options within the Static Mesh Editor

As seen in *Figure 2.2*, the UV map of the toy tank model is comprised of several islands. These are areas that contain connected vertices that correspond to different parts of the 3D mesh—the body, the tracks, the barrel, the antenna, and all of the polygons that make up the tank. We want to treat some of those areas differently, applying unique textures depending on what we would like to see on those parts, such as on the main body or the barrel. To do so, we will need to create masks that cover the selected parts we want to treat differently. The project includes one such texture designed specifically for this toy tank asset, which we'll use in the following steps. If you want to use your own 3D models, make sure that their UVs are properly unwrapped and that you create a similar mask that covers the areas you want to separate.

> **Tip**
>
> You can always adjust the models provided with this project by exporting them out of Unreal and importing them into your favorite **Digital Content Creation** (**DCC**) software package. To do that, right-click on the asset within the Content Browser and select **Asset Actions | Export**.

4. Open our newly created material and create a **Texture Sample** node within it (right-click and then select **Texture Sample**, or hold down the *T* key and click anywhere inside the graph).

5. Assign the **T_TankMasks** texture (located within **Content | Assets | Chapter 02 | 02_01**) to the previous **Texture Sample** node (or your own masking image if working with your own assets).

6. Create two **Constant3Vector** nodes and select whichever color you want under their color selection wheel. Make sure they are different from one another, though!

7. Next, create a **Lerp** node—that strange word is short for *linear interpolation*, and it lets us blend between different assets according to the masks that we connect to the **Alpha** pin.

8. Connect the **R** (red) channel of the **T_TankMasks** asset to the **Alpha** pin of the new **Lerp** node.

9. Connect each of the **Constant3Vectors** to the **A** and **B** pins of the **Lerp** node.

10. Create another **Lerp** node and a third **Constant3Vector**.

11. Connect the **B** (blue) channel of the **T_TankMasks** texture to the **Alpha** pin of the new **Lerp** node, and the new **Constant3Vector** to the **B** pin.

12. Connect the output of the **Lerp** node we created in *step 7* (the first of the two we should have by now) to the **A** slot of the new **Lerp** node.

13. Assign the material to the tank in the main scene. The material graph should look something like this at the moment:

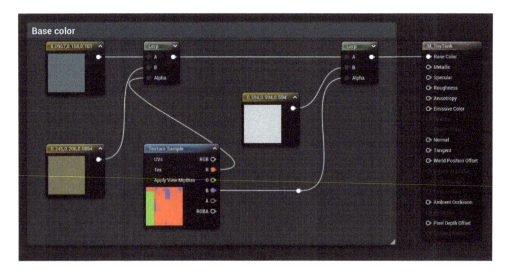

Figure 2.3 – The current state of the material graph

We've managed to isolate certain areas of the toy tank model thanks to the use of the previous mask, which is one of our main goals for this recipe. We now need to apply this technique to the **Metallic** and **Roughness** parameters of the material, just so we can control those independently.

14. Copy all the previous nodes and paste them twice—we'll need one copy to connect to the **Roughness** slot and a different one that will drive the **Metallic** attribute.

15. In the new copies, replace the **Constant3Vector** nodes with simple **Constant** ones. We don't need an RGB value to specify the **Metallic** and **Roughness** properties, so that's why we only need standard **Constant** nodes this time around.

16. Assign custom values to the new **Constant** nodes you defined in the previous step. Remember what we already know about these parameters: a value of 0 in the **Roughness** slot means that the material has very clear reflections, while a value of 1 means the exact opposite. Similarly, a value of 1 connected to the **Metallic** node means that the material is indeed metal, while 0 determines that it is not. I've chosen a value of 0 for the first two **Constant** nodes and a value of 1 for the third one. Let's now review these new sections of the graph:

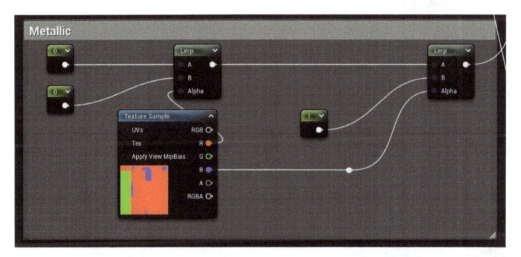

Figure 2.4 – A look at the new Metallic section of the material, which is a duplicate of the Roughness one

Finally, think about tidying things up by creating comments, which is a way to group the different sections of the material graph together. This is done by selecting all of the nodes that you want to group and pressing the *C* key on your keyboard. This keeps things organized, which is very important no matter whether you work with others or whether you revisit your own work. You saw this in action in the previous screenshot with the colored outlines around the nodes. Having said that, let's take a final look at the model, calling it a day!

Figure 2.5 – The toy tank with the new material applied to it

How it works...

In essence, the textures that we've used as masks are images that contain a black-and-white picture stored in each of the files' RGB channels. This is something that might not be that well known, but the files that store the images that we can see on our computers are actually made out of three or four different channels—one containing the information for the red pixels, another for the green ones, and yet one more for the blue tones, with an extra optional one called **Alpha** where certain file types store the transparency values. Those channels mimic the composition of modern flat panel displays, which operate by adjusting the intensity received by the three lights (red, green, and blue) used to represent a pixel.

Seeing as masks only contain black-and-white information, one technique that many artists use is to encode that data in each of the channels present in a given texture. This is what we saw in this recipe when we used the **T_TankMasks** texture, an image that contains different black-and-white values in each of its RGB channels. We can store up to three or four masks in each picture using this technique—one per available channel, depending on the file type. You can see an example of this in the following screenshot:

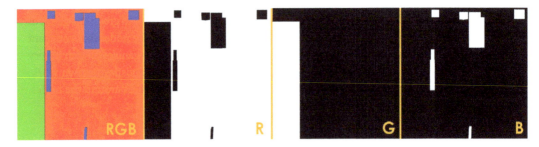

Figure 2.6 – The texture used for masking purposes and the information stored in each of the RGB channels

The **Lerp** node, which we've also seen in this recipe, manages to blend two inputs (called **A** and **B**) according to the values that we provide in its **Alpha** input pin. This last pin accepts any value situated in the 0 to 1 range and is used to determine how much of the two **A** and **B** input pins are shown: a value of 0 means that only the information provided to the **A** pin is used, while a value of 1 has the opposite effect—only the **B** pin is shown. A more interesting effect happens when we provide a value in between, such as 0.25, which would mean that both the information in the **A** and in the **B** pin is used—75% of **A** and 25% of **B** in particular. This is also the reason why we used the previous mask with this node: as it contained only black-and-white values in each of its RGB channels, we could use that information to selectively apply different effects to certain parts of the models.

Using masks to drive the appearance of a material is often preferable to using multiple materials on a model. This is because of the way that the rendering pipeline works behind the scenes—without getting too technical, we could say that each new material that a model has makes it more expensive to render. It's best to be aware of this whenever we work on larger projects!

See also

We've seen how to use masks to drive the appearance of a material in this recipe, but what if we want to achieve the same results using colors instead? This is a technique that has seen widespread use in other 3D programs, and even though it's not as simple or computationally cheap as the masking solution we've just seen in Unreal, it is a handy feature that you might want to explore in the future. The only requirement is to have a color texture where each individual shade is used to mask the area we want to operate on, as you can see in the following screenshot:

Figure 2.7 – Another texture used for masking, where each color is meant to separate different materials

The idea is to isolate a color from the mask in order to drive the appearance of a material. If this technique interests you, you can read more about it here: https://answers.unrealengine. com/questions/191185/how-to-mask-a-singlecolor.html.

Finally, there are more examples of masks being used in other assets that you might want to check. We can find some within the **Sample Content** node, inside the following folder: **Starter Content | Props | Materials**. Shaders such as **M_Door** or **M_Lamp** have been created according to the masking methodology explained in this chapter. On top of that, why not try to come up with your own masks for the 3D models that you use? That is a great way to look at UVs, 3D models, image-editing software, and Unreal. Be sure to try it out!

Instancing a material

We started this chapter by setting up the material graph used by a small wooden toy tank prop in the previous recipe. This is something that we'll be doing multiple times throughout the rest of the book, no matter whether we are working with big or small objects— after all, creating materials is what this book is all about! Having said that, there are occasions when we don't need to recreate the entire material graph, especially when all we want to do is change certain parameters (such as the textures being used or the constant values that we've set). It is then that we can reach for the assets known as **Material Instances**.

This new type of asset allows us to quickly modify the appearance of the parent material by adjusting the parameters that we decide to expose. On top of that, instances remove the need for a shader to be compiled every time we make a change to it, as that responsibility falls to the parent material instead. This is great in terms of saving time when making small modifications, especially on those complex materials that take a while to finish compiling.

In this recipe, we will learn how to correctly set up a parent material so that we can use it to create Material Instances—doing things such as exposing certain values so that they can be edited later on. Let's take a look at that.

Getting ready

We are going to continue working on the scene we adjusted in the previous recipe. This means that, as always, there won't be a lot that you need in order to follow, as we'll be taking a simple model with a material applied to it and tweaking it so that the new shader that we end up using is an instance of the original one. Without further ado, let's get to it!

How to do it...

Let's start by reviewing the asset we created in the previous recipe—it's a pretty standard one, containing attributes that affect the base color, the roughness, and the metallic properties of the material. All those parts are very similar in terms of the nodes that we've used to create them, so let's focus on the **Metallic** property once again:

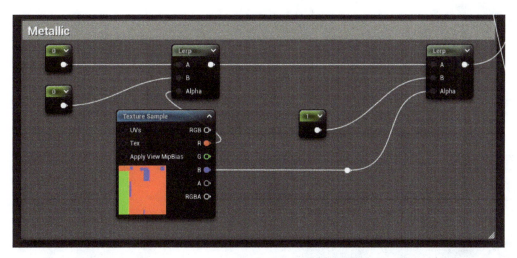

Figure 2.8 – A look at the different nodes that affect the Metallic attribute of the material

Upon reviewing the screenshot, we can see that there are basically two things happening in the graph—we are using Constant nodes to adjust the **Metallic** property of the shader, and we are also employing texture masks to determine the parts where the Constant values get used within our models. The same logic applies elsewhere in the material graph for those nodes that affect the **Roughness** and **Base Color** properties, as they are almost identical copies of the previous example.

Even though the previous set of nodes is quite standard when it comes to the way we create materials in Unreal, opening every material and adjusting each setting one at a time can prove a bit cumbersome and time-consuming—especially when we also have to click **Compile** and **Save** every single time we make a change. To alleviate this, we are going to start using parameters and Material Instances, elements that will allow us to more quickly modify a given asset. Let's see how we can do this:

1. Open the material that is currently being applied to the toy tank in the middle of the scene. It should be **M_ ToyTank_ Parameterized_ Start** (though feel free to use one similar to what we used in the previous recipe).

2. Select the **Constant** nodes that live within the **Metallic** material expression comment (the ones seen in *Figure 2.8*), right-click with your mouse, and select the **Convert to Parameter** option. This will create a **Scalar Parameter** node, a type of node used to define a numerical value that can be later modified when we create a Material Instance.

> **Tip**
>
> You can also create scalar parameters by right-clicking anywhere within the material graph and searching for **Scalar Parameter**. Remember that you can also look for them in the **Palette** panel!

3.   With that done, it's now time to give our new parameters a name. Judging from the way the mask is dividing the model, I've gone with the following criteria: **Tire Metalness**, **Body Metalness**, and **Cannon Metalness**. You can rename them by right-clicking on each of the nodes and selecting the **Rename** option.

4.   Convert the **Constant** nodes that can be found within the **Roughness** and **Base Color** sections into **Scalar Parameter** nodes and give them appropriate names, just as we did before. This is how the **Base Color** nodes look after doing that:

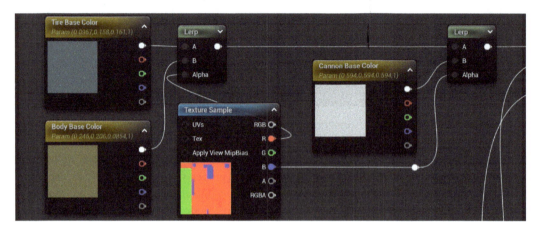

Figure 2.9 – The appearance of the Base Color section within the material graph on which we are working

---

Important note

The effects of the **Convert to Parameter** option are different depending on the type of variable that we want to change: simple **Constant** nodes (like the ones found in the **Roughness** and **Metallic** sections of the material) will be converted into the scalar parameters we've already seen, while the **Constant3Vectors** will get transformed into **Vector Parameters**. We'll still be able to modify them in the same way, the only difference being the type of data that they demand.

Once all of this is done, we should be left with a material graph that looks like what we previously had, but one where we are using parameters instead of Constants. This is a key feature that will play a major role in the next steps, as we are about to see.

5.   Locate the material we've been operating on within the Content Browser and right-click on it. Select the **Create Material Instance** option and give it a name. I've gone with **MI_ToyTank_Parameterized** since *"MI"* is a common prefix for this type of asset.

6.   Double-click on the newly created Material Instance to adjust it. Unlike their parents, instances let us expose the parameters we previously created and adjust them—either before or during runtime, and without the need to recompile any shaders.

7. Tweak the parameters that we have exposed to something different than their defaults. To do so, be sure to first check the boxes of those you want to change. As a reference, and just to see that we are on the same page, here is a reference screenshot of the settings I've chosen:

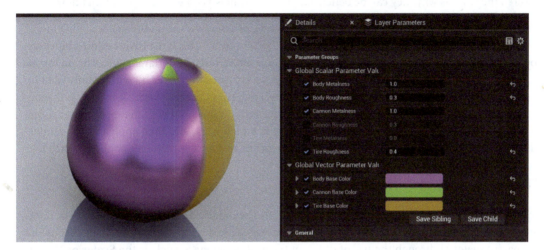

Figure 2.10 – A look at the editor view for Material Instances and the settings I've chosen

> **Tip**
> Take a look at the *See also* section to discover how to organize the previous parameters in a different way, beyond the standard **Scalar Parameter** and **Vector Parameter** categories that the engine has created for us.

8. With your first Material Instance created, apply it to the model and look at the results!

Figure 2.11 – The final look of the toy tank using the new Material Instance we have just created

How it works...

We set our sights on creating a Material Instance in this recipe, which we have done, but we haven't covered the reasons why they can be beneficial to our workflow yet. Let's take care of that now.

First and foremost, we have to understand how this type of asset falls within the material pipeline. If we think of a pyramid, a basic material will sit at the bottom layer—this is the basic building block on top of which the rest of what we are about to be talking about rests. Material Instances lie on top of that: once we have a basic parent material set up, we can create multiple instances if we only want to modify parameters that are already part of that first material. For example, if we have two toy tanks and we want to give them different colors, we could create a master material with that property exposed as a variable and two Material Instances that affect it, which we would then apply to each model. That is better in terms of performance compared to having two master materials applied to each toy.

There's more...

Something else to note is that Material Instances can also be modified at runtime, which can't be done using master materials. The properties that can be tweaked are the ones that we decide to expose in the parent material through the use of different types of parameters, such as the **Scalar** or **Vector** types that we have seen in this recipe. These materials that can be modified during gameplay receive the distinctive name of **Material Instance Dynamic (MID)**, as even though they are also an instance of a master material, they are created differently (at runtime!). Let's take a quick look at how to create one of these materials next:

1. Create an Actor Blueprint.

2. Add a **Static Mesh** component to it and assign the **SM_ToyTank** model as its default value (or the model that you want to use for this example!).

3. In the **Construction Script** node, add the **Create Dynamic Material Instance** node and hook the **Static Mesh** node to its **Target** input pin. Select the parent material that you'd like to modify in the **Source Material** drop-down menu and store this material as a variable. You can see this sequence in the next screenshot:

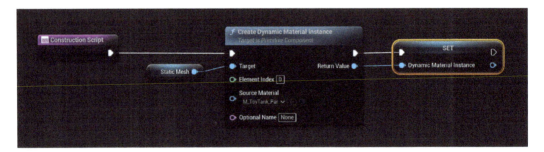

Figure 2.12 – The composition of the Construction Script node

4. Head over to the **Event Graph** window and use a **Set Vector Parameter Value** node in conjunction with a **Custom Event** node to drive the changing of the material parameter that you'd like to affect. You can set the event to be called from the editor if you want to see its effects in action without having to hit **Play**. Also, remember to type the exact parameter you'd like to change and assign a value to it! Take a look at these instructions here:

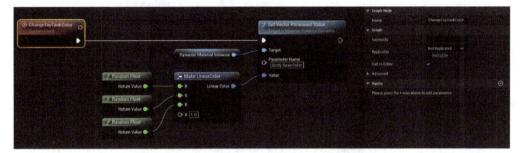

Figure 2.13 – Node sequence in the Event Graph

I'll leave the previous set of steps as a challenge for you, as it includes certain topics that we haven't covered yet, such as the creation of Blueprints. In spite of that, I've left a pre-made asset for you to explore within the Unreal Engine project included with this book. Look for the **BP_Changing Color Tank** Blueprint and open it up to see how it works in greater detail!

See also

Something that we didn't do in this recipe was to group the different parameters we created. You might remember that the properties we tweaked when we worked on the Material Instances were automatically bundled into two different categories: *vector parameter values* and *scalar parameter values*. The names are only representative of the type they belong to, and not what they are affecting. That is something that we can, fortunately, modify by going back to the parent material, so let's open its material graph to see how to do so.

Once in there, we can select each of the parameters we created and assign them to a group by simply typing the desired group name inside the **Group Editable** textbox within each of the parameters' **Details** panels. Choosing a name different from the default **None** will automatically create a new group, which you can reuse across other variables by simply repeating the exact same word. This is a good way to keep things tidy, so be sure to give this technique a go!

On top of this, I also wanted to make you aware of other types of parameters available within Unreal. So far, we've only seen two different kinds: the **Scalar** and **Vector** categories. There are many more that cover other types of variables, such as the **Texture Sample 2D** parameter used to expose texture samples to the user. That one and others can help you create even more complex effects, and we'll see some of them in future chapters.

Finally, let me leave you with the official documentation dedicated to Material Instances in case you want to take a further look at that:

- `https://docs.unrealengine.com/5.0/en-US/instanced-materials-in-unreal-engine/`

- `https://docs.unrealengine.com/5.0/en-US/creating-and-using-material-instances-in-unreal-engine/`

Reading the official docs is always a good way to learn the ins and outs of the engine, so make sure to go over said website in case you find any other topics that are of interest!

Texturing a small prop

It's time we looked at how to properly work with textures within a material, something that we'll do now thanks to the small toy tank prop we've used in the last two recipes. The word *properly* is the key element in the previous sentence, for even though we've worked with images in the past, we haven't really manipulated them inside the editor or seen how we can adjust them without leaving the editor. It's time we did that, so let's take care of it in this recipe!

Getting ready

You probably know the drill by now—just like in previous recipes, there's not a lot that you'll actually need in order to follow along. Most of the assets that we are going to be using can be acquired through the **Starter Content** asset pack, except for a few that we provide alongside this book's Unreal Engine 5 project.

As you are about to see, we'll continue using our famous toy tank model and material, so you can use either those or your own assets; we are dealing with fairly simple assets here, and the important bit is the actual logic that we are going to create within the materials.

If you want to along with me, feel free to open the level named `02_03_Start` just so that we are looking at the same scene.

How to do it...

So far, all we've used in our toy tank material are simple colors—and even though they look good, we are going to make the jump to realistic textures this time. Let's start:

1. Duplicate the material we created in the first recipe of this chapter, **M_ToyTank**. This is going to be useful, as we don't want to reinvent the wheel so much as expand a system that already works. Having everything separated with masks and neatly organized will speed things up greatly this time. Name the new asset whatever you like—I've gone with **M_ToyTank_Textured**.

2. Open the material graph by double-clicking on the newly created asset and focus on the **Base Color** section of the material. Look for the **Constant3Vector** node that is driving the appearance of the main body of the tank (the one using a brownish shade) and delete it.

3. Create a new **Texture Sample** node and assign the **T_Wood_Pine_D** texture to it (this asset comes with the **Starter Content** asset pack, so we should already have access to it).

4. Connect the **RGB** output pin of the new **Texture Sample** node to the **B** input pin of the **Lerp** node, to which the recently deleted **Constant3Vector** node used to be connected:

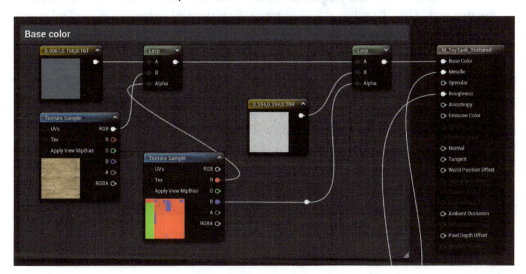

Figure 2.14 – The now-adjusted material graph

This will drive the appearance of the main body of the tank.

> **Tip**
>
> There's a shortcut for the **Texture Sample** node: the letter *T* on your keyboard. Hold it and left-click anywhere within the material graph to create one.

There are a couple of extra things that we'll be doing with this texture next—first, we can adjust the tiling so that the wood veins are a bit closer to each other once we apply the material to the model. Second, we might want to rotate it just to modify the current direction of the screenshot. Even though this is an aesthetic choice, it's useful to get familiar with these nodes, as they are often quite useful when we need to use them.

5. While holding down the *U* key on your keyboard, left-click in an empty space of the material graph to add a **Texture Coordinate** node. Assign a value of **3** to both its **U Tiling** and **V Tiling** settings, or adjust it until you are happy with the result.

6. Next, create a **Custom Rotator** node and place it immediately after the previous **Texture Coordinate** node.

7. With the previous node in place, plug the output of the **Texture Coordinate** node into the **UVs (V2)** input pin of the **Custom Rotator** node.

8. Connect the **Rotated Values** output node of the **Custom Rotator** node to the **UVs** input pin of the **Texture Sample** node used to display the wood texture.

9. Create a **Constant** node and plug it into the **Rotation Angle (0-1) (S)** node of the **Custom Rotator** node. Give it a value of **0.25** to rotate the texture 90 degrees (you can find more information regarding why a value of 0.25 equals 90 degrees in the *How it works…* section).

Here is what the previous sequence of nodes should look like:

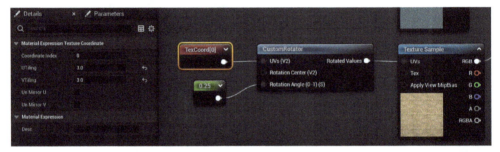

Figure 2.15 – A look at the previous set of nodes

Let's now continue to add a little extra variation to the other parts of the material. I'd like to focus on the other two masked elements next—the tires and the metallic sections—and make them a little bit more interesting visually speaking.

10. We can start by changing the color of the **Constant3Vector** node that we are using to drive the color of the toy tank's tires to something that better represents that material (rubber). We've been using a blueish color so far, so something closer to black would probably be better.

11. After that, create a **Multiply** node and place it right after the previous **Constant3Vector** node. This is a very common type of node that does what its name implies: it outputs the result obtained after multiplying the inputs that reach this pin. One way to create it is by simply holding down the *M* key of the keyboard and left-clicking anywhere in the material graph.

12. With the previous **Multiply** node created, connect its **A** input pin to the output of the same **Constant3Vector** node driving the color of the rubber tires.

13. As for the **B** input pin, assign a value of **0.5** to it, something that you can do through the **Details** panel of the **Multiply** node without needing to create a new **Constant** node.

14. Interpolate between the default value of the **Constant3Vector** node for the rubber tires and the result of the previous multiplication by creating a **Lerp** node. You can do this by connecting the output of said **Constant3Vector** node to the **A** input pin of a new **Lerp** node and the result of the **Multiply** node to the **B** input pin. Don't worry about the **Alpha** parameter for now—we'll take care of that next.

15. Create a **Texture Sample** node and assign the **T_Smoked_Tiled_D** asset as its default value.

16. Next, create and connect a **Texture Coordinate** node to the previous **Texture Sample** node and assign a higher value than the default 1 to both its **U Tiling** and **V Tiling** parameters; I've gone with **3**, just so that we can see the effect this will have more clearly in the future.

17. Proceed to drag a wire out of the output pin of the **Texture Sample** node for the new black-and-white smoke texture and create a **Cheap Contrast** node.

18. We'll need to feed the previous **Cheap Contrast** node with some values, which we can do by creating a **Constant** node and connecting it to the **Contrast (S)** input pin of the previous node. This will increase the difference between the dark and white areas of the texture it is affecting, making the final effect more obvious. You can leave the default value of **1** untouched, as that will already give us the desired effect.

19. Connect the result of the **Cheap Contrast** node to the **Alpha** input pin of the **Lerp** node we created in *step 14* and connect the output of that to the original **Lerp** node which is being driven by the **T_TankMasks** texture mask.

Here is a screenshot that summarizes the previous set of steps:

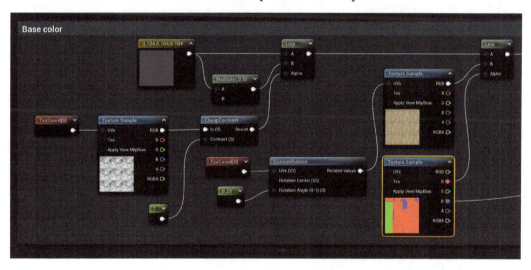

Figure 2.16 – The modified color for the toy tank tires

We've managed to introduce a little bit of color variation thanks to the use of the previous smoke texture as a mask. This is something that we'll come back to in future recipes, as it's a very useful technique when trying to create non-repetitive materials.

> **Tip**
>
> Clicking on the teapot icon in the Material Editor viewport will let you use whichever mesh you have selected on the Content Browser as the visible asset. This is useful for previewing changes without having to move back and forth between different viewports.

Finally, let's introduce some extra changes to the metallic parts of the model.

20. To do that, head over to the **Roughness** section of the material graph.

21. Create a **Texture Sample** parameter and assign the **T_MacroVariation** texture to it. This will serve as the **Alpha** pin for the new **Lerp** node we are about to create.

22. Add two simple **Constant** nodes and give them two different values. Keep in mind that these will affect the roughness of the metallic parts when choosing the values, so values lower than 0.3 will work well when making those parts reflective.

23. Place a new **Lerp** node and plug the previous two new **Constant** nodes into the **A** and **B** input pins.

24. Connect the red channel of the **Texture Sample** node we created in *step 21* to the **Alpha** input pin of the previous **Lerp** node.

> **Important note**
>
> The reason why we are connecting the red channel of the previous **Texture Sample** node to the **Alpha** input pin of the **Lerp** node is simply that it offers a grayscale value which we can use in our favor to drive the mixture of the different roughness values.

25. Finally, replace the initial Constant that was driving the roughness value of the metallic parts with the node network we created in the three previous steps. The graph should now look something like this:

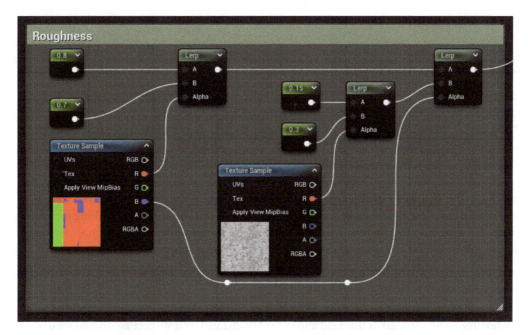

Figure 2.17 – The updated roughness section of our material

And after doing that, let's now... oh, wait; I think we can call it a day! After all of those changes have been made, we will be left with a nice new material that is a more realistic version of the shader we had previously created. Furthermore, everything we've done constitutes the basics of setting up a real material in Unreal Engine 5. Combining textures with math operations, blending nodes according to different masks, and taking advantage of different assets to create specific effects are everyday tasks that many artists working with the engine have to face. And now, you know how to do that as well! Let's check out the results before moving on:

Figure 2.18 – The final look of the textured toy tank material

How it works...

We've had the opportunity to look at several new nodes in this recipe, so let's take a moment to review what they do before moving forward.

One of the first ones we used was the **Texture Coordinate** one, which affects the scaling of the textures that we apply to our models. The **Details** panel gives us access to a couple of parameters within it, **U Tiling** and **V Tiling**. Modifying those settings will affect the size of the textures that are affected by this node, with higher values making the images appear smaller. In reality, the size of the textures doesn't really change, as their resolution stays the same: what changes is the amount of UV space that they occupy. A value of **1** for both the **U Tiling** and the **V Tiling** parameters means that any textures connected to the **Texture Coordinate** node occupy the entirety of the UV space, while a value of **2** means that the same texture is repeated twice over the same extent. Decoupling the **U Tiling** parameter from the **V Tiling** parameter lets us affect the repetition pattern independently on those two axes, giving us a bit more freedom when adjusting the look of our materials.

> **Important note**
>
> There's a third parameter that we can tweak within the **Texture Coordinate** node— **Coordinate Index**. The value specified in this field affects the UV channel we are affecting, should we use more than one.

On that note, it's probably important to highlight the importance of well-laid-out UVs. Seeing as how many of the nodes that we've used directly affect the UV space, it only makes sense that the models with which we work contain correctly unwrapped UV maps. Errors here will most likely impact the look of any textures that you try to apply, so make sure to check for mistakes in this area.

Another node that we used to modify the textures used in this recipe was the **Custom Rotator** node. This one adjusts the rotation of the images you are using, so it's not very difficult to conceptualize. The trickiest setting to wrap our heads around is probably the Constant that drives the **Rotation Angle (0-1) (S)** parameter, which dictates the degrees by which we rotate a given texture. We need to feed a value within the 0 to 1 range, 0 being no rotation and 1 mapping to 360 degrees. A value of 0.25, the one we used in this recipe, corresponds exactly to 90 degrees following a simple rule of three.

Something that we can also affect using the **Custom Rotator** node are the UVs of the model, just as we did with the **Texture Coordinate** node before (in fact, we plugged the latter into the **UVs (V2)** input pin of the former). This allows us to increase or decrease the repetition of the texture before we rotate it. Finally, the **Rotation Centre (V2)** parameter allows us to determine the center of the rotation. This works according to the UV space, so a value of (0,0) will see the texture rotating around its bottom-left corner, a value of (0.5,0.5) will rotate the image around its center, and a value of (1,1) will do the same around its upper-right corner.

The third node we will discuss is the **Multiply** one. It is rather simple in terms of what it does: it multiplies two different values. The two elements need to be of a similar nature —we can only multiply two **Constant3Vector** nodes, two **Constant2Vector** nodes, two **Texture Sample** nodes, and so on. The exception to this rule is the simple **Constant** nodes, which can always be combined with any other input. The **Multiply** node is often used when adjusting the brightness of a given value, which is what we've done in this recipe.

Finally, the last new node we saw was **Cheap Contrast**. This is useful for adjusting the contrast of what we hook into it, with the default value of 0 not adjusting the input at all. We can use a constant value to adjust the intensity of the operation, with higher values increasing the difference between the black and the white levels. This particular node works with grayscale values—should you wish to perform the same operation in a colored image, note that there's another **CheapContrast_RGB** node that does just that.

> **Tip**
>
> A useful technique that lets you evaluate the impact of the nodes you place within the material graph is the **Start Previewing Node** option. Right-click on the node that you want to preview and choose that option to start displaying the result of the node network you've chosen up to that point. The results will become visible in the viewport.

See also

I want to leave you with a couple of links that provide more information regarding **Multiply**, **Cheap Contrast**, and some other related nodes:

- `https://docs.unrealengine.com/4.27/en-US/RenderingAndGraphics/Materials/Functions/Reference/ImageAdjustment/`

- `https://docs.unrealengine.com/4.27/en-US/RenderingAndGraphics/Materials/ExpressionReference/Math/`

Adding Fresnel and Detail Texturing nodes

We started to use textures extensively in the previous recipe, and we also had the opportunity to talk about certain useful nodes, such as **Cheap Contrast** and **Custom Rotator**. Just as with those two, Unreal includes several more that have been created to suit the needs of 3D artists, with the goal of sometimes improving the look of our models or creating specific effects in a smart way. Whatever the case, learning about them is sure to improve the look of our scenes.

We'll be looking at some of those useful nodes in this recipe, paying special attention to the **Fresnel** one, which we'll use to achieve a velvety effect across the 3D model of a tablecloth—something that would be more difficult to achieve without it. Let's jump straight into it!

Getting ready

The scene we'll be using in this recipe is a bit different from the ones we've used before. Rather than a toy tank, I thought it would be good to change the setting and have something on which we can demonstrate the new techniques we are about to see, so we'll use a simple cloth to highlight these effects. Please open 02_04_Start if you want to follow along with me.

If you want to use your own assets, know that we'll be creating a velvet-like material in this recipe, so it would be great to have an object onto which you can apply that.

How to do it...

We are about to take a simple material and enhance it using some of the material nodes available in Unreal, which will help us—with very little effort on our side—to improve its final appearance.

1. First of all, let's start by loading the 02_04_Start level and opening the material that is being applied to the tablecloth, named **M_TableCloth_Start**, as we are going to be working on that one. As you can see, it only has a single **Constant3Vector** node modifying the **Base Color** property of the material, and that's where we are going to start working.

2. Next, create a **Texture Sample** node and assign the **T_TableCloth_D** texture to it.

3. Add a **Texture Coordinate** node and connect it to the input pin of the previous **Texture Sample** node. Set a value of **20** for both its **U Tiling** and **V Tiling** properties.

4. Proceed to include a **Constant3Vector** node and assign a soft red color to it, something similar but lighter to the color of the texture chosen for the existing **Texture Sample** node. I've gone with the following RGB values: **R = 0.90, G = 0.44**, and **B = 0.44**.

5. Moving on, create a **Lerp** node and place it after both the original **Texture Sample** node and the new **Constant3Vector** node.

6. Connect the previous **Texture Sample** node to the **A** input pin of the new **Lerp** node, and wire the output of the **Constant3Vector** node to the **B** input pin.

7. Right-click anywhere within the material graph and type Fresnel to create that node. Then, connect the output of the **Fresnel** node to the **Alpha** input pin of the previous **Lerp** node.

   This new node tries to mimic the way light reflects out of the objects it interacts with, which varies according to the viewing angle at which we see those surfaces—with areas parallel to our viewing angle having clearer reflections.

Having said that, let's look at the next screenshot, which shows the material graph at this point:

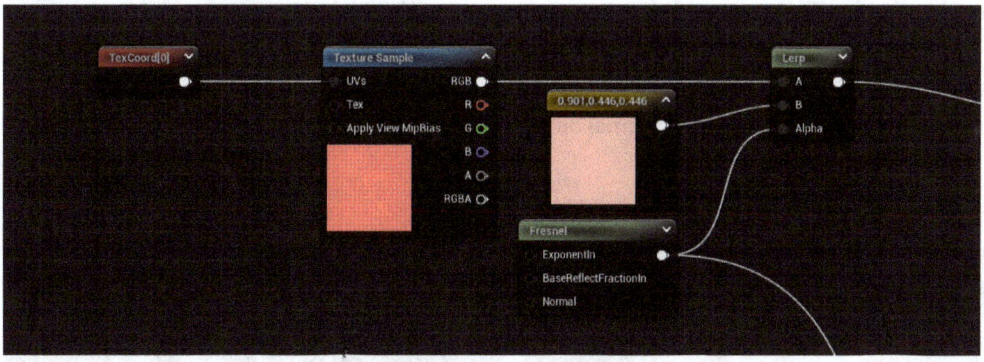

Figure 2.19 – A look at the current state of the material graph

8. Select the **Fresnel** node and head over to the **Details** panel. Set the **Exponent** parameter to something lower than the default, which will make the effect of the **Fresnel** function more apparent. I've gone with 2.5, which seems to work fine for our purposes.

    Next up, we are going to start using another new different node: the **Detail Texturing** one. This node allows us to use two different textures to enhance the look of a material. Its usefulness resides in the ability to create highly detailed models without the need for large texture assets. Of course, there might be other cases where this is also useful beyond the example I've just mentioned. Let's see how to set it up.

9. Right-click within the material graph and type `Detail Texturing`. You'll see this new node appear.

10. In order to work with this node, we are going to need some extra ones as well, so create three **Constant** nodes and two **Texture Sample** nodes.

11. Proceed to connect the **Constant** nodes to the **Scale**, **Diffuse Intensity**, and **Normal Intensity** input pins. We'll worry about the values later.

12. Set **T_ground_Moss_D** and **T_Ground_Moss_N** as the assets in the **Texture Sample** nodes we created (they are both part of the **Starter Content** asset pack).

Important note

Using the **Detail Texturing** node presents a little quirk, in that the previous two **Texture Sample** nodes that we created will need to be converted into what is known as a **Texture Object** node More about this node type in the *How it works...* section.

---

13. Right-click on the two **Texture Sample** nodes and select **Convert to Texture Object**.

14. Give the first **Constant** node—the one connected to the **Scale** pin—a value of **20**, and assign a value of **1** to the other **Constant** nodes.

15. Connect the **Diffuse** output pin of **Detail Texturing** to the **Base Color** input pin in the main material node, and hook the **Normal** output of the **Detail Texturing** node to the **Normal** input pin of our material.

This is what the material graph should look like now:

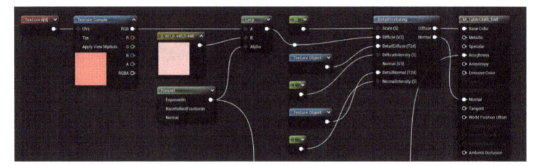

Figure 2.20 – The material graph with the new nodes we've just added

Before we finish, let's continue working a bit longer and reuse the **Fresnel** node once again to drive the **Roughness** parameter of the material. The idea is to use it to specify the roughness values, making the parts that are facing the camera have less clear reflections than the parts that are facing away from it.

16. Create two **Constant** nodes and give them different values—I've gone with **0.65** and **0.35** this time. The first of those values will make the areas where it gets applied appear rougher, which will happen when we look directly at the surface by virtue of using the **Fresnel** node. Following the same principle, the second one will make the areas that face away from us have clearer reflections.

17. Add a **Lerp** node and connect the previous **Constant** nodes to its **A** and **B** pins. The rougher value (**0.65**) should go into the **A** input pin, while the second one should be fed into the **B** pin.

18. Drag another pin from our original **Fresnel** node and hook it to the **Alpha** input pin of the new **Lerp** node.

19. Connect the output of the **Lerp** node to the **Roughness** parameter in the main material node.

The final graph should look something along these lines:

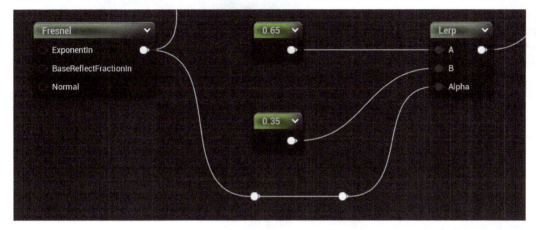

Figure 2.21 – The final addition to the material graph

Finally, all we need to do is to click on the **Apply** and **Save** buttons and assign the material to our model in the main viewport. Take a look at what we've created and try to play around with the material, disconnecting the **Fresnel** node and seeing what its inclusion achieves. It is a subtle effect, but it allows us to achieve slightly more convincing results that can trick the eye into thinking that what it is seeing is real. You can see this in the following screenshot when looking at the cloth: the top of the model appears more washed out because it is reflecting more light, while the part that faces the camera appears redder:

Figure 2.22 – The final look of the cloth material

How it works...

Let's take a minute to understand how the **Fresnel** node works, as it can be tricky to grasp. This effect assigns a different value to each pixel of an object depending on the direction that the normal at that point is facing. Surface normals perpendicular to the camera (those pointing away from it) receive a value of 1, while the ones that look directly at it get a value of 0. This is the result of a dot product operation between the surface normal and the direction of the camera that results in a falloff effect we can adjust with the three input parameters present in that node:

- The first of them, **ExponentIn**, controls the falloff effect, with higher values pushing it toward the areas that are perpendicular to the camera.

- The second, **BaseReflectFrctionIn**, specifies "*the fraction of specular reflection when the surface is viewed from straight on*," as stated by Unreal. This basically specifies how the areas that are facing the camera behave, so using a value of 1 here will nullify the effect of the node.

- The third, **Normal**, lets us provide a normal texture to adjust the apparent geometry of the model the material is affecting.

Detail Texturing is also a handy node to get familiar with, as seen in this example. But even though it's quite easy to use, we could have achieved a similar effect by implementing a larger set of nodes that consider the distance to a given object and swap textures according to that. In fact, that is the logic that you'll find if you double-click on the node itself, which isn't technically a node but a **Material Function**. Doing that will open its material graph, letting you take a sneak peek at the logic driving this function. As we see in there, it contains a bunch of nodes that take care of creating the effect that we end up seeing playing its part in the material we created. Since these are normal nodes, you can copy and paste them into your own material. Doing that would remove the need to use the **Detail Texturing** node itself, but this also highlights why we use Material Functions in the first place: they are a nice way of reusing the same logic across different materials and keeping things organized at the same time.

Finally, I also wanted to touch base on the **Texture Object** node we encountered in this recipe. There can be a bit of confusion around this node, as it is apparently similar to the **Texture Sample** one—after all, we used both types to specify a texture, which we then used in our material. The difference is that **Texture Object** nodes are the format in which we send textures to a Material Function, which we can't do using **Texture Sample** nodes. If you remember, we used **Texture Object** nodes alongside the **Detail Texturing** node, which was a Material Function itself—hence the need to use **Texture Object** nodes. We also sent the information of the **Texture Sample** node used for the color, but the **Detail Texturing** node sampled that as Vector3 information and not as a **Texture** node. You can think of **Texture Object** nodes as references to a given texture, while the **Texture Sample** node represents the values of that given texture (its RGB information as well as the **Alpha** input pin and each of the texture's channels).

See also

You can find more documentation about the **Fresnel** and **Detail Texturing** nodes through Epic's official documents:

- `https://docs.unrealengine.com/4.27/en-US/RenderingAndGraphics/Materials/HowTo/Fresnel/`

- `https://docs.unrealengine.com/4.27/en-US/RenderingAndGraphics/Materials/HowTo/DetailTexturing/`

Creating semi-procedural materials

Thus far, we've only worked with materials that we have applied to relatively small 3D models, where a single texture or color was enough to make them look good. However, that isn't the only type of asset that we'll encounter in a real-life project. Sometimes we will have to deal with bigger meshes, making the texturing process not as straightforward as we've seen so far. In those circumstances, we are forced to think creatively and find ways to make that asset look good, as we won't be able to create high enough resolution textures that can cover the entirety of the object.

Thankfully, Unreal provides us with a very robust Material Editor and several ways of tackling this issue, as we are about to see next through the use of semi-procedural material creation techniques. This approach to the material-creation process relies on standard effects that aim to achieve a procedural look—that is, one that seems to follow mathematical functions in its operation rather than relying on existing textures. This is done with the intention of alleviating the problems inherent to using those assets, such as the appearance of obvious repetition patterns. This is what we'll be doing in this recipe.

Getting ready

We are about to use several different assets to semi-procedurally texture a 3D model, but all the resources that you will need to use come bundled within the **Starter Content** asset pack in Unreal Engine 5. Be sure to include it if you want to follow along using the same textures and models I'll be using, but don't worry if you want to use your own, as everything we are about to do can be done using very simple models.

As always, feel free to open the scene named `02_05_ Start` if you are working on the Unreal project provided alongside this book.

How to do it...

Let's start this recipe by loading the `02_05_ Start` level. Looking at the object at the center of the screen, you'll see a large plane that acts as a ground floor, which has been placed there just so that we can see the problems that arise when using textures to shade a large-scale surface—as evidenced by the next screenshot:

Figure 2.23 – A close-up look at the concrete floor next to a faraway shot showing
the repetition of the same pattern over the entirety of the surface

As you can see, the first image in the screenshot could actually be quite a nice one, as we have a concrete floor that looks convincing. However, the illusion starts to fall apart once the camera gets farther from the ground plane. Even though it might be difficult to notice at first, the repetition of the texture across the surface starts to show up. This is what we are going to try to fix in this recipe: creating materials that look good both up close and far from the camera, all thanks to semi-procedural material creation techniques. Let's dive right in:

1. Open the material being applied to the plane, **M_SemiProceduralConcrete_Start**. You should see a lonely **Texture Sample** node (**T_Concrete_Poured_D**) being driven by a **Texture Coordinate** node, which is currently adjusting the tiling. This will serve as our starting point.

2. Add another **Texture Sample** node and set the **T_Rock_Marble_Polished_D** texture from the **Starter Content** asset pack as its value. We are going to blend between these first two images thanks to the use of a third one, which will serve as a mask.

   Using multiple similar assets can be key to creating semi-procedural content. Blending between two or more textures into a random pattern helps alleviate the visible tiling across big surfaces.

3. Connect the output of the existing **Texture Coordinate** node to the **UVs** input pin of the previous **Texture Sample** node we created.

   The next step will be to create a **Lerp** node, which we can use to blend between the first two images. We want to use a texture that has a certain randomness to it, and the **Starter Content** asset pack includes one named **T_MacroVariation** that could potentially work. However, we'll need to adjust it a little bit, as it currently has a lot of gray values that wouldn't work too well for masking purposes. Using it in its current state would have us blend between two textures simultaneously when we really want to use one or the other based on the previous texture. Let's adjust the texture to achieve that effect.

4. Create a new **Texture Sample** node and assign the **T_MacroVariation** asset to it.

5. Include a **Cheap Contrast** node after the previous **Texture Sample** node and hook its input pin to the output of the said node.

6. Add a **Constant** node and connect it to the **Contrast** slot in the **Cheap Contrast** node. Give it a value of 0.5 so that we achieve the desired effect we mentioned before.

Here is a screenshot of what we should have achieved through this process:

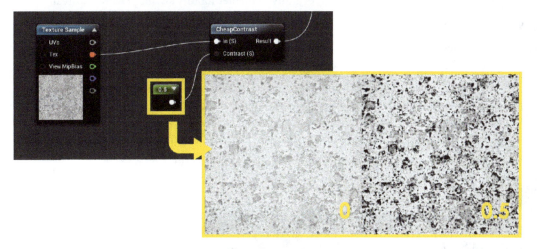

Figure 2.24 – The previous set of nodes and their result

7. Create a **Lerp** node, which we'll use in the next step to combine the first two **Texture Sample** nodes according to the mask we created in the previous step.

8. Connect the output nodes of the first two texture assets (**T_Rock_Marble_Polished_D** and **T_Concrete_Poured_D**) to the **A** and **B** input pins of the **Lerp** node and connect the **Alpha** input pin to the output pin of the **Cheap Contrast** node. The graph should look something like this:

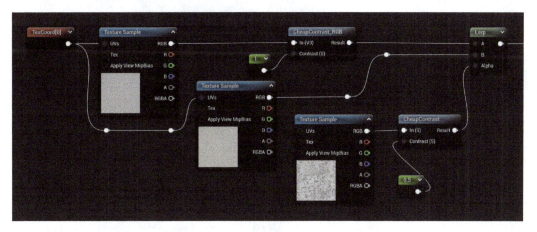

Figure 2.25 – The look of the material graph so far

We've managed to add variety and randomness to our material thanks to the nodes we created in the previous steps. Despite that, there's still room for improvement, as we can take the previous approach one step further. We will now include a third texture, which will help further randomize the shader, as well as give us the opportunity to further modify an image within Unreal.

9. Create a new couple of nodes—a **Texture Coordinate** node and a **Texture Sample** one.

10. Assign the **T_Rock_Sandstone_D** asset as the default value of the **Texture Sample** node, and give the **Texture Coordinate** node a value of 15 in both the **U Tiling** and **V Tiling** fields.

11. Continue by adding a **Desaturation** node and a **Constant** node right after the previous **Texture Sample** node. Give the **Constant** node a value of 0.8 and connect it to the **Fraction** input pin of the **Desaturation** node, whose job is to desaturate or remove the color from any texture we provide it, creating a black-and-white version of the same asset instead.

12. Connect the output of the **Texture Sample** node to the **Desaturation** node.

This is what we should be looking at:

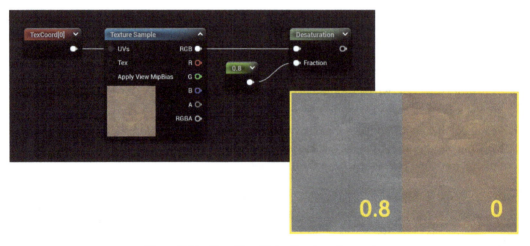

Figure 2.26 – The effect of the Desaturation node

Following these steps has allowed us to create a texture that can work as concrete, but that really wasn't intended as such. Nice! Let's now create the final blend mask.

13. Start by adding a new **Texture Sample** node and assign the **T_MacroVariation** texture as its default value.

14. Continue by including a **Texture Coordinate** node and give it a value of 2 in both its **U Tiling** and **V Tiling** parameters.

15. Next, add a **Custom Rotator** node and connect its **UVs** input pin to the output of the previous **Texture Coordinate** one.

16. Create a **Constant** node and feed it to the **Rotation Angle** input pin of the **Custom Rotator** node. Give it a value of **0.167**, which will rotate the texture by 60 degrees.

Important note

You might be wondering why a value of 0.167 translates to a rotation of 60 degrees. This is for the same reason we saw a couple of recipes ago—the **Custom Rotator** node maps the 0-to-360-degree range to another one that goes from 0 to 1. This makes 0.167 roughly 60 degrees: 60.12 degrees if we want to be precise!

---

17. Hook the **Custom Rotator** node into the **UVs** input pin of the **Texture Sample** node.

18. Create a new **Lerp** node and feed the previous **Texture Sample** node into the **Alpha** input pin.

19. Connect the **A** pin to the output of the original **Lerp** node and connect the **B** pin to the output of the **Desaturation** node.

20. Wire the output of the previous **Lerp** node to the **Base Color** input pin in the main material node. We've now finished creating random variations for this material!

One final look at the plane in the scene with the new material applied to it should give us the following result:

Figure 2.27 – The final look of the material

The final shader looks good both up close and far away from the camera. This is happening thanks to the techniques we've used, which help reduce the repetition across our 3D models quite dramatically. Be sure to remember this technique, as I'm sure it will be quite handy in the future!

How it works...

The technique we used in this recipe is quite straightforward—just introduce as much randomness as you can until the eye gets tricked into thinking that there's no single texture repeating endlessly. Even though this principle can be simple to grasp, it is by no means simple—humans are very good at recognizing patterns, and as the surfaces we work on increase in size, so does the challenge.

Choosing the right textures for the job is part of the solution—be sure to blend gently between different layers, and don't overuse the same masks all the time. Doing so could result in our brains figuring out what's going on. This is a method that has to be used gently.

Focusing now on the new nodes that we used in this recipe, we probably need to highlight the **Desaturation** one. It is very similar to what you might have encountered in other image-editing programs: it takes an image and gradually takes the "color" away from it. I know these are all very technical terms—not!—but that is the basic idea. The Unreal node does the same thing: we start from an RGB image and we use the **Fraction** input to adjust the desaturation of the texture. The higher the value, the more desaturated the result. Simple enough!

Something else that I wanted to mention before we move on is the real procedural noise patterns available in Unreal. In this recipe, we created what can be considered semi-procedural materials through the use of several black-and-white gradient-noise textures. Those are limited in that they too repeat themselves, just like the images we hooked to the **Base Color** property—it was only through smart use of them that we achieved good enough results. Thankfully, we have access to the **Noise** node in Unreal, which gets rid of that limitation altogether. You can see some examples here:

  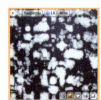 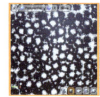 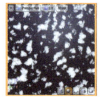

Figure 2.28 – The Noise node and several of its available patterns

This node is exactly what we are looking for to create fully procedural materials. As you can see, it creates random patterns, which we can control through the settings in the **Details** panel, and the resulting maps are similar to the masks we used in this recipe.

The reason why we didn't take advantage of this asset is that it is quite taxing with regard to performance, so it is often used to create nice-looking materials that are then baked. We'll be looking at material-baking techniques in a later chapter, so be sure to check it out!

See also

You can find an extensive article on the **Noise** node at the following link:

```
https://www.unrealengine.com/es-ES/tech-blog/getting-the-most-out-
of-noise-in-ue4
```

It's also a great resource if you want to learn more about procedural content creation, so be sure to give it a go!

Blending textures based on our distance from them

We are now going to learn how to blend between a couple of different textures according to how far we are from the model to which they are being applied. Even though it's great to have a complex material that works both when the camera is close to the 3D model and when we are far from it, such complexity operating on a model that only occupies a small percentage of our screens can be a bit too much. This is especially true if we can achieve the same effect with a much lower-resolution texture.

With that goal in mind, we are going to learn how to tackle those situations in the next few pages. Let's jump right in!

Getting ready

If we look back at the previous recipe, you might remember that the semi-procedural concrete material we created made use of several nodes and textures. This serves us well to prove the point that we want to make in the next few pages: achieving a similar look using lower-resolution images and less complex graphs. We will thus start our journey with a very similar scene to what we have already used, named 02_06_Start.

Right before we start, though, know that I've created a new texture for you, named **T_DistantConcrete_D**, which we'll be using in this recipe. The curious thing about this asset is that it's a baked texture from the original, more complex material we used in the *Creating semi-procedural materials* recipe. We'll learn how to create these simplified assets later in the book when we deal with optimization techniques.

How to do it...

Let's jump right into the thick of it and learn how to blend between textures based on the distance between the camera and the objects to which these images are being applied:

1. Let's start by creating a new material and giving it a name. I've chosen **M_DistanceBasedConcrete**, as it roughly matches what we want to do with it. Apply it to the plane in the center of the level. This is the humble start of our journey!

2. Continue by adding two **Texture Sample** nodes, and choose the **T_Concrete_Poured_D** and the **T_DistantConcrete_D** images as their default values.

3. Next, include a **Texture Coordinate** node and plug it into the first of the two **Texture Sample** nodes. Assign a value of **20** to its two parameters, **U Tiling** and **V Tiling**.

4. Place a **Lerp** node in the graph and connect its **A** and **B** input pins to the previous **Texture Sample** nodes. This is how the graph should look up to this point:

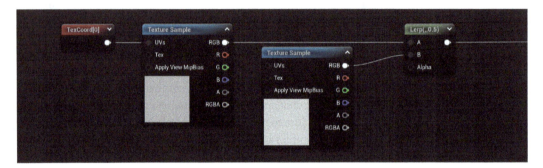

Figure 2.29 – The material graph so far

Everything we've done so far was mixing two textures, one of which we already used in the previous recipe. The idea is to create something very similar to what we previously had, but more efficient. The part that we are now going to tackle is distance-based calculations, as we'll see next.

5. Start by creating a **World Position** node, something that can be done by right-clicking anywhere within the material graph and typing its name. This new node allows us to access the world coordinates of the model to which the material is being applied, enabling us to create interesting effects such as tiling textures without considering the UVs of a model.

> **Important note**
> The **World Position** node can have several prefixes (**Absolute, Camera Relative…**) based on the mode we specify for it in the **Details** panel. We want to be working with its default value—the **Absolute** one.

6. Create a **Camera Position** node. Even though that's the name that appears in the visual representation of that function, the name that will show up when you look for it will be **Camera Position WS** (*WS* meaning *World Space*).

7. Include a **Distance** node after the previous two, which will give us the value of the separation between the camera and the object that has this material applied to it. You can find this node under the **Utility** section. Connect the **Absolute World Position** node to its **A** input pin, and the **Camera Position** one to the **B** input.

8. Next, add a **Divide** node and a **Constant** node. You can find the first node by just typing that name, or alternatively by holding down the *D* key on your keyboard and clicking in an empty space of the material graph. Incidentally, **Divide** is a new node that allows us to perform that mathematical operation on the two inputs that we need to provide it with.

9. Give the **Constant** node a value of something such as 256. This will drive the distance from the camera at which the swapping of the textures will happen. Higher numbers mean that it will be further from the camera, but this is something that has to be interactively tested to narrow down the precise sweet spot at which you want the transition to happen.

10. Add a **Power** node after the previous **Divide** node and connect the result of the latter to the **Base** input pin of the former. Just as with the **Divide** node, this function performs the homonymous mathematical operation according to the values that we provide it with.

11. Create another **Constant** node, which we'll use to feed the **Exp** pin of the **Power** node. The higher the number, the softer the transition between the two textures. Sensible numbers can be anything between 1 to 10 – I've chosen **4** this time.

12. Include a **Clamp** node into the mix at the end, right after the final **Power** node, and connect both. Leave the values at their default, with **0** as the minimum and **1** as the maximum.

13. Connect the output pin of the previous **Clamp** node to the **Lerp** node that we used to blend between the two textures acting as the **Base Color** node, the one we created in *step 4*.

The graph we've created should now look something like this:

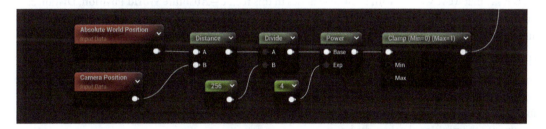

Figure 2.30 – The logic behind the distance-blending technique used in this material

We should now have a material that can effectively blend between two different assets according to the distance at which the model is from the camera. Of course, this approach could be expanded upon and made so that we are not just blending between two textures, but between as many as we want. However, one of the benefits of using this technique is to reduce the rendering cost of materials that were previously too expensive to render—so, leaving things relatively simple will help work in our favor this time.

Let me leave you with a screenshot that highlights the results both when the camera is up close and when it is far away from our level's main ground plane:

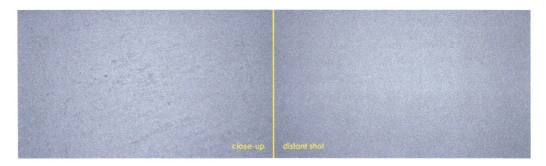

Figure 2.31 – Comparison shots between the close-up and distant versions of the material

As you can see, the differences aren't that big, but we are using fewer textures in the new material compared to the previous one. That's a win right there!

> **Tip**
>
> Try to play around with the textures used for the base color and the distance values used in the **Constant** node created in *steps 8* and *9* to see how the effect changes. Using bright colors will help highlight what we are doing, so give that a go in case you have trouble visualizing the effect!

How it works...

We can say that we've worked with a few new nodes in this recipe—**Absolute World Position**, **Distance**, **Camera Position**, **Divide**, **Power**, **Clamp**… We did go over all of them a few pages ago, but it's now time to delve a little bit deeper into how they operate.

The first one we saw, the **Absolute World Position** node, gives us the location of the vertices we are looking at in world space. Pairing it with the **Camera Position** and **Distance** nodes we used immediately after allowed us to know the distance between our eyes and the object onto which we applied the material—the former giving us the value for the camera location and the latter being used to compare the positional values we have.

We divided the result of the **Distance** node by a Constant using the **Divide** node, a mathematical function similar to the **Multiply** node that we've already seen. We did this in order to adjust the result we got out of the **Distance** node—to have a value that was easier to work with.

We then used the **Power** node to drive the mathematical operation, which allowed us to create a softer transition between the two textures that we were trying to blend. In essence, the **Power** node simply executes said operation with the data that we provide it. Finally, the **Clamp** node took the value we fed to it and made sure that it was contained within the 0 to 1 range that we specified for it—so, any higher values than the unit would have been reduced to one, while lower values than zero would have been increased to zero.

All the previous operations are mathematical in nature, and even though there are not that many nodes involved, it would be nice if there were a way for us to visualize the values of those operations. That would be helpful with regard to knowing the values that we should use—you can imagine that I didn't just get the values that we used in this recipe out of thin air, but rather after careful consideration.

There are some extra nodes that can be helpful in that regard, all grouped within the **Debug** category. Elements such as **DebugScalarValues**, **DebugFloat2Values**, and others will help you visualize the values being computed in different parts of the graph. To use them, simply connect them to the part of the graph that you want to analyze and wire their output to the **Base Color** input pin node of the material. Here is an example of using the **DebugScalarValues** node at the end of the distance blend section of the material we've worked on in this recipe:

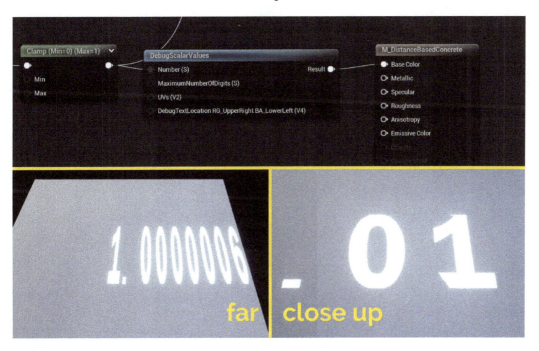

Figure 2.32 – The effects of using debug nodes to visualize parts of the graph

See also

You can find extra information about some of the nodes we've used, such as **Absolute World Position**, in Epic Games' official documentation:

`https://docs.unrealengine.com/en-US/Engine/Rendering/Materials/ExpressionReference/Coordinates`

There are some extra nodes in there that might catch your eye, so make sure to check them out!

3

Making Translucent Objects

The previous chapter saw us dealing with multiple different materials, and even though they were all unique, they also shared a couple of important common characteristics: making use of **Opaque Blend Mode** and **Default Lit Shading Model**. This limits us a bit in terms of the effects that we can create, as not all of the objects that we see around us can be described using those settings.

It is now time for us to explore other materials that differ from the previous ones in at least one of those two attributes – so, get ready to talk about things such as translucency, refraction, and subsurface scattering. Knowing about those topics will help us create important materials such as glass or water, and this is the place where we will start looking at those types of effects.

To summarize, these are the recipes that we'll be covering next:

- Creating a translucent glass
- Displaying holograms
- Working with subsurface scattering
- Using refraction in a body of water
- Faking caustics with a Light Function
- Animating a sea shader

You can see some of the things we'll be creating in the following figure:

Figure 3.1 – A look at some of the materials we'll be tackling in this chapter

Technical requirements

As mentioned previously, this chapter will see us working with multiple material effects that are a bit more complex than the ones we've seen so far. In light of that, we'll start to use some custom models, textures, and masks that will make our lives easier when it comes to demonstrating the different techniques we are about to see. Not wanting to leave you behind, we've made them available through the following link in case you want to download them: `https://packt.link/20u7B`.

You'll be able to find the complete project for this book there, granting you access to the same assets we'll be using in the next few pages. All of those resources have been provided with the hopes that they can be used in case you are not able to author them, as we need access to specific software packages to create them (3D modeling tools, image editing software, and more). Having said that, feel free to use your own creations if you already know how to create those, as it might make things even more interesting for you.

Beyond that, everything else we need has already been covered – a copy of the Unreal Engine 5 editor and a lot of passion. Already have those at hand? Then let's jump into this chapter!

Creating a translucent glass

As indicated by the title of this chapter, we are going to dedicate the next few pages to studying, creating, and understanding some of the different types of translucent objects that can be created within Unreal. We are going to start with what is probably one of the most common translucent objects that I can think of: glasses. They are easy to create, and we will be able to explore different Shading Models and Blend Modes through them. Let's jump right in and see what they are all about.

Getting ready

The look of the glass material we'll be creating in this recipe can be achieved without the need for almost any additional assets – everything required is available as part of the Material Editor's standard nodes. Having said this, the Unreal project we've been using so far includes the 3D model of a glass onto which we can apply said material, so feel free to use that resource if you want to. I've also included a custom gradient texture that will make the result look a bit better, which is meant to work alongside the aforementioned model.

All of this is included with this book's Unreal Engine project, and you can open the `03_01_Start` level (from the download link in the *Technical requirements* section) if you want to follow along and work on the same map. If you're working on your own, try to have a cleanly unwrapped 3D model where it makes sense to apply a glass material and be prepared to create a custom texture if you want to improve its final look in a similar way to what you'll see by the end of this recipe.

How to do it...

As we are going to be using a different type of material compared to the ones we have created thus far in this book, we'll need to focus on two different important places. The first one is where most of the logic is coded, the so-called **Main Material graph**, a place with which we are already familiar. The second one is the **Details** panel for the **Main Material** node itself, where we can find the options that define how a shader behaves concerning the light that hits it (something known as the **Shading Model** – more on this in the *How it works...* section).

Let's start by creating the material and looking at the **Details** panel:

1. Create a new material, meant to be used on the glass model located at the center of the screen, and apply it to that element. This is what we'll be working on over the next few pages. I've named this material **M_Glass**, but feel free to give it your own name!

2. Open the Material Editor and select the **Main Material** node. Focus on the **Details** panel, as we are going to modify several settings there.

3. The first parameter we want to change is **Blend Mode**. Select the **Translucent** option instead of the default **Opaque** option.

4. Scroll down to the **Translucency** category. There are two parameters that we want to change here: tick the checkbox next to **Screen Space Reflection**, and change **Lighting Mode** to **Surface Translucency Volume**.

5. Expand the **Advanced** section within the **Translucency** category and change the value of **Translucency Pass** so that it reads **Before DOF**.

6. Search within the **Usage** category for the checkbox next to **Used with Static Lighting**. We want to ensure we check this one if we use that type of mobility in our lights.

    Implementing all of the previous steps has effectively changed the way the surface we apply the material to reacts to the light that hits it, and how it deals with other things such as reflections, transparency, or depth-of-field calculations. Be sure to check out the *How it works...* section to learn more about those topics, as they are quite important to master when it comes to translucent materials.

    Next up is creating the material's logic. We'll start gently, playing with parameters that we've already used, only to introduce new concepts, such as view-dependent opacity and refraction later on.

7. Create a **Vector** parameter, which we'll use to feed the **Base Color** node of the material. This will act as the main color of the glass, so choose something with a blueish tint (and remember to also give it a name).

> **Important note**
>
> Remember vector parameters? We've used them in the past to create Material Instances, and that is what we are going to be doing in this recipe as well. This will enable us to quickly change the material without needing to recompile it, something vital when it comes to translucent objects, as they are more expensive to render.

8. Throw a **Scalar** parameter into the mix and give it a value of **0.9**. Call it **Metalness**.

9. Plug the previous node into the **Metallic** input pint of the material. Even though assigning a metallic value to a glass might seem like a questionable thing to do, it makes the final image a bit more vibrant and closer to the real object.

    So far, so good! The next bit that we are going to be doing is crucial to making a translucent object. Since we are dealing with glass, we need to tackle reflections, the refraction of light, and the opacity of the material itself. All of these are going to be view-dependent – reflections on the glass are most obvious when our view is parallel to the surface of the object. That being the case, we'll need to call our old friend the Fresnel function to help us with this task.

10. Right-click within the material graph and start typing `Fresnel`. We've used this node before, but there is a variant that we now want to choose – if only to spice things up a little bit. Its name is **Fresnel_Function**, which will appear once we finish typing. It is a bit more complex to set up than the one we already know, but it also gives you more control over the different parameters it contains in case you want to play around with them. So, select **Fresnel_Function**.

11. Add a **Scalar Parameter** right before **Fresnel_Function** and name it **Power**, since that's the input pin we'll be connecting it to on the newly created function. Give it a value of **3**, as seen in the following figure:

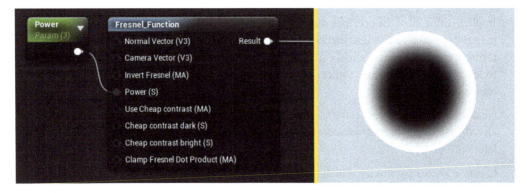

Figure 3.2 – The new Fresnel Function and its effect

The previous two nodes will allow us to drive the appearance of the material in terms of those three parameters we mentioned: the **Reflection**, **Refraction**, and **Opacity** parameters. We'll be using the previous Fresnel function to drive the linear interpolation of two values in those categories, so let's create six scalar parameters for that purpose (two per category).

12. Add two scalar parameters to control the reflections and give them appropriate names, such as **Frontal Roughness** and **Side Roughness**.

13. After that, create a **Lerp node** and plug each of the previous parameters into the A and the B input pins (A for the front reflection, B for the side one).

14. Repeat *steps 12* and *13*, creating two other scalar parameters and a **Lerp** node, but give them names such as **Frontal Opacity** and **Side Opacity**.

15. Repeat the same sequence for a third time, naming the new parameters **Frontal Refraction** and **Side Refraction**, creating a **Lerp** node after them once again.

In total, we should have three sets of two scalar parameters, each with a corresponding Lerp node after them, as shown in the following screenshot:

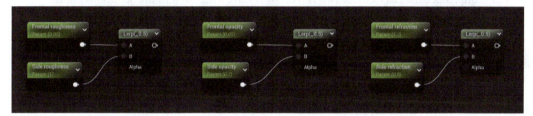

Figure 3.3 – The different scalar parameters and their values

Apart from creating those parameters, remember to assign them some values! We didn't do that before as we'll be able to interactively change them once we create a Material Instance, but we might as well set some default ones now:

- **0.05** for **Frontal Roughness**, 1 for **Side Roughness**
- **0.05** for **Frontal Opacity**, 0.2 for **Side Opacity**
- **1.2** for **Frontal Refraction**, 0.8 for **Side Refraction**

16. Connect the **Result** output node of **Fresnel_Function** to the **Alpha** pins of the last three **Lerp** nodes we created.

17. Plug the output pin of said **Lerp** nodes into the **Roughness**, **Opacity**, and **Refraction** input pins of the **Main Material** node.

This is what the material should now look like if applied to the glass in the scene:

Figure 3.4 – The current look of the glass material

Looking sharp, right? But we can still make it look even better! Something that I like to adjust is the opacity of the upper edges of the glass, the part of the real-life object where we place our lips whenever we drink from it. That area is usually more opaque than the rest of the body. This is an optic effect, one that is difficult to mimic inside the engine. However, we can still fake it, which we'll do next:

18. Create a **Texture Sample** node close to the scalar parameters controlling the opacity, and assign the texture named **T_Glass_OpacityMask** to it.

    T_Glass_OpacityMask is a custom texture that we've created for the model that we are using in this recipe, which will mask the rim of the glass. But what if you are using your own models? You'll have to create a mask of your own, which can be easily done in a program such as Photoshop or Gimp. Since the last of those two is free, we'll leave a link to it in the *See also* section.

19. Create another **Scalar Parameter**, set its value to **1**, and name it **Rim Opacity**.

20. Add a **Lerp** node after the two new ones we've just created and plug **Rim Opacity** into its **B** input pin. Connect the **Opacity** mask to the **Alpha** pin, and the output of the original **Opacity Lerp** node to pin **A**.

The previous sequence should look something like this:

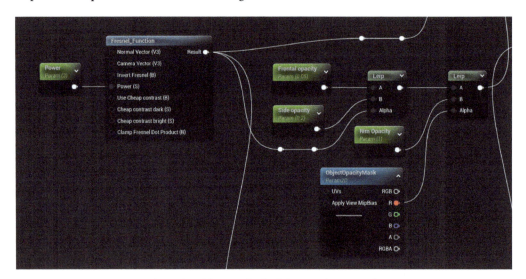

Figure 3.5 – The new look of the opacity controls

Once we implement those steps, we'll end up with a glass that has a much opaquer edge, which should look nicer than before. These types of adjustments are often made to materials such as glass, where we can't just rely on the engine's implementation of translucency and we need an artist to tweak the scene a little bit. Let's take a final look at the scene before calling it a day!

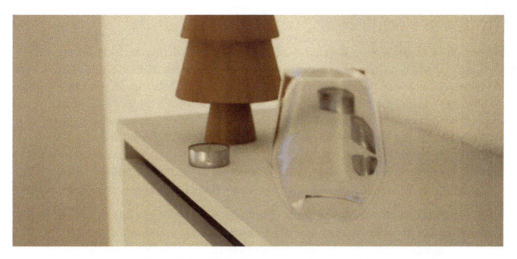

Figure 3.6 – The final look of the glass material

How it works...

Glass! It can be a complicated thing to get right in real-time renderers, and even though we've come a long way in recent years, it still has its hurdles. Because of that, we'll need to talk about these issues to fully understand what we've just created.

Different types of translucency

Most of the issues that we might experience when working with translucent materials often boil down to how this effect is tackled by the engine. Whenever we have an object that makes use of this type of material, such as the glass in this recipe, Unreal needs to know how it affects the objects that are behind it. Not only that, but it also needs to know whether the image has to be distorted, how much light can pass through its surface, or the type of shadows that the object is casting. Some of these problems are solved by adopting smart rendering methods that can provide affordable solutions to hard computational problems, which can sometimes introduce little errors for the sake of performance.

One such example of this balance between realism and performance is the implementation of refraction within the engine. Unreal has introduced two different solutions to the problem: one that's physically based, which uses the index of refraction to calculate the effect (which is the one we've used in this recipe), and another more artistic one, named **Pixel Normal Offset**. Even though the first one relies on real-world measurable values, the second one is sometimes better to achieve the desired effect. This is because the refraction method takes the scene color into account to calculate the distortion, which can cause undesired artifacts, especially in big translucent objects. We'll take a look at this other method for calculating refraction in an upcoming recipe in this very chapter, *Using refraction in a body of water*.

The Translucent Blend Mode

With regards to how these types of materials work, we need to go back to some of the settings that we used in this recipe and explain them a bit further. Translucent materials make use of the **Translucent Blend Mode**, one of the first parameters we tweaked at the start of this recipe. Blend Modes tell the engine how "*the output of the current Material will blend over what is already being drawn in the background*," according to Unreal's official docs. The one we've seen the most so far is the **Opaque** one, which was the **Blend Mode** that was used by all of the materials that we created in the previous chapter. This is probably the easiest one to render as it blocks everything behind the models where the material is applied. Assets that have a translucent material applied to them need to combine their color with that of the other models located behind them, making them more expensive to render.

Translucent materials present other peculiarities if we compare them with the opaque ones we already know. Something that might have caught your attention was the need to specify the use of screen space reflections through a custom checkbox that we needed to tick early in this recipe. The reason why we needed to do so was simply that this feature is not supported by default by this type of material, something that isn't the case in opaque ones. Furthermore, we also specified the moment when we wanted the effect to render when we changed the value of the **Translucency Pass** menu from the default **After DOF** to our chosen **Before DOF**. This consideration is an important one, especially if we are creating cinematic shots that make use of that photographic technique to blur out elements in the background. Switching from one option to the other will affect the moment when **Translucency Pass** gets rendered, and sticking to the default option here will cause problems with this effect, making it unreliable with these types of objects.

To finish up, I want to focus on a couple of extra topics that we've covered in this recipe before moving on.

Benefits of Material Instances in translucent materials

The first thing I'd like to mention is the already-familiar Material Instance asset – a tool that can save us quite a bit of time when working with translucent materials. You might have noticed that the material took a while to update itself every time we changed any of its settings. Creating an instance out of the material once we set it up can greatly speed up this process – this way, we can modify every exposed parameter on the go without having to wait for our shader to recompile.

Values used for the refraction

The second item we should talk about is the **Refraction** parameters we have chosen as our default values. We chose **1.2** and **0.8** as the **Frontal** and the **Side Refraction** values, respectively, and even though the real-life **index of refraction** (**IOR**) for glass is closer to 1.52, those looked better in the viewport. Something that also works well and that you can try is to combine the following nodes: two constants, one with a value of 1 and another one with the real IOR value, hooked into a **Lerp** node being driven by a Fresnel one. This is usually closer to the real-life appearance than using just the real

IOR value. All in all, these are all quirks of the refraction implementation: be sure to play around with different values until you achieve your desired look.

See also

There are lots of useful and interesting official docs that cover this topic, so be sure to read them if you are interested:

- **Transparency**: `https://docs.unrealengine.com/5.0/en-US/using-transparency-in-unreal-engine-materials/`

- **Index of refraction**: `https://docs.unrealengine.com/5.0/en-US/using-refraction-in-unreal-engine/`

- **Getting Gimp**: `https://www.gimp.org/`

Displaying holograms

Now that we are a bit more familiar with translucent materials, why don't we spice things up a little and see what other cool effects we can create with them? One that comes to mind is **holograms** – the ever-so-sci-fi, cool-looking effect that we can see in futuristic movies. This can be quite an interesting technique to implement, as not only are we going to have to deal with transparent surfaces, but also with animated textures and light-emitting properties.

All in all, we'll also use this opportunity to introduce a few new features within the Material Editor, which will come in handy not just in this recipe, but in any animated material we want to create in the future. Let's see what they are all about!

Getting ready

You'll probably remember from previous recipes that we always provide you with the same assets you see on the pages of this book. This is, of course, so you can easily follow along. This time won't be any different, but you'll probably start to see that, sometimes, we won't even need any type of models or textures to tackle a specific topic. As we get more and more proficient with the engine, we'll sometimes start to make use of procedural and mathematical nodes in the material creation process. This allows us to expose some very powerful material creation techniques, as well as free ourselves from the need to use resolution-dependent textures.

Having said this, this recipe makes use of a custom model that will act as the shape of the hologram, as well as a custom texture based on that object. The texture was created by taking the UVs of said 3D asset into account, so keep that in mind if you are planning to use your own assets. Finally, feel free to open the `03_02_Start` level if you want to follow along using the same resources we'll be providing.

How to do it...

Let's start by looking at the scene with which we are going to be working:

Figure 3.7 – The initial look of the scene

The tree at the center of the screen is the 3D model on which we are going to be working – it's similar to one we've seen in the past, specifically in the *Adding Fresnel and Detail Texturing nodes* recipe in the previous chapter, but tweaked just enough so that it could make it into a sci-fi movie. You'll be able to see that it has two material slots if you select it: the first one with a material called **M_ChristmasTree_Base** and the second one named **M_ChristmasTree_Hologram**. This is the one we'll modify, so make sure you open it! Let's get started:

1. Open the second material being applied to the 3D model of the tree (named **M_ChristmasTree_Hologram**), or create a new one and apply it to the second material slot of said static mesh.

2. Select the main material and head over to the **Details** panel. Look for the **Blend Mode** and **Shading Model** sections, and change their default values from **Opaque** to **Translucent** and from **Default Lit** to **Unlit**, respectively. This will make any models onto which the material is applied unaffected by lighting. We'll learn more about this in the *How it works...* section.

> **Important note**
>
> The reason why we are applying two unique materials to the same model is that they each make use of a different **Shading Model** and **Blend Mode**. Unlike other attributes, such as the metalness or the roughness we've seen before, these properties can not be masked apart in a single material.

The previous step ensures that the material won't be affected by any lighting in the scene, which is quite common for light-emitting materials such as this hologram. Changing **Blend Mode** to **Translucent** has also enabled the see-through quality that most sci-fi holograms have. All in all, we are now ready to start focusing on the look of our material.

3. Start by creating a **Texture Sample**, a **Constant**, and a **Cheap Contrast** node.

4. Select the **T_ChristmasTree_D** asset in the **Details** panel of **Texture Sample**. We will use this asset as a reminder of what the prop once looked like with a different shader.

5. Plug the **Red** channel of **Texture Sample** into the **In (S)** input pin of the **Cheap Contrast** node. This means that we are using a black-and-white version of the image, as we don't want the color one.

6. Assign a value of **0.2** to the **Constant** node and plug it into the **Contrast (S)** pin of the **Cheap Contrast** node. This will increase the contrast between the light and dark areas of the picture.

7. Add a **Multiply** node and a **Constant** node, and assign a value of **2** to that last node.

8. Connect the output pin of the **Cheap Contrast** node to input pin **A** of the previous **Multiply** node, and the output of the previous **Constant** node to input pin **B**.

9. Right-click on the **Multiply** node and select the **Start Previewing Node** option. This will let you see what the graph will look like up until that point, as depicted in the following screenshot:

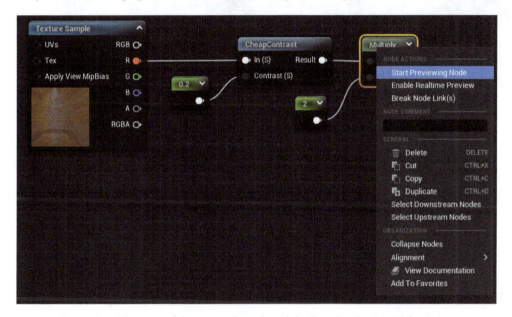

Figure 3.8 – The state of the material graph and the Start Previewing Node option

The previous set of nodes allowed us to have a black-and-white version of the wooden Christmas tree that we can use later in the graph. This is mostly thanks to the use of the **Cheap Contrast** node, while the **Multiply** and **Constant** nodes placed at the end crank up the intensity of the white areas even further.

With that out of the way, let's focus on adjusting the color of the effect by introducing the archetypical blue holographic tint.

10. Create a new **Multiply** node and place it after the previous sequence.

11. Add a new **Constant 3** vector and give it a blueish value. I've gone with the following values: **0.52**, **0.55**, and **0.74**.

12. Plug the result of the **Constant 3** vector into pin **B** of the new **Multiply** node, and hook pin **A** to the output of the **Multiply** node we created in *step 7*.

13. Let's add a third **Multiply** node to control the overall brightness of what we've created. I've connected the **A** pin to the output of the second **Multiply** node and given the **B** pin a value of **0.5**, without creating any **Constants** this time.

Let's look at all of the nodes we have created so far:

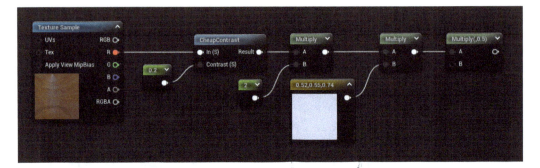

Figure 3.9 – The first part of the material graph

Everything we've done so far has left us with a washed-out, blueish wood-ghost material, which we'll use at a later stage. If this were a cookbook dedicated to food, we would put what we've made aside to focus on other things. For organizational purposes, let's call what we've created *Part A*. The next part we are going to be looking at is the creation of a wireframe overlay, which will give us a very sci-fi look – we'll call that *Part B*, as we'll need to merge both upon completion.

14. Create a **Texture Sample** node and assign the **T_ChristmasTree_UVs** asset to it.

15. Add a **Constant 3** node and give it a blueish color, brighter than the one we created in *step 11*. RGB values of **0.045**, **0.16**, and **0.79** would be a good starting point.

16. Multiply both assets using a **Multiply** node.

17. Create another **Multiply** node after the previous one and connect its **A** input pin to the output of the first **Multiply** node. The **B** value will decide the brightness of the wireframe, so feel free to give it a high value. I've gone with **50**; you can see the result here:

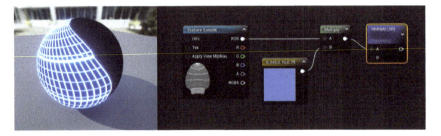

Figure 3.10 – The previous set of nodes and their material preview

The previous set of nodes is what we will call *Part B*, and we are using them to show the wireframe of the model. Now that we have it, let's learn how to blend both sections.

18. Create a **Lerp** node and place it after both previous parts, after all of the existing nodes.

19. Connect the result of *Part A* (remember, the output of the last **Multiply** node of the wood ghost effect, which we created in *step 13*) to the A pin of the new **Lerp** node. Do the same with pin **B** and the last **Multiply** node of *Part B* (the other **Multiply** node at the end of the part controlling the wireframe effect, which we created in *step 17*).

20. Add a new **Texture Sample** node and select the **T_ChristmasTree_UVs** asset as its value. Connect its **Alpha** output pin to the **Alpha** input of the **Lerp** node. You can also reuse the texture sample we created in *step 14* instead of creating a new one if you prefer to do that.

Tip

You can double-click on a given wire to create a reroute node, which you can then move around to better organize the material graph and all the wires that exist within it. This is especially useful as the complexity of your graph grows, so be sure to keep that in mind in case you ever need it.

Doing this will see us almost finished with most of the nodes that we are going to be connecting to the **Emissive Color** input pin of the **Main Material** node. However, we are not finished with this section yet, as there's a small addition we can implement. Some holographic implementations have a Fresnel quality to them – that is, they are brighter around the edges of the model than they are at the center of it. Let's implement that.

21. Create a **Multiply** node and assign a high value to the **B** constant, something like **10**. Place it immediately after the last **Lerp** node we created in *step 18* and connect the output of said node to the **A** input pin of this new **Multiply** one.

22. Introduce a **Lerp** node after this last **Multiply** one and connect its **A** input pin to the output of the previous **Lerp** node. Connect its **B** input pin to the previous **Multiply** node.

23. Add a **Fresnel** node to work as the **Alpha** pin for the last **Lerp** node. If you click on it and look at the **Details** panel, you'll be able to modify its default values. I've chosen **8** as its **Exponent** and **0.01** for its **Base Reflect Fraction** value.

We've just achieved what we set out to do just before the previous numbered points: we made the areas that face away from the camera brighter than the ones that face it directly. We have the Fresnel node to thank for that, as it's the one driving the interpolation between the base and the brighter input we fed the **Lerp** node. Before moving on, let's take a second to review all of the nodes that we've placed so far to ensure that we are on the same page:

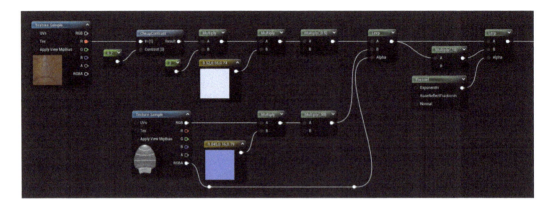

Figure 3.11 – A look at the state of the material graph

With those last changes in place, we can finally say that our material looks like a proper hologram! However, a nice addition to the current setup would be the inclusion of some varying levels of opacity. Most holograms we see in movies are translucent objects, as we can see what's behind them. Furthermore, holograms are also known to flicker and get distorted over time – this is often done to emulate how a hypothetical technology that allows the holograms to even exist would work. Think of it like an old TV effect, where the scan lines get updated from one side of the screen to the other every frame. That's what we'll be trying to replicate next!

24. Head over to an empty area of the material graph and add a **Texture Sample**. Select the **T_ ChristmasTree_Scanlines** texture as its value, which we'll use to drive the opacity.

The previous texture can't be applied as it currently is to the toy Christmas tree 3D model as its contents don't align with the UVs of the asset. Let's fix that by creating a **BoundingBoxBased_ 0-1_ UVW** node, which will allow us to project any given texture according to a bounding box centered around the asset onto which the material is being applied. This is a different way of applying textures, where we bypass the UVs of the model in favor of certain custom projections.

25. Create the aforementioned node by right-clicking and typing its name. Place the node a little bit before the previous **Texture Sample** since we'll be driving its UVs through it.

26. Create a **MakeFloat2** node after the previous **BoundingBox** one.

27. Connect the **R** value of the **BoundingBox** node to the **X** input pin of the **MakeFloat2** node, and the **B** value to the **Y** one. This will make a **float 2** vector out of the XZ plane that defines the bounding box, allowing us to map the height of the object.

28. Add a **One Minus** node after the **MakeFloat2** one, which will invert the values that we had so far – white turning black and vice versa. Hook its input pin to the output pin of the **MakeFloat 2** node.

29. Include a **Multiply** node that will control the scale of the texture affected by the previous set of nodes. Connect its **A** pin to the output of the **One Minus** node and assign a value to pin **B** – I've gone with **3** in this case.

30. Create a **Panner** node and plug the result of the previous **Multiply** node into its **Coordinate** input pin.

31. Select the previous **Panner** node and focus on the **Details** panel. As we want the scan lines to be moving upward, leave **Speed X** set to **0** and increase the value of the **Speed Y** setting. I've chosen **0.025** to give it a nice, slow effect.

32. Connect the output of the **Panner** node to the **UVs** input pin of the **Texture Sample** node we created in *step 24*.

This is what the previous part of the graph should look like, which we'll call *Part C* for future reference:

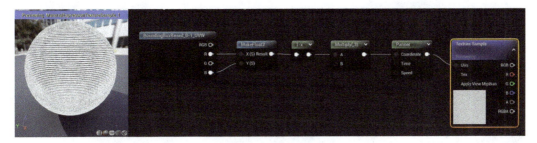

Figure 3.12 – A look at Part C of the material graph, with the last node being previewed

> **Important note**
>
> The **Panner** node we used before is an interesting resource that comes in handy whenever we want to animate a texture as it allows us to move the UV space (pan it) to create the effect of a moving image. More on this in the *How it works...* section!

Something we can do now is add a little bit of variation to this scanline effect, just before we finally go ahead and plug the results into the opacity channel of the **Main Material** node.

33. Copy the graph we've just created and paste it immediately under the previous one. We should have two identical bits of visual scripting code, one on top of the other, which we can refer to as *Part C* (the part that we already had) and *Part D* (the duplicate).

34. Head over to the **Multiply** node of the copied graph section (*Part D*) and change its current value of **3** to something like **6**. This will make the scan-line effect in this section smaller than the previous one.

35. Combine both effects by creating an **Add** node and hooking its **A** and **B** input pins to the output of the final **Texture Samples** node in *Part C* and *D*.

36. Add two **Constant 3** vectors after the last **Add** node. Give the first a value of **0.2** in each of its channels (R, G, and B), and a value of **1** on each channel for the second one.

37. Create a **Lerp** node and connect its **A** input pin to the first of the previous **Constant 3** vectors, and pin **B** to the second one.

38. Use the output of the **Add** node from *step 35* as the **Alpha** value of the previous **Lerp** node.

39. Finally, connect the output of the previous **Add** node to the **Opacity** input pin of the **Main Material** node, and hook the output of the **Lerp** node we created in *step 22* (the one that combines *Parts A* and *B* of the material) to the **Emissive Color** one.

Completing all of the previous steps will give us the final look that we were trying to achieve: a nice holographic material based on a small wooden tree prop. Enjoy looking at it in the following screenshot!

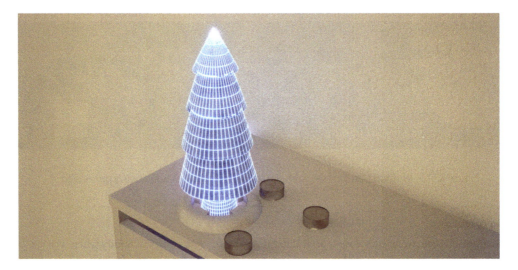

Figure 3.13 – The final look of the holographic material

How it works...

Having encountered a few new concepts in this recipe, let's take a bit of time to go over the theory behind them.

The Unlit Shading Model

The holographic material we've been dealing with made use of the **Translucent Blend Mode**, something we already worked on when we created the glass material in the previous recipe. In addition, we've also started to explore the new **Unlit Shading Model**, something completely new to us at this point. Even though we are just starting to see this new option, it might be easier to understand the way it works by just comparing it to the other type we've used in the past – the **Default Lit** one.

In essence, a **Shading Model** is a function used to describe how light affects the surfaces of the objects within Unreal. So far, we've explored the **Default Lit** one, as mentioned previously. That one is meant to be used on those types of objects that reflect light in a similar way to opaque ones, such as wood, iron, or concrete. There are other ones, such as **Subsurface Profile**, which we'll see in the next recipe, which are very helpful when working with models that have part of the light that reaches them absorbed by its surface and scattered underneath it.

The new nodes

Regarding the new nodes that we saw in this recipe, the first of the bunch was **Bounding Box Based _ 0-1_ UVW**, which kickstarts a new methodology for texturing assets that we haven't seen before. It's fair to say that we have relied on the UVs of an object to project textures onto them. This meant that the UVs of a given model had to be preprepared in an external editor, carefully curated before importing the object into Unreal, and considered before creating the textures for the asset. This workflow has the benefit of being able to perfectly map each polygon of the original object to a coordinate of the 2D UV space, letting us create custom textures that work great on those models and that can be tailored-made for them. It also has some cons though, such as the need to use specific textures for the assets that make use of this workflow and not being able to reuse them on other objects – with the added problem that presents, such as the extended footprint that those custom assets leave on our hard drives.

The **Bounding Box UVW** node tackles the texturing process differently. Instead of leaving the UV creation to the artist, it automatically creates a new mapping projection based on the shape of a box that covers the entirety of the object in its interior – hence the *Bounding Box* prefix. The projection happens along three axes (*U*, *V*, and *W*, which translate to *X*, *Y*, and *Z*), so it tends to work well on "boxy" objects, or on those assets where their faces align well with that projection. Limitations start to show as soon as we introduce curves, cavities, or continuous faces that change throughout the shape of the object, as those areas won't be evenly mapped due to the incompatibility between the projection and their shape.

It's also worth mentioning that we don't need to use the three axes as we can map along just a couple, as we did in this recipe. This works quite well for achieving certain effects, such as the scanline we saw in this recipe, and reusing assets across our projects. Choosing a single plane of projection from that Bounding Box system can even work well on rounded objects, such as the wooden Christmas tree we've been using, and the possibilities are usually bigger than what we might initially think when dealing with this node.

Beyond that, we've also had the chance to use the **Make Float 2** node, a utility that allows us to combine single values into a duo. On the other hand, the **One Minus** node allows us to subtract whatever we feed it from one. This translates to inverting the value when working with grayscale values: a white one will become black, and a black one will become a white one instead. It works on all of the channels that we input, so it also accepts Constant 2, 3, and 4 vectors. It's important to remember that it doesn't technically invert the values as it only subtracts what we feed it from 1, something important to keep in mind when working with values that have multiple channels.

Finally, another very interesting node we worked with in this recipe was the **Panner** function. This allows us to animate textures, something crucial when trying to create multiple different effects: from the scanline one we saw in this recipe to running water, dynamic smoke, or anything else that can benefit from a bit of movement. We can pan those textures along the *U* or the *V* axis, or even both directions according to a specific speed that we can control independently for each channel. On top of that, the **Time** input grants us control throughout the effect in case we want to narrow it down or limit it to a specific range rather than having it continuously happening.

Emissive materials

The emissive property of our materials is a powerful resource when we want to fake lights or create them. Even though this is not the standard procedure to add a new light, setting up a material to make use of the emissive node can replace them under certain conditions. So, what options do we have when working with these types of assets?

We have two basic choices: we can use the emissive property to give the impression that our material is casting light, or we can set it up so that it affects the world around it by emitting photons. The way the engine deals with the faking part is by using the emissive node of the material to affect its contribution to the bloom effect and give the impression that you are looking at a bright light when working with static lighting.

If we don't want this technique to just be an illusion, but to affect the shadows in the scene, we need to select the model that the emissive material is being applied to and look at its **Details** panel. Under the **Lighting** tab, we should be able to find an option named **Use Emissive for Static Lighting**. Turn that on, click on the **Build** button to calculate the shadow maps, and you're good to go!

Everything we've covered before regarding emissive materials is only useful when working with static lighting and not relying on Lumen, as we saw in the *Using static lighting in our projects* recipe back in *Chapter 1*. We can now say that these types of shaders contribute to lighting and global illumination thanks to the arrival of Lumen in Unreal Engine 5, which sees *"emissive materials propagate light through Lumen's Final Gather with no additional performance cost,"* according to Epic's official documentation on this technique.

See also

You can find more information on the **Unlit Shading Model** and the emissive material input pin in Epic Games' official documentation:

- https://docs.unrealengine.com/4.27/en-US/RenderingAndGraphics/Materials/MaterialProperties/LightingModels/

- https://docs.unrealengine.com/4.26/en-US/RenderingAndGraphics/Materials/HowTo/EmissiveGlow/

Working with subsurface scattering

The new material we'll explore in this recipe is going to be wax. Wax! So, what is so interesting about it, you might wonder? Isn't it just another opaque material, something we already covered before? Well, no! It's a good example of the new **Shading Model** we'll be using next, the **SSS** one – short for **Sub Surface Scattering**. This is different from the other ones we've seen so far, as light neither bounces completely off the objects onto which materials that use this **Shading Model** are applied nor does it go through them; instead, light penetrates their surface just enough that we can see its effects in the actual exterior, changing the color of the final pixels that are being rendered. It's an interesting effect that affects multiple objects, such as the candle made out of wax we'll see next to snow, skin, or thin opaque curtains. Let's see how it works!

Getting ready

You can follow this recipe by opening the `03_03_Start` level provided with the Unreal project included alongside this book. As always, all of the assets that you'll be seeing in the next few pages can be found there – but what if you want to use your own?

You'll need three things, the first one being a basic scene setup with some lights in it so that we can see the materials that we are about to introduce. The second is a type of model where the subsurface effect can be seen, such as a candle or an ear – the effect will be more obvious when it's applied to an object where it would also be visible in real life. Take the following screenshot of the model that we'll be using as an example:

Figure 3.14 – Render of the candle model we'll be using throughout this recipe

As you would expect, the thinner areas of that candle are the places where the effect will be the most visible. You might want to consider that when choosing your objects.

The third element we'll need to be aware of if we want to work with the subsurface scattering effect is a project-wide setting that we need to adjust: we'll need to have our project use the Default Shadow Maps instead of the Virtual Shadow Maps that are new in this fifth version of Unreal. We'll learn how to adjust this parameter once we start the recipe, but the key idea to take away from this fact is that Virtual Shadow Maps are still in beta access and that they don't support all of the features of the engine, such as the subsurface profile effect. Despite that, they are the way the engine is going to move forward, as Virtual Shadow Maps enable much better shadows across the board compared to what was possible in the past, and many more features will start to work as soon as they exit their beta phase.

How to do it...

Unreal offers several possibilities in terms of enabling the subsurface scattering effect – two main ones available through the **Subsurface Profile** and **Subsurface** Shading Models, and some other more specific ones, such as the **Preintegrated skin** method. In this section, we'll be exploring the first of those implementations, but we'll talk about the others in the *How it works...* section. With that said, let's start by adjusting the Shadow Maps used in the project so that we can visualize the effect, as mentioned in the *Getting started* section:

1. Assuming we are starting from the `03_03_Start` level, select **Post Process Effect Volume** from the **Details** panel and search for the **Global Illumination** property. Create one such actor if you are working on your own level.

2. Look for the **Global Illumination** category and change the **Method** property from the default **Lumen** to the option called **None**.

3. Next, head over to the **Reflection** category and assign a value of **None** to the **Method** property, just like we did in the **Global Illumination** section.

    Implementing the previous changes will see us reverting to the previous rendering method of illuminating our scenes, just like we saw back in the *Using static lighting in our projects* recipe in *Chapter 1*. The next thing we'll do is adjust the types of Shadow Maps used in the project.

4. Head over to **Edit | Project Settings** and look for the **Shadow Map Method** option within the **Shadows** section of the **Engine – Rendering** category. Change from the default **Virtual Shadow Map (beta)** option to the **Shadow Maps** one.

    That's everything we need to do to enable the subsurface effect we are about to explore. With that done, let's create the material that we'll use in this recipe.

5. Create a new material and assign it to the **SM_Candle** model you placed in the scene, or to the model that you are using if you're working on your own level. I've named the material **M_CandleWax_SSS**, but feel free to choose your own name!

6. Double-click on said material to open the Material Editor.

7. Drag and drop some of the textures included with the project into the material we've just created. You should be using three, named **T_ Candle_ ColorAndOp**, **T_ Candle_ AORM**, and **T_ Candle_ Normals**, which can be found by going to **Content | Assets | Chapter01 | 01_07**. Connect them to the appropriate pins in the **Main Material** node.

You might remember these textures back from the last recipe of *Chapter 1*, called *Checking the cost of our materials*, where they were simply being used in a similar material as the one we are creating now to demonstrate the cost of using a shader with subsurface scattering properties. We'll sometimes use assets in a recipe that were previously used in another, so simply search for their name in the Content Browser of the Unreal Engine project provided alongside this book when you're in doubt as to where to find them.

> **Important note**
>
> Instead of using the textures I've mentioned previously, feel free to create your own or use simple constants to drive the appearance of those basic properties of the material. The important stuff is what comes next: setting up the subsurface properties of the shader.

Before moving any further, this is what the material graph should look like at this point:

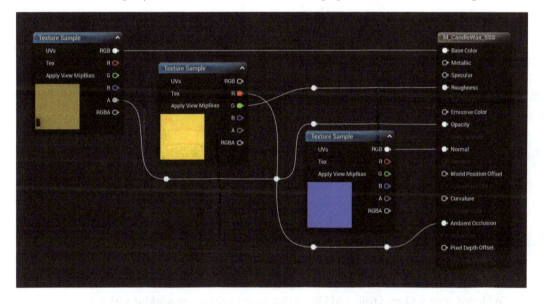

Figure 3.15 – The three texture samples we should be using at this stage of the recipe

With that out of the way, it is now time to create the object that we'll be using to drive the subsurface properties of the material: the **Subsurface Profile** asset. We are going to need it very soon, so let's create it now.

8. Go back to the Content Browser and create a new **Subsurface Profile** object. You can find that class by right-clicking and looking for it within **Create Advanced Asset category | Materials | Subsurface Profile**. I've named mine **SSP_Wax**.

Now that we have created it, let's put some time aside to properly set it up and assign it to the material that will be using it. Subsurface profiles are similar to data sheets in that we are going to define some values within them, properties such as the **Surface Albedo** color or **Mean Free Path Color** that will define how the effect looks.

Important note

Not having a **Subsurface Profile** object defined for our material means that Unreal will assign the default one to it, which has been tailored to the visualization of the human skin of a Caucasian male. Remember to always make and assign one for your materials if you want them to look right!

---

9. Head back to the Material Editor for the **M_CandleWax_SSS** material and, without having anything selected, look at the **Details** panel and focus your attention on the **Shading Model** category. The default value should be **Default Lit**, but we want to change that to **Subsurface Profile**.

10. Assign the **SSP_Wax** asset as the object reference within the **Subsurface Profile** property. You can find this option right in the **Details** panel, as shown in the following screenshot:

Figure 3.16 – Location of the subsurface properties within the Details panel of the main material node

With that done, we have finally told Unreal that the material we are working with is going present subsurface scattering properties. The next steps are going to deal with tweaking that effect until it looks like what we are after. To achieve that, we'll need to head back to the **Subsurface Profile** object that we created and adjust its parameters. Let's do that now!

> **Tip**
>
> It's good to have the **Subsurface Profile** object applied to the material before tweaking its properties, as we are about to do, since this will enable us to check how they affect the model in real time.

11. Open the previously created **Subsurface Profile** object by double-clicking on it within the Content Browser. Having it open at the same time as the main viewport will help you see how the candle material changes in real time.

12. The first parameter we want to change is **World Unit Scale**. This specifies the ratio between the world units and those used by Unreal to the **Subsurface Profile** asset. Unreal's default units are hard-coded to centimeters, and the model of the candle has also been created with those units in mind. Given that fact, we have a 1:1 ratio, so let's just use that value for this setting.

13. For the effect to be visible, we need to force the light to travel through the object quite a bit. To do so, adjust the **Mean Free Path Distance** value to something greater than its default value, such as **20**. We'll tweak this property later, but it's good to make it visible at this stage.

14. The next field we'll adjust is going to be **Surface Albedo**. Let's choose a similar tone to the one used for the **Base Color** property, so let's go for a dark yellowish/orange tone.

15. Now, let's tweak the **Mean Free Path Color** property. This property controls the distance that light travels across the RGB channels, working in tandem with the parameter we tweaked in *step 13*. Set a value of **White (1,1,1)**, which will have the light travel the maximum amount in all three channels. Adjust this setting if you want certain tones to travel farther than others.

16. Something else we can tweak is **Tint**, especially if we want to achieve more dramatic results. I've chosen a light blue color, making the material look redder – as the RGB values are 0,31 **Red**, 0,75 **Green**, and 1 **Blue**, red tones get absorbed less, becoming more visible.

    Other options within the **Subsurface Profile** asset are things such as **Extinction Scale**, which controls the absorption scale of the material. **Normal Scale** modulates how much of the Normal input affects the transmission of the subsurface effect. The **Roughness 0** and **Roughness 1** values, along with the **Lobe Mix** property, control the roughness values that you'll see on the areas that display the subsurface scattering effect. Play around with those values until you are happy with the results, and take a look at the following screenshot, which shows the settings that I've used:

Figure 3.17 – The settings I've used to set up the Subsurface Profile asset

With all of those changes in place, there's one further interesting element that we can look at before moving on: using the **Opacity** channel within the material graph. Through this, we can change how the subsurface effect works across the entirety of the model. Going back to our candle example, we want the scattering effect to happen on the main body but not on the wick. We can set this up by plugging a mask into the **Opacity** input pin of the **Main Material** node, as shown here:

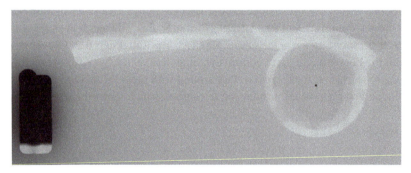

Figure 3.18 – The Alpha channel of T_Candle_ColorAndOp used to drive the Opacity parameter

With all of those changes implemented in the material, we can now say that we've achieved our objective: creating a candle made out of wax. We'll leave it here for the moment, but not before we look at the result:

Figure 3.19 – The final look of the scene

How it works...

Seeing as we've touched on a few topics in this recipe, let's quickly go over two of the most important ones we've covered: the **Subsurface Profile Shading Model** and Texture Packing.

The Subsurface Profile Shading Model

In the previous section, we looked at the implementation of the **Subsurface Profile Shading Model** inside UE5. This is quite a powerful rendering technique, one that makes a huge difference when we are dealing with materials that exhibit properties such as wax or skin. But how does it work under the hood? First of all, we need to understand that this is not the only way we can enable the subsurface effect in our materials. In total, four Shading Models can do so:

- The Standard Subsurface Model
- The Preintegrated Skin Model
- The Subsurface Profile Model
- The Two-Sided Foliage Model

Each of these is geared toward achieving a specific goal. For now, let's focus on the one we know, which is aimed for use in high-end projects.

The **Subsurface Profile Shading Model** works as a screen space technique, similar to what we saw with ambient occlusion or screen space reflections back in *Chapter 1*. This is a different implementation from what the other three methods offer and this is the key difference we need to be aware of when working with it. While the standard **Subsurface Shading Model** is cheaper and can run faster overall, using the **Subsurface Profile Shading Model** offers several advantages in terms of quality. The different settings we can modify, some of which we've already seen in the previous pages – such as **Surface Albedo** and **Mean Free Path Color** – help us to realistically define the effect that is applied throughout the material. Other settings, such as the ones found under the **Transmission** tab within the **Subsurface Profile** asset, help define how light scatters from the back of an object.

I would also like to point out the difference in the functionality of some of the nodes that make up a **Subsurface Profile** material. If you look at the **Main Material** node in the material graph, you'll see that the **Opacity** input pin is no longer grayed out, unlike what happened when we were using a **Default Lit** type of shader. That input is now available for us to use. However, it can be a bit counter-intuitive as to what it does – it affects the strength of the subsurface effect in our model, as we saw in this recipe, and not how opaque the 3D mesh is going to be.

Something else that we need to note is the limitation of the **Metallic** property. If we plug anything into that slot, the subsurface effect will disappear; this is because that material channel has now been repurposed in this **Shading Model** to accommodate the data from the **Subsurface Profile** asset. Keep that in mind!

Texture packing

You might remember that the material for the candle that we created was driven by several textures placed within the material graph. We used three main ones: **T_ Candle_ColorAndOp**, **T_ Candle_ AORM**, and **T_ Candle_ Normals**. I'd like to talk a bit more about the first two, as we are doing something smart with them that we haven't seen before.

As you already know, some material attributes can be driven by grayscale values. That being the case, we find ourselves in a situation where we can use the RGB channels (as well as the **Alpha** one) of a texture to drive certain material inputs. That's exactly what we are doing with the **T_Candle_AORM** asset; we store the **Ambient Occlusion**, **Roughness**, and **Metallic** textures into each of that image's channels (hence the AORM suffix, an acronym for the words Ambient Occlusion, Roughness, and Metallic). You can see this in the following screenshot:

 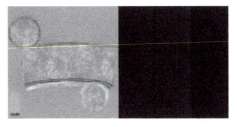

Figure 3.20 – The T_Candle_AORM texture and the information contained in each of its channels

See also

Epic Games is continuously updating its documentation on this topic, so be sure to check it out if you want to learn more: `https://docs.unrealengine.com/5.0/en-US/subsurface-profile-shading-model-in-unreal-engine/`.

Using refraction in a body of water

Water is one of the most cherished materials in real-time visualizations, as shown by the continuous advances made to the materials that have tried to depict it over the years. We've come a long way from the days of early 3D graphics when simple planes and animated textures were used to represent it. Even though we can still use those older tricks, thankfully, we have more techniques at our disposal that can help us achieve richer results. Things such as refraction, displacement maps, tessellation, and caustics can add a bit of extra flair to our projects, and we are going to be looking at that first effect in this recipe.

Getting ready

More so than in previous recipes, this one will require us to set up the scene in a way that allows the effects that we are about to implement to shine. The first thing we'll need is a plane onto which we can apply the water material we'll be creating. Beyond that, we'll also need something to act as a container for the water – I'm going to use a 3D model of a bathtub, so something similar will do. Having something opaque wrapping around the water plane is quite important as we'll be using nodes that detect the intersection of several geometries. Those two things are the basics, but we'll probably want to add a small prop just so that we can check the refraction within the body of water – a rubber duck can work, even though we'll be using a shampoo bottle!

As always, we'll be providing you with everything you need if you want to use the same assets. Open the `03_ 04_ Start` level, where you'll find everything you need to follow along.

Finally, remember to revert the change we made to the Shadow Maps in the previous recipe – go back to using Virtual Shadow Maps instead of the regular ones by adjusting the **Shadow Map Method** option within the **Shadows** section of the **Engine – Rendering** category, all within the Project Settings.

How to do it...

Since we are about to create a new type of translucent material, it's always good to think about what we want to achieve before we begin. We've dealt with different types of translucent surfaces in the past, but this time, things won't be as straightforward as before. The reason for that lies within the specific asset that we want to create: water.

Unlike glass, water is a material that is rarely found resting perfectly still. Its condition as a fluid means that it is almost always in motion, and that introduces certain conditions that we need to consider, such as how the normals of the material are going to be animated, how the direction of those affect the color of the material, and how the opacity changes across the body of water.

We will explore all of those conditions next:

1. Create a new material and name it whatever you like – I've gone with **M_BathtubWater**, as that's what we'll be creating.

2. Change **Blend Mode** to **Translucent** and check the **Screen Space Reflections** checkbox a little further down the **Details** panel, in the **Translucency** section.

3. Change **Lighting Mode** to **Surface Translucency Volume** and modify the value of the **Translucency Pass** option so that it reads **Before DOF**.

4. Scroll down to the **Usage** category and tick the checkbox next to **Used with Static Lighting**.

    So far, these are the same steps we implemented when we created the glass material three recipes ago. The first bespoke additions we are going to introduce deal with the opacity and color values of our material. Starting with the first one, we'll introduce a gradual loss of transparency as the depth of the water increases. We'll replicate this using a combination of the **Scene Depth** and **Pixel Depth** nodes, which will help us with those calculations.

5. Start by adding a **Scene Depth** node to the material. This will let us sample the depth of the scene behind the object onto which we'll apply the water shade. This node is only available when working with translucent materials as it gives us information about objects that are behind the model onto which this effect is being applied.

6. Create a **Pixel Depth** node and place it close to the previous one. This node gives us the distance between the surface of the object onto which the material is being applied and the camera, and not the distance to what's behind it.

7. Include a **Subtract** node and place it right after the previous two nodes.

8. Connect the output of the **Scene Depth** node into the **A** input pin of the **Subtract** one; do the same for **Pixel Depth** and the **B** input pin.

    Performing the previous operation will give us the results we are after – a value that depicts the distance between the 3D model of the body of water and the objects behind it. We can use that information to drive effects such as the opacity of the water, albeit not in its current state: the subtraction operation throws very large numbers, which represent the distance between the location of the pixels in world space and the position of the camera that is viewing them. We need to normalize those values – that is, fit them within the 0 to 1 bracket needed to operate on nodes such as the **Lerp** node. Let's take care of that now:

9. Create a **Scalar Parameter** and name it something along the lines of **Depth Adjustment Value**. This node will be tasked with reducing the result of the subtraction operation.

10. Place a **Multiply** node after the previous **Scalar Parameter** and **Subtract** ones, and connect its input pin to the outputs of those two.

11. Assign a value of **0.03** to **Depth Value Adjustment Scalar Parameter**. This is a number that works well given the scene that we are working with, but please note that you might need to choose a different figure if you're working with your own assets.

Completing the previous steps will enable us to create a mask that works well in terms of showing the depth of the scene behind our translucent material. Let's quickly connect the output of the previous **Multiply** node to the **Base Color** input pin of the material and ensure that the scene looks like this:

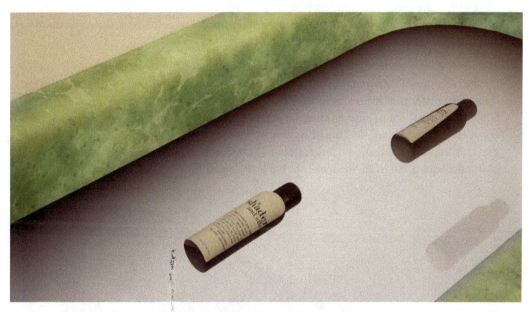

Figure 3.21 – The effect that the depth logic should have when applied to the water model in our scene

With that done, we can now say that we can use the depth of the scene to drive the appearance of our material. Useful as this is when working with liquids such as water, we can still go a step further and apply another interesting visual effect that we can sometimes see at the edge of these substances: their opacity and color change right at the point where they meet another object. This is something that we can also achieve within Unreal, so we'll learn how to do that next.

12. Create a **Depth Fade** node and place it somewhere after the ones we created previously. This function generates a gradient mask based on the proximity of our model to others. It's a great tool whenever we want to isolate those intersecting areas.

13. Add a **Scalar Parameter** and name it something along the lines of **Fade Distance**. We'll use this node to control the distance covered by the gradient generated with the previous **Depth Fade** function. Assign a value of **6** to it and connect its output pin to the **Fade Distance** input of the **Depth Fade** node.

14. Create a **One Minus** node and place it immediately after the **Depth Fade** one, and connect the input pin of the latter to the output of the former. This node allows us to invert the color of the gradient we are currently creating, something necessary to make it work with the depth calculations we completed earlier.

15. Include a **Power** and another **Scalar Parameter** after the previous **One Minus** node. Assign a value of **8** to the parameter, connect it to the **Exp** input pin of **Power**, and give it a name; I've gone with **Depth Fade Smooth Exponent** as we'll control the smoothness of the edge detection gradient through it.

The previous set of nodes should have allowed us to identify the area of the model that is close to the intersection with other 3D objects. We'll use that information to adjust the opacity and the color of the material, as many waterbodies show an increased darkening of those areas in contact. Something important to note at this point is also the inversion of the gradient we would have obtained had we not used the **One Minus** node. We'll need to combine this new mask with the previous, depth-based one, so it's important to get the values right before we attempt to merge both calculations. All in all, the scene should look something like the following at this point:

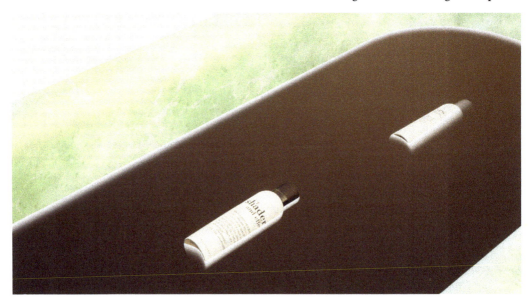

Figure 3.22 – The edge detection mask we have just created

With both the edge and the depth masks ready, it is time to combine them. We'll pay a second visit to the **Depth Fade** node's effect, given how the area where we need to merge both masks is right at the meeting point of the water model and the different objects that intersect it. This second **Depth Fade** node will have to be different from the first one as the goal now is to create a soft gradient that bridges the values of both masks right where they meet.

16. Create a second **Depth Fade** node.

17. Include a new **Scalar Parameter** and name it something along the lines of **Depth Fade Blend Distance**. I've assigned a value of **3** to it, which controls the distance covered by the depth gradient around the models that intersect the object where we are applying our material.

18. Connect **Scalar Parameter** to the **Fade Distance** input pin of the previous **Depth Fade** node.

19. Throw another **One Minus** node immediately after the **Depth Fade** one and connect the output of the latter to the input of the former.

We should now be looking at a third mask that will hopefully look something like this:

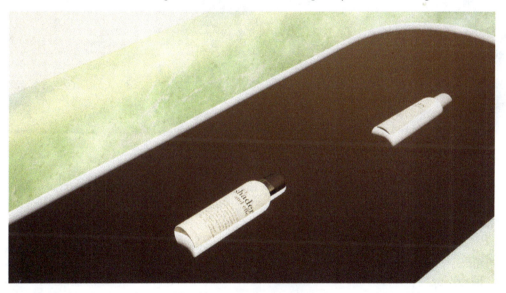

Figure 3.23 – The mask we'll use to blend between the depth calculations and the edge detection ones

The previous mask should be very similar to the second one we created, which depicts the intersection areas of the model. However, this new one should be sharper than the previous one as we are only interested in having a black and white mask that covers the edges of the model.

> **Tip**
>
> Remember to connect different parts of the material to the **Base Color** input pin of the shader, as that will allow you to visualize what you are creating on the screen – something vital in this recipe!

With that in place, we are now in a position where we can blend everything we've created thus far. Let's tackle that now.

20. Create a **Lerp** node and place it after everything we created previously.

21. Connect its **A** input pin to the output of the **Multiply** node we created in *step 10*.

22. Hook the output of the **Power** node we created in *step 15* to the **B** input pin of the new **Lerp** node.

23. As for the **Alpha** input pin, connect the output of the **One Minus** node from *step 19* to it.

Completing the previous steps will have granted us the ability to create depth and edge-aware effects within our body of water. This is both a great achievement and a fantastic opportunity to review the visual code we should have created thus far:

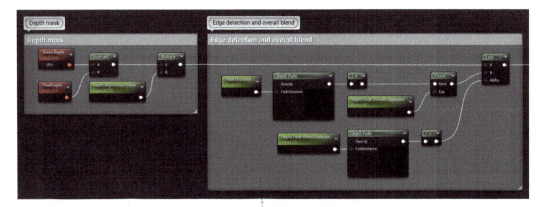

Figure 3.24 – The current look of the material graph

With those nodes in place, we can finally say that we are ready to start driving both the opacity and the color of the water. Let's not waste another minute and jump straight into tackling the color values.

24. Create a couple of **Vector Parameter** nodes to drive the color of the water. An apt name for these two nodes would be **Shallow Water Color** and **Deep Water Color**, as that's what they'll be driving.

25. Assign values to the previous two nodes that you feel should represent the water well. I've gone with values of **0.241**, **0.31**, and **0.316** to depict the shallow areas and values of **0.109**, **0.208**, and **0.178** for the deeper parts.

26. Throw a **Lerp** node after the two Vector Parameters and connect the one controlling the shallow areas to the **A** input pin. The color driving the deeper parts should be connected to the **B** input pin, while the output of the **Lerp** node we created in *step 20* (the one mixing the depth masks) should be connected to the **Alpha** input pin of this latest **Lerp** node.

27. Connect the output of the previous **Lerp** node to the **Base Color** input pin of the material.

28. The previous set of actions focused on the **Base Color** property of the material; now, let's focus on its opacity.

29. Create a couple of scalar parameters and assign them the following names: **Shallow Water Opacity** and **Deep Water Opacity**.

30. Assign a value of **0.2** to the first of the two new parameters and a value of **0.6** to the other one.

31. Create a new **Lerp** node and, just like we did with the color blending one, connect its **A** input pin to the output of the **Shallow Water Opacity** parameter. Hook the **B** input pin to the other parameter driving the opacity of the deeper parts and connect the **Alpha** input pin to the output of the **Lerp** node we created in *step 20*, just like we did for the color property.

32. Hook the output of the previous **Lerp** node to the **Opacity** property of our material.

    Implementing that last step will have seen us tackle both the **Base Color** and **Opacity** properties of the material. That's not the end of the story, though, as we still need to take care of parameters such as roughness and refraction.

    Starting with the first of those two, the roughness property is something that we know varies according to the viewing angle, something that we've already talked about in previous recipes. We'll need to know whether the material is facing toward the camera or away from it, but seeing as we are dealing with water, we are also going to complicate things a bit further and introduce small, animated waves to the equation, just to increase the realism of the final material. We'll tackle all of those challenges next.

33. Create a **Texture Sample** node and assign the **T_Water_N** asset to it. This image is part of the Starter Content, so make sure you add it to your project if you are working with a scene different from the one available alongside this book.

34. Create a **Texture Coordinate** node and place it to the left of the previous sample. Leave the default values untouched.

35. Add a **Scalar Parameter** and assign a value of **0.1** to it, and name it something like **Water Scale**. This node will be in control of the size of the texture we'll be using to drive the appearance of the water waves.

36. Place a **Multiply** node after the previous **Texture Coordinate** and **Scalar Parameter** and connect the outputs of those into the **A** and **B** input pins of the new **Multiply** node. This will easily allow us to control the size of the waves if we decide to create a Material Instance later.

37. Create a **Panner** node and use the output of the previous **Multiply** node to drive its **Coordinate** input pin.

38. Create a couple of **Scalar Parameter** nodes to control the speed at which the Panner effect works. Name them **Water X Speed** and **Water Y Speed**, and assign a value of **0** to the first node and a value of **0.001** to the second one. This will make the UVs be in the *Y* direction, as we'll see later.

39. Create an **Append** node and connect the previous two scalar parameters to its **A** and **B** input pins.

40. Connect the output of the previous node to the **Speed** input pin of the **Panner** function.

41. Finally, connect the output of the **Panner** function to the **UVs input** property of **Texture Sample**.

42. Implementing the previous set of actions will allow us to animate a texture, which we've chosen to be the normal map of a waterbody. Seeing as this isn't an asset that we have created ourselves, we probably want to implement a way to control its intensity next.

43. Create a **Component Mask** node and place it immediately after the **Texture Sample** node we created in *step 32*. This node allows us to isolate the texture channels that we choose from a given texture – and seeing as the intensity of a normal map is stored in its red and green channels, tick the **R** and **G** checkboxes in the **Details** panel to select them.

44. Add a new **Scalar Parameter** and name it something along the lines of **Normal Intensity**, as this is the task that it will be controlling. Set its value to **0.8**.

45. Connect both of the previous nodes through the **Multiply** one.

46. Create another **Append** node, which we'll use to recombine the adjusted red and green channels and the original blue one from the normal texture. Hook its **A** input pin to the output of the **Multiply** node and draw a wire from the **B** output pin of the **Texture Sample** node to the **B** input pin of this new **Append** node.

The final **Append** node is the last of the adjustments that we'll make to the normal texture since, by now, we should have achieved the goal of animating it. Let's take a look at that part of the graph just to make sure that we are on the same page:

Figure 3.25 – A look at the section of the graph controlling the normal map and its animation parameters

We can now use the information generated through the previous set of nodes to drive the Roughness and Normal inputs of the main material. The latter seems quite obvious given what we've just created – using a normal map to drive its homonymous material parameter shouldn't be a breakthrough at this point! – but we need to touch base on how we are going to affect the roughness property. It won't take us too long, though: we'll simply assign different values to the parts of the material that face the viewport as opposed to the ones that look away from it, just as we've done in the past. The difference this time is that we can't simply use the **Fresnel** node as we did back in *Chapter 2*, as we also need to consider the values of the normals assigned to the material.

47. Create a **Fresnel** node and place it somewhere after the last **Append** node.

48. Connect the output of the last **Append** node we created (the one from *step 44*) to the **Normal** input pin of the Fresnel function. This will allow us to use the animated normal information to drive the effects achieved through the use of the Fresnel technique.

49. Create a couple of scalar parameters and assign them names such as **Camera-facing roughness** and **Perpendicular Roughness**. Set the value of the first node to **0.025** and the second to **0.2**.

50. Create a **Lerp** node and connect the output of the **Fresnel** node to its **Alpha** input pin. Hook the result of the **Camera-facing roughness** parameter to the **A** input pin and connect the other one to the last available input.

51. Connect the output of the **Lerp** node to the **Roughness** input pin of the material.

The last five steps have taken care of adjusting the roughness parameter of our water material. With that out of the way, we can finally move on to the last setting that we need to tweak: the refractive property of the shader. This is where things get interesting, as we are about to use a different technique compared to the one we saw when we created the glass shader in the first recipe of this chapter. This time, instead of working with the standard **Index of Refraction Translucency** mode, we'll rely on a technique called **Pixel Normal Offset**, which offers better results when working with large flat surfaces such as waterbodies or windows. Let's see how we can implement it.

52. Select the main material node and focus on the **Details** panel. Scroll down to the **Refraction** section and change the value of **Refraction Mode** from the default, **Index of Refraction**, to **Pixel Normal Offset**.

53. Connect the output of the **Append** node we created back in *step 44* to the **Normal** input of the material.

54. Create a new **Scalar Parameter** and simply name it **Refraction**. Assign it a value of **1.25** and connect its output pin to the **Refraction** input of the material.

Undertaking that last step will see us fulfill all the goals that we set out to complete. The last thing that we need to do is apply the material to the body of water present in our scene, and maybe create a Material Instance so that we can play around with some of the scalar parameters we created. That is probably a good idea, if only just so that we can visualize in real time how those different settings affect the look of the material without the need to continuously recompile the parent shader. In any case, make sure that you take one last look at the scene and enjoy the fruits of your labor!

Figure 3.26 – A final look at the scene

How it works...

We had the opportunity to take a look at a multitude of different new nodes in this recipe, among which we can find the **Scene Depth** node, its sibling the **Pixel Depth** function, **Depth Fade**, **Append**, **Component Mask**, and **Pixel Normal Offset**… So many that we need to take a further look at how they all work!

Let's start with the first one we used: **Scene Depth**. As mentioned previously, the functionality provided by this node can only be used in materials that make use of **Translucent Blend Mode**, and that is because of the information that we get when using it. The data that we can gather through it represents the distance between the opaque surfaces located beyond the translucent surface we are working with and the viewport. You can think of those values as the coordinates of the last visible elements in the scene.

In contrast, the **Pixel Depth** node provides us with similar data, with the key difference that the distance we are offered this time is between the viewport and the pixels depicting the object onto which we are applying the material that contains the **Pixel Depth** node. This means that we get the distance between the camera and the object the material is being applied to, regardless of whether it uses the **Opaque** or **Translucent Blend Mode** property. This can, in turn, explain why we can only use the **Scene Depth** node on translucent objects, as the information provided by that function would be the same as the information gathered with the **Pixel Depth** node for this type of object.

Another depth-related function is the **Depth Fade** one, which allows us to retrieve the areas near the intersection with other objects. This can be useful to hide the seams between them or for other purposes, such as the one we saw in this recipe. All in all, it's another way of controlling and masking certain parts of our models dynamically, one that doesn't require static texture masks to be used and one that can change with gameplay.

On top of this, it's only natural that we needed to use an **Append** node at some point given how many different parameters we've used in this recipe. You might remember that we used it immediately after creating the two scalar parameters that controlled the speed of the water in the *X* and *Y* directions. We needed it there to combine those single values into a vector 2 variable, which is what the **Panner** node demands as input for its **Speed** parameter. In essence, the **Append** function allows us to create a higher magnitude vector, one that contains more channels than the inputs we are feeding it.

Something related to this is the **Component Mask** node. This node works in a relatively opposing way to the previous **Append** one: instead of combining simpler vectors into a single entity, it allows us to choose specific channels out of its required multi-channel input. That can be great for us when we only need to modify specific texture channels, as we did in the previous pages.

The last aspect I'd like to go into a bit more detail is the **Pixel Normal Offset** refraction mode. We used this technique instead of the traditional **Index of Refraction** mode we saw in the first recipe of this chapter because it tends to work better on large and/or flat surfaces. Even though we might think that the IOR method is closer to real-life refraction mechanics, the implementation that Unreal has chosen for this system is not always the best option when trying to mimic reality. Without getting too technical, their IOR implementation makes some assumptions that can introduce artifacts in the type of models that we mentioned previously. To counter this, Epic has introduced the **Pixel Normal Offset Refraction** method as an alternative.

This new refraction model works by using the vertex normal data to calculate the refraction offset that we see. This is possible by computing the difference between the per-pixel normal against that of the vertex one, offering the results that you see onscreen, which match real-world scenarios better than the IOR method.

See also

As always, make sure you read Epic Games' official documentation if you want to learn more about these topics: `https://docs.unrealengine.com/5.0/en-US/refraction-using-pixel-normal-offset-in-unreal-engine/`.

Faking caustics with a Light Function

Nailing down the right look for a water surface can greatly increase the realism of our scenes, especially when dealing with tricky materials such as the ones that make use of translucency. Getting those nice reflections to show up, the refraction working in the right way, and the animation to feel believable is something we tackled in the previous recipe. On top of that, we can add an extra effect that is often seen in certain bodies of water: caustics.

This technique, which tries to mimic how the refracted envelope of light rays that the surface of a water body is projecting onto another object works, is difficult to calculate in real-time renderers. That being the case, we usually rely on approaches that try to fake the effect rather than realistically show it. We'll be exploring one such method in the following pages.

Getting ready

Since this recipe is going to follow in the footsteps of the previous one, there's nothing new that you'll need this time around. All of the same considerations we took into account some pages ago still apply here: we'll need a plane that can act as a body of water and a model to act as its container. With regards to the new bits that we'll be using, all of them are provided as part of the Starter Content, so make sure you include that!

If you want to follow along using the same scene and assets, make sure you open the level called 03_05_Start and buckle yourself for some sweet Unreal Engine magic.

See you in the next section!

How to do it...

As mentioned previously, caustics are usually faked within the engine rather than computed in real time as there's no current method that allows for that operation to run at a high refresh rate. As such, we'll fake them with something called a Light Function, which is a type of asset that is applied to lights in UE5. With that in mind, let's place the light that will support this technique:

1. Add a **Spotlight** to the scene. This type of light is quite helpful when dealing with caustics since it has a couple of parameters (specifically, **Outer Cone Angle** and **Inner Cone Angle**) that allow us to have a nice area where the light isn't fully on or off.

2.   Place this Spotlight above the water plane and tilt it, as seen in the following screenshot:

Figure 3.27 – Suggestion regarding the placement of our Spotlight above the water surface

3.   Now, let's focus our attention on the **Details** panel. Set the **Mobility** parameter of our new spotlight to **Movable**.

4.   Play a little bit with the other values for the selected light – tweak its **Intensity** and adjust its **Inner Cone Angle** and **Outer Cone Angle**. I've set the first of those parameters to read **2.3** candelas and used **20** degrees for **Inner Cone Angle** and **42** for **Outer Cone Angle**.

5.   Look a bit further down in the **Light** category of our **Spotlight** and expand the **Advanced** panel, which contains another expandable area called **Lighting Channels**. Uncheck **Channel 0**, which should be on by default, and check **Channel 1**.

6.   Select the **SM_BathTub** actor and enable the **Channel 1** option within the **Lighting Channels** expandable area, which you'll be able to find by expanding the **Advanced** section of the **Lighting** category within the actor's **Details** panel. Make sure that both **Channel 0** and **Channel 1** are enabled this time, unlike what we did with the previous spotlight.

> **Important note**
>
> Assigning a specific Lighting Channel to a mesh means that said model will only be affected by lights that make use of the same Lighting Channel. Keep that in mind when you want to create special effects that only affect certain surfaces.

With that out of the way, we can finally dedicate our efforts to building the Light Function through which we are going to emulate the caustic effect. We'll start this process by heading over to the Content Browser and creating a new material.

7. Create a new material and perform the usual steps – give it a name (I've chosen **M_WaterCaustics**) and save it.

8. Open the new asset in the Material Editor and head over to the **Details** panel. Change **Material Domain** to **Light Function**, which will enable us to describe a material that can work as a light.

9. Assign the new material to our **Spotlight**. To do so, head back to the main editor and select the spotlight. Look at its **Details** panel – there's a category there called **Light Function** that you'll need to adjust by setting it to use the new material we have just created.

   Having the material applied to the light will allow us to see the changes we make. With that done, let's start modifying it!

10. Create a **Texture Coordinate** node, as we need one of these whenever we want to alter the size or the position of our textures.

11. Add a **Multiply** node after the previous **Texture Coordinate** node and connect the output of that one to the **A** input pin of the new node.

12. Create a **Scalar Parameter** somewhere below the **Texture Coordinate** node. We will use this to change the tiling of the **Texture Sample** node we'll create later instead of relying on the **Texture Coordinate** node, simply because scalar parameters can be tweaked in Material Instances, unlike **Texture Coordinate** nodes. Set its default value to **1** and name it something along the lines of **Texture Tiling**.

13. Connect the **Scalar Parameter** node to the **B** input pin of the previous **Multiply** node.

   For organizational purposes, let's refer to the previous set of nodes simply as *Part A*. This will help us identify this part of the graph whenever we need to come back to it at a later stage.

14. Create a couple of scalar parameters; name one of them **X-Axis Speed 01** and the other one **Y-Axis Speed 01**. Feel free to name them whatever you want, but know that this is what we'll be using them for: to define the speed at which a future texture reference is going to move. Set their values to **0.01** and **0.025**, respectively.

15. Create a **MakeFloat2** node. As this is a new node, know that you can do so if you right-click anywhere within the material graph and start to type that name.

16. Connect the **Scalar Parameter** node that affects the horizontal speed to the input pin named **X (S)** of the **MakeFloat2** node and the one that affects the vertical speed to the pin labeled **Y (S)**.

17. Drag a wire from the **Result** output pin of the **MakeFloat2** node and create a **Multiply** function at the end of it. It will automatically connect itself to pin **A** and leave pin **B** with the default value of **1**, which is exactly what we want at this point.

   We can call these last few nodes *Part B*. With these two parts created, we can create the next bit, which is going to tie both groups together.

18. Add a **Panner** node.

19. Connect the output of the **Multiply** node at the end of *Part A* to the **Coordinate** input pin of the new **Panner** node, and then connect the output of *Part B* to the **Speed** input pin of the same **Panner** node.

This is what the material graph should like by now:

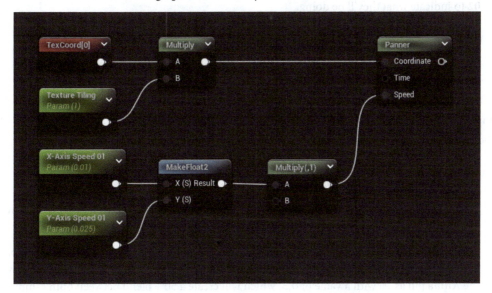

Figure 3.28 – The current state of the material graph

> **Tip**
> Feed a 2D vector to the **Speed** input pin of a **Panner** node! That's why we used the **MakeFloat2** node — to control the **Speed** input pin in the X and Y axes independently. Had we only used a **Constant**, we would be tweaking both speeds with just one value, resulting in a diagonal movement.

20. Drag a wire out of the output pin of the **Panner** node and create a **Texture Sample** node, which is going to be affected by everything we've created before it.

21. Select the **T_Water_M** texture as the value for our new **Texture Sample** node. This asset is part of the Starter Content, so be sure you include it if you haven't done so already.

Every step we've performed so far has left us with a network of nodes that we can use to drive the appearance of the Light Function we are creating. We are already in a good position when it comes to achieving the looks that water caustics are supposed to have – simply connect the previous **Texture Sample** node to the **Emissive Color** property of our material and apply it to the spotlight in our scene to see what that achieves. Despite that, we still have room for improvement, especially when it comes to making the effect even more subtle. We'll take care of that next.

22. Make a copy of all the nodes that we have created so far and place those duplicates somewhere below the originals.

23. Change the names of the duplicate scalar parameters located in the new section of the graph we have just created. I've gone with **Texture Tiling 02**, **X-Axis Speed 02**, and **Y-Axis Speed 02** to indicate what they'll be doing.

24. Modify the values of the duplicate scalar parameters. Making them similar but different from the original ones will work well when we eventually mix them as we intend to create a slightly different pattern. I've gone with the following values: **1.5** for the tiling, **-0.008** for the horizontal speed, and **-0.0125** for the vertical one.

> **Important note**
> The values in those fields can change drastically, depending on the type of surface that you are working on and how its UVs are laid out. Feel free to experiment with them if you are using your own assets.

Having an almost identical copy of the original network of nodes will let us mix both, creating a nice, subtle effect. With that as our objective, let's proceed with the mixing part.

25. Add a **Lerp** node after all of the existing ones. Connect the output pin of the first **Texture Sample** node to its **A** input pin and connect the output of the second sampler to the other **B** input. Leave the **Alpha** pin as-is, with a value of **0.5**, which will create a 50% blend between both samples.

26. Include a **Multiply** node right after the last **Lerp** node.

27. Create another **Scalar Parameter** and name it **Overall Intensity**. We'll use it to control the strength of the Light Function. Assign a value of **1** to it.

28. Connect the output of the **Lerp** node into the **A** input pin of the **Multiply** one, as well as the output of the last **Scalar Parameter** to the **B** input pin.

29. Plug the result of the **Multiply** node into the **Emissive Color** property of our **Main Material** node.

Even though we are almost finished, we can still add one more **Scalar Parameter** to control the overall speed of the effect. Do you remember that we left an unconnected pin on the **Multiply** node located after the **MakeFloat2** one? We'll take care of that now.

30. Create a **Scalar Parameter**. Assign a value of **1** to it and place it right at the beginning of the material graph. Name it something along the lines of **Overall Speed**.

31. Hook the output pin of the previous **Scalar Parameter** to the **B** input pins of both of the **Multiply** nodes located right after the **MakeFloat2** nodes.

32. Finally, compile and save your work.

With all of those changes in place, this is what our scene should now look like:

Figure 3.29 – The final look of our scene

That will do it for this recipe. See you in the next one!

How it works...

Light Functions are great! You already knew that, but the way we've used them in this recipe highlights Unreal's versatility. Think about it: we've used a light. However, this light doesn't cast any shadows, and the light it emits only affects certain surfaces – a direct result of setting it to work on a specific lighting channel that's removed from most of the objects in the level. Pretty uncommon behavior as far as lights go!

To achieve that, one of the things we did was change the Material Domain used by the material acting as the Light Function. The function of that parameter is quite straightforward: we need to adjust it according to the type of actor that is going to use the shader. We can choose from several options, ranging from **Surface** to **User Interface**, with **Deferred Decals**, **Light Functions**, and **Post Process** as the other options. Each of those types indicates the type of objects or entities that might use the shader: if we want to use a material within a UI element, we'll need to choose the **User Interface** option for the Material Domain, which will be different from the one used by a material that is meant to be applied on a 3D model (which use the standard **Surface** for this).

See also

You can find more information regarding Material Domains and Light Functions in Unreal's official docs:

- `https://docs.unrealengine.com/4.27/en-US/BuildingWorlds/ LightingAndShadows/LightFunctions/`

- `https://docs.unrealengine.com/4.26/en-US/Resources/ ContentExamples/MaterialProperties/1_5/`

Animating a sea shader

Even though we've worked with water in the past two recipes, I couldn't pass up the opportunity to talk about large-scale ocean shaders in UE5. I'll admit it: this is one of my favorite subjects in computer rendering, and achieving great results is completely possible thanks to the methods we are about to see. However, this is not going to be a mere expansion of any of the previous topics. Instead, we are going to continue to learn new techniques and apply new concepts throughout the following pages. So, buckle up – there's no time to rest in our journey!

Getting ready

There are going to be some big visual changes in this recipe but, despite this, everything you'll need will be either provided by the project bundled with this book, included in the Starter Content, or part of the engine. Since we are going to be working on a large ocean material, it makes sense to move away from the familiar interior scene we've been working on to a large outdoor environment. The level we'll be working on is called `03_06_ Start`, so make sure you open it if you want to follow along using the same instructions we'll be covering in the next few pages.

But what if you want to apply what we'll be learning to your own projects, you wonder? In that case, here are the basic building blocks you'll need: an evenly subdivided plane, and the **BP_LightStudio** blueprint that is a part of the Starter Content. We'll need a plane that has enough subdivisions so that we can fake the motion of the waves and a light to see it. That's it!

How to do it...

Working with oceans usually means dealing with large-scale surfaces, and as we've discovered in the past, that circumstance comes with its own set of challenges. Having textures that work well when viewed up close and from afar can prove difficult, but that is something that we've already tackled in the past when we created a concrete surface back in *Chapter 2*. What we didn't do back then was create an animated material that looked nice at those two different scales, something that will require us to tackle things a bit differently this time around. On top of that, we'll also make sure we include other new and interesting techniques, such as adding sea foam, depending on the height of the waves that we create. There are plenty of new things to discover, so let's jump straight into them:

1. With your scene already set up – either by opening the one we provided or after creating your own – create a new material in the Content Browser. I've named mine **M_OceanWater**, which is self-explanatory!

    As one of our main worries will be to hide any repetition that may become obvious in our material, we'll start by defining two sets of randomly-generated waves – one smaller and one larger in scale. We'll call this first bit *Part A*, dedicated to the creation of the small waves.

2. Create a **Material Function Call**, either by holding the *F* key on your keyboard and left-clicking in an empty space of the material graph or by right-clicking and typing its name.

3. With the new node selected, look at the **Details** panel and expand the drop-down menu that allows us to choose a Material Function. You won't see too many options available at this stage, mostly because we've only created one Material Function so far. Don't worry, though: click on the cog icon that appears when the drop-down menu is expanded and check the **Show Engine Content** option in there.

4. Choose the **Motion_4WayChaos_Normal** Material Function. This handy resource will create a randomized animated motion based on a normal texture that we need to provide as input. Let's proceed and set up its input parameters.

5. Create an **Absolute World Position** node. To retrieve this node, simply type the words `World Position`, as that is the real name of this function. The *"Absolute"* prefix appears by default when we create the node, as that is the default subtype that the function is set up to use.

6. Throw two scalar parameters into the mix. The first one, which I've named **Water Scale**, will be used as a general control for the scale of our ocean. The second one, named **Small Wave Scale**, will, in turn, control the specific scale of the smaller waves. Set their values to **1** and **256**, respectively.

7. Add a **Multiply** node and connect its **A** and **B** input pins to the previous two scalar parameters.

8. Next, create a **Divide** node and place it after the **Absolute World Position** node. Connect its **A** input pin to the output of the **World Position** node and wire its **B** input pin to the output of the previous **Multiply** node.

9. Place a **Component Mask** after the **Divide** node. Select the **Red** and **Green** channels in the **Details** panel.

10. Connect the output of that network of nodes to the **Coordinates/UVs** input pin of the original Material Function.

    This first part that is feeding into the **UVs** input pin of the **Four-Way Chaos** function defines how this node will project itself onto the surfaces we apply this material to. On the one hand, the **World Position** node gives us a projection that doesn't rely on the UVs of the model. We affect the scale of this projection through the two scalar parameters we created, and we finally mask the red and green channels as we need a two-coordinate vector to feed into the final input pin. With the coordinates under control, let's affect the **Speed** and **Texture** parameters.

11. Create a couple of extra scalar parameters call them **Small Wave Speed** and **Water Speed**. This is the same approach we took when we created the previous scalar parameters that affected the scale of the small waves – except this time, it's the speed we are controlling! Assign the default values of **0.2** and **1.5**, respectively.

12. Next, add a **Multiply** node and connect the previous scalar parameters to it. After that, connect its output to the **Speed** input pin of the **Four-Way Chaos** function.

    Include a **Texture Object Parameter**, assign the **T_ Water_ N** texture to it, and connect it to the **Texture** input node of our previous function. I've gone with the name **Normal Texture Object** as the name of the new parameter.

    We've created the different scalar and texture object parameters to eventually create a Material Instance that lets us adjust all of the previous settings more easily. With that done, we can say that we are almost finished with this section of the graph, which for reference will be called *Part A*. We still need to create a couple more nodes before moving on to a different section, so let's do that now.

13. Create two **Constant3Vector** nodes. Assign a value of **0,0,0** to the first one and set the other to a value of **1,1,0.2** instead.

14. Create a **Scalar Parameter** and name it **Small Wave Amplifier**. Set its value to **0.5**.

15. **Lerp** between the first two Constant3Vectors and use the previous **Scalar Parameter** as the value driving the **Alpha** channel.

16. Add a **Multiply** node after the **Four-Way Chaos** function and connect the output pin of the latter to the **A** input pin of the former. Remember to also connect its **B** input pin to the output of the previous **Lerp** node.

    Implementing all of these steps will see us finished with *Part A* of the material, which controls the scale, speed, and magnitude of the small-scale waves:

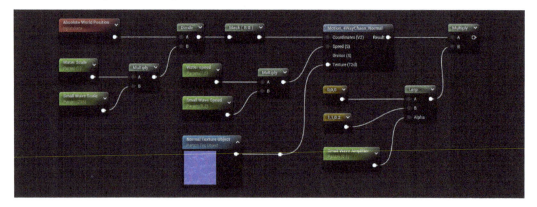

Figure 3.30 – The current look of the material graph, which we'll refer to as Part A

Having this section labeled as *Part A* will enable us to quickly duplicate the contents we've just created and generate a second set of waves. This copy will be larger in scale, which will drive the name of the scalar parameters we'll use there. We'll call this next section of the graph *Part B*.

17. Copy all of the nodes that we have created up until this point (what we've come to call *Part A*) and paste them a little bit further down the material graph. This next section is going to be called *Part B* and it will control the scale, speed, and amplitude of the large-scale waves.

18. Rename the copies of the **Small Wave Scale**, **Small Wave Speed**, and **Small Wave Amplifier Scalar** parameters to something that indicates the role of the new ones. I've gone with **Large Wave Scale**, **Large Wave Speed**, and **Large Wave Amplifier** this time around.

19. Change the values of the **Large Wave Scale**, **Large Wave Speed**, and **Large Wave Amplifier** parameters. The new values should be something like **1024**, **1.5**, and **0.9**, respectively. Also, change the values of the Constant3Vectors that feed into the **Lerp** node being driven by the **Large Wave Amplifier** parameter. The new values should be **1,1,3** and **1,1,0.1** in the **A** and **B** pins, respectively.

20. As a small addition, include a **Rotator** node between your **Component Mask** and the **Four-Way Chaos** function in this *Part B* of the material. This will spice things up a little bit as we would otherwise end up with just a larger-scale version of the previous texture.

21. Feed the **Time** input pin of this rotator with a simple constant, which I've given a value of **1.342**.

This is what *Part B* should look like:

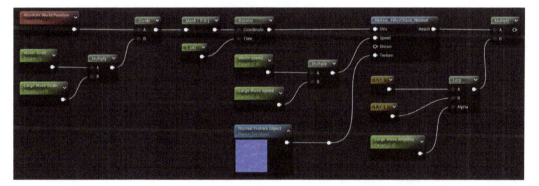

Figure 3.31 – Part B of the material graph

The previous changes have left us with a functional *Part B* of the graph, identical to the other one except for the names of the different scalar parameters and the values that we've applied. This means that we now have two functions that are creating two patterns of waves of different sizes, animated at different speeds, and that have different normal intensities. You can see the results for yourself by connecting the output of each part to the **Base Color** input of the material, as shown in the following screenshot:

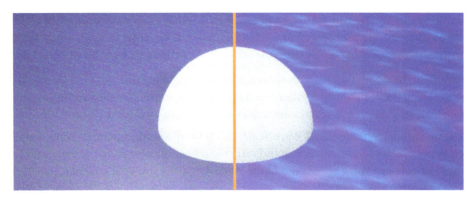

Figure 3.32 – A side-by-side view of the small and large-scale wave patterns

As you can see, we've managed to create a couple of different animated wave patterns. The reason why we want to have two is so that we can blend between them, to remove the repetition that we would see if we only used one. This will be useful to drive other parameters of the material, as we'll see later in this recipe, but something we can already do at this stage is blend both parts and use that to drive the **Normal** property of the material.

22. Create an **Add** node, place it after both *Parts A* and *B*, and connect the output of those two sections to the input pins of this new node.

23. Connect the output pin of the previous **Add** node to the **Normal** input pin of the **Main Material** node.

    Those actions have taken care of defining the **Normal** property of the material. However, we still need to address some of the other parameters, such as **Base Color**, **Displacement**, and **Roughness**. Seeing as this will be the first time we will deal with **Displacement**, we'll focus on that property next.

24. Copy what we've come to call *Part B* of the material and paste it somewhere below in the material graph. We'll call this new section *Part C*, which we'll modify now.

25. Delete the **Multiply** and **Lerp** nodes located after the **4 Way Chaos** function, along with the two Constant3Vectors and the scalar value connected to the **Lerp** node.

26. Select the **Motion_4WayChaos_Normal** Material Function and change it to the one called **Motion_4WayChaos**. Be careful when selecting this new resource as its name is very similar to the previous one being used.

27. Next, change the texture being fed to **Texture Object Parameter** to the one named T_Water_M. This is a heightmap version of the previous one, also included with the Starter Content. While you're at it, change the name of this parameter to **Large Wave Height**.

28. Add a **Power** node after the previous function and connect the output of the **Motion_4WayChaos** node to its **Base** input pin.

Completing all the previous steps has seen us finish dealing with the new *Part C* of our material, which will be driving the displacement of the models that have this shader applied to them. *Parts B* and *C* work in a very similar way; they define the scale and speed of the large waves. However, while *Part B* focused on calculating the normals, *Part C* deals with calculating the height of those waves, which is going to be useful to drive the displacement of our model, as well as the location of the foam in the shader – an effect we'll introduce later on.

At this stage, we are going to branch off *Part C*, as we'll need this to drive the location of the sea foam and the displacement of the material. The displacement can be used almost as-is, so let's focus on that for the time being.

29. Add a **Scalar Parameter** and name it **Luminance Bias**, which we'll use to control the intensity of the height map driving the displacement. Give it a value of **0**. This will control how high or low the displacement effect will be in our world.

30. Create an **Add** node and connect both the output of the **Luminance Bias** node and that of the **Power** node from *step 28* to its input pins.

31. Mask the **Red** channel with a **Component Mask**. This will give us a grayscale texture that we want to store in a different channel – the blue one in particular, or the *Z* value as it is sometimes known.

32. Create an **Append** node and a **Constant2Vector**. Give the vector a value of **(0,0)** and connect it to the **A** input pin of the **Append** node. Connect the previous mask we created to input pin **B**. This will effectively create a three-component vector and store the height map as the *Z* value.

33. Include a **Multiply** node to increase the intensity – connect its **A** input to the previous **Append** node and set the **B** value to **5**.

34. Create another **Multiply** node and a **Scalar Parameter**. This latest setting will control the displacement of the material, so name it just that and give it an initial value of something like **10**. Connect the new **Scalar Parameter** and the output of the previous **Multiply** node to the inputs of the new one.

35. Right-click and create a **Vertex Normal** vector by typing its name into the search box. This will give us the world space value of the vertex normals our material is applied to, which will let us displace the model.

36. Add the output of the **Vertex Normal** vector to the previous **Multiply** node by creating an **Add** node, and connect its output to the **World Position Offset** input pin of the material:

Figure 3.33 – The previous set of nodes we've created that drive the displacement property of the material

With that out of the way, let's tackle the **Base Color** property of the material. The first step in that journey is going to be creating a sea foam texture. Let's do that now.

37. Copy *Part C* and paste it between *Parts B* and *C*. Create space in between them if you don't have any right now.

38. Rename some of the scalar parameters; we should have a **Seafoam Scale** instead of the previous **Large Wave Scale**, and **Seafoam Speed** instead of **Large Wave Speed**.

39. Set the **Water_d** texture as the default value for **Texture Object Parameter** and change its name to **Sea Foam Texture**.

Before we continue, we need to make sure that we only show the sea foam texture where it makes sense – at the crest of the large waves. Let's learn how we can take care of that.

40. Remove the **Rotator** node (and the **Constant** node feeding into its **Time** input pin) located between **Component Mask** and the **Motion_4WayChaos** Material Function, and place an **Add** node there instead. This **Component Mask** should be fed into one of the input pins of the **Add** node, whereas the output of the **Add** node needs to be connected to the Material Function again. Remove the **Power** node located after the **Motion_4WayChaos** Material Function as well.

We should now have an empty input pin on that last **Add** node. We'll come back to this in a few steps, so keep this in the back of your head.

41. Head back to *Part B*, the section controlling the normals of the large waves, and drag another wire out of the Material Function (**Motion_4WayChaos_Normal**). Create a **Component Mask** and select the **Red** and **Green** channels.

42. Continue to drag another wire out of the output pin of the new mask we created in the previous step and create a **One Minus** node.

43. Create a **Multiply** node.

44. Add a **Scalar Parameter** and name it **Foam Distortion**. Hook it into the **B** input pin of the previous **Multiply** node and assign a value of **0,2** to it.

45. Connect the output of the **One Minus** node to the **A** input pin of the **Multiply** node we created in *step 44*.

46. Connect the result of the **Multiply** node to the remaining available input pin of the **Add** node we created in *step 40*.

47. Connect the output of the **Add** node we created in *step 40* to the **UVs** input pin of the **Motion_4WayChaos** Material Function.

We'll refer to *steps 38* to *47* as *Part D*. Those steps have enabled us to include a small sea foam texture that will be driven in part by the large waves. We will use this to modify the appearance of our material. It is now time to tackle the **Base Color** property of the shader, a process that will involve several different steps. The first of them will see us trying to mimic the effect that we see when we look at the ocean: the color seems to change according to the position and movement of the waves, where surfaces that are pointing toward us seem darker than those that are parallel to our vision. Before that, though, let's look at *Part D* of the material graph:

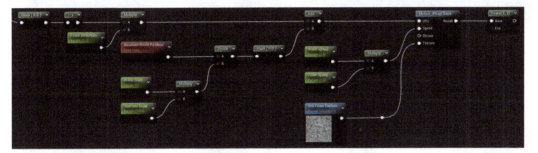

Figure 3.34 – Part D of the material graph

48. Drag a wire from the **Add** node we created in *step 22* (the one that combines *Parts A* and *B*) and create a **Normalize** node at the end of it. This expression takes the vector values we give it and returns a unit vector to us (a vector with a length of 1).

49. Add a **Transform Vector** node after the previous **Normalize** one (simply type `Transform` to find it). This function converts the input tangent space values we feed it into world space ones, something that we need at this point since those are the values we got from the previous operations that lead us to this point.

50. Create a **Fresnel** node and connect the previous **Transform Vector** node to its **Normal** input pin. With the node still selected, set **Exponent** to **5** and **Base Reflect Fraction** to **0.04**.

51. Include a **Clamp** node after the previous Fresnel and leave the **Min** and **Max** values untouched.

Let's refer to the previous part of the graph as *Part E* for future reference. All of those nodes have given us a mask that treats the areas of the material that aren't facing the camera differently from those that are and, what's even more useful, they are also taking the waves into account. We finally have something that can help us drive the appearance of the different parts of the ocean shader. Next, we'll tackle the **Base Color** property of the shader, which we'll call *Part F*.

52. Create two vector parameters. Name the first one **Water Camera Facing Color** and the second one **Water Side Color**. Assign values that you think will work well in such a material – maybe a deep blue in those areas that face the camera, and a lighter version on those that face away from it. You can also give them wildly different colors if you want to see the effect in action in a more obvious way.

53. **Lerp** between the previous two vector parameters by creating one such node and using the output of *Part E* as the **Alpha** value.

Doing this will allow us to see the effects that those changes have on our scene. Let's take a look at them:

Figure 3.35 – The appearance of the ocean shader thus far

At this point, all we need to do is merge the preceding colors with the sea foam texture we previously selected. To do so, we still need to implement a little bit of code… all aboard for this last part!

54. Go back to the **Power** node located at the end of *Part C* (where we calculated **Large Wave Height**, the one we created in *step 28*) and drag a wire from it to create a new **Power** node at the end.

55. Create a **Scalar Parameter**, which we'll use to control how much foam we can see. I've named it **Seafoam Height Power** and given it a value of **2**. Connect it to the **Exp** pin of the previous **Power** node.

56. Create a **Multiply** node and connect its **A** input pin to the output of the previous **Power** node.

57. Create another **Scalar Parameter**, which we'll use to control the tolerance at which the sea foam is placed. I've named it **Seafoam Height Multiply** and given it a value of **2,048**. Connect it to pin **B** of the previous **Multiply** node.

58. Create a **Component Mask** after the previous **Multiply** node and select the **Red** channel as its only option.

59. Limit the scope of the output of the previous **Component Mask** to the 0 to 1 range by creating a **Clamp** node. This is the graph we should be looking at:

Figure 3.36 – A look at the previous set of nodes

We'll refer to the previous set of nodes as *Part G*. They control the height at which the foam is going to be positioned, so they are bound to come into play soon.

60. Drag a wire out of the output pin of the **Lerp** node we created in *Part F*, the one that interpolates between the colors of the water, and create another such node at the end of that wire, which should automatically connect itself to the output of the previous **Lerp** node.

61. Connect its **B** input pin to the output of the **Motion_4WayChaos** Material Function from *Part D*, where we created the sea foam.

62. Create a **Component Mask** with the **Green** channel selected after that same **Motion_4WayChaos** function at the end of *Part D*. Connect the output of that mask to the **Alpha** pin of the **Lerp** node we created in *step 60*.

The previous steps have enabled us to interpolate between the selected color of the water and the sea foam texture, all according to the motion of the large waves. We also need to introduce a final requirement so that the foam only shows on the highest parts of the large waves, which we calculated in *Part G*.

63. Drag another wire from the **Lerp** node located in *Part F* and create a new **Lerp**.

64. Connect the output of the **Lerp** node we created in *step 60* to the **B** input pin of the new Lerp node. The **Alpha** pin should be hooked to the output of the **Clamp** mode from *Part G*, the one controlling the position of the seafoam.

65. Hook the output of this final **Lerp** node into the **Base Color** property of the material node.

The only thing we need to do now is create a couple of constants to define the **Metallic**, **Specular**, and **Roughness** properties of the material. We can create three such scalar parameters and assign a value of 0.1 to define the metalness, 1 for the specular (to boost those reflections), and 0.01 to the **Roughness** slot, since we want to see the details reflected in the surface of our water body. Make sure you create a Material Instance after doing that and tweak the values until you are happy with the result. Here's what the scene should look like by the end of this journey:

Figure 3.37 – A final look at the ocean material we created in this recipe

How it works...

We've done a lot in this recipe! I'm sure things will take a while to settle and become clear, but let's see whether we can expedite that process with the next few sentences. At the end of the day, it's always useful to think about any complex materials in terms of what we are doing instead of the individual steps that we are taking. If we break things down into smaller pieces, such as the different parts we've mentioned in this recipe, things can become clear. Let's review our work!

Parts A and *B* are quite straightforward: we created two animated wave patterns, one small and another large in scale, thanks to a Material Function that the engine includes. We also introduced a small variation to the large wave pattern by introducing a rotator node. *Part C*, which is almost a copy of *Part B*, focused on calculating the height of the large waves, a valuable piece of information that we used to drive the displacement of the model onto which we applied the material.

Beyond that, we used *Part D* to create the sea foam texture. Unlike the previous sections, which were completely independent, *Part D* relies on the large-scale waves as we only want the foam to appear on their crests. Similarly, *Parts E* and *F* take the values of the normals calculated in *Parts A* and *B* to drive the appearance of the base color property of the material, as we want the parts of the water that face away from the camera to look different to the ones that directly face it. In particular, *Part E* focuses on defining the mask that *Part F* then uses to drive the different colors used by the **Base Color** property. Finally, *Part H* tackles the creation of the mask, which allows us to place the sea foam texture only on the areas where we want it to appear.

That's pretty much it! I know it's still a lot to take in, so my suggestion here is to go back to the Material Editor and try to replicate some of the techniques we've used on smaller, more manageable materials. Doing so will enable you to make sense of the individual bits that make up this big material, so don't lose the chance to perform a final review on your own. On that note, don't forget about the ability to preview certain parts of the graph by selecting the **Start Previewing Node** option available when we right-click on a given node. This feature should enable us to take a closer look at what we are creating without the need to implement more complicated workarounds. Make sure you give it a go!

See also

One of the things we used the most in this recipe was the **4 Way Chaos** Material Function. Even though we've used several useful nodes in the past, functions are kind of new to us, and going back to Unreal Engine's official docs can be a great way to learn more about them: `https://docs.unrealengine.com/4.27/en-US/RenderingAndGraphics/Materials/Functions/`.

Additionally, we all know that this last recipe has been a bit on the long side, but the different parts we've tackled are very important and quite advanced stuff! As an assignment, try to test yourself and go over the different steps we've reproduced by creating a similar material once again, introducing your own tweaks. You'll gain a lot of knowledge and hands-on experience when doing so!

4

Playing with Nanite, Lumen, and Other UE5 Goodies

Having arrived at the fourth chapter of this book, we can safely say that we've already had the chance to go over a few of the different rendering techniques available in Unreal. Putting that experience aside for the moment, it's time for us to delve a little bit deeper into the new technologies that have arrived alongside the latest version of the engine: Nanite, Lumen, and hardware-accelerated ray tracing effects.

With that in mind, our goal for this chapter is going to be exploring these new rendering features. We'll study each in their own dedicated recipe, including Nanite, Quixel Megascans, Lumen, and software and hardware ray tracing—while also looking at some extra ones, such as more realistic thin glass surfaces.

Here is what we'll be learning next:

- Taking advantage of Nanite and Quixel Megascans assets
- Using software and hardware ray tracing
- Revisiting screen-space and planar reflections
- Creating an arch viz scene with realistic-looking glass and virtual textures
- Varnishing wood through the Clear Coat Shading Model

And here is a sneak peek at some of the content with which we'll work:

Figure 4.1 – Some of the models and materials we'll be studying in this chapter

Technical requirements

As with the previous chapters, most of the things that you'll need to tackle the recipes in this chapter are included within Unreal Engine 5. Just as before, we also make the Unreal Engine project available for download through the following link: `https://packt.link/20u7B`

Something that we need to highlight is the need for special hardware in an upcoming recipe. One of the topics that we'll cover deals with hardware ray tracing, a feature that requires a graphics card with ray tracing acceleration capabilities: any of the RTX series from Nvidia, the Radeon RX line from AMD, or the Arc category from Intel will do.

Despite that, we'll also take a look at software ray tracing for those users that lack those hardware-accelerated features, so don't worry if that's you—we've got you covered!

Taking advantage of Nanite and Quixel Megascans assets

The fifth installment of Unreal Engine came bundled with a couple of surprises—the inclusion of the Quixel Megascans library and Nanite being two of them. The first is a repository that contains hundreds of high-quality photo-scanned assets, which mostly consist of 3D models, decals, and materials. The second one is none other than the engine's virtualized geometry system, a feature that allows us to render extremely detailed models without breaking a sweat.

We'll be looking at these two new additions to the engine through Quixel Bridge, a native app that runs within Unreal that allows us to view, download, and integrate any of the Megascans content directly into our projects. Let's explore how to take advantage of them in the next few pages.

Getting ready

Something that we'll do in this recipe is build a level using the available Quixel Megascans assets. Given how we are going to download different models from that library, you don't need to worry about creating any of your own to tackle this recipe. On top of that, you can also open the level named `04_01_Start` if you want to use the same map I'll be working on over the following pages.

Even though we don't need to have a very complex scene ready for this recipe, we do need to make sure that we have both an Epic Games account and a Quixel account. This is a requirement if we want to use **Quixel Bridge** within the engine, the native Unreal Engine 5 extension browser that lets you find and integrate any content from the Quixel Megascans library directly within the engine. Let's take a look at how to do this before we start the recipe:

1. Create a Quixel account by visiting `https://quixel.com/`.
2. Open Unreal Engine 5 and navigate to **Window** | **Quixel Bridge**.

3.   Once there, look in the upper-right corner and click on the **User Profile** symbol.

4.   Sign in using your Epic Games account. Once you reach this point, or before attempting to download any of the Megascans assets, you will be prompted to link your Epic Games and Quixel accounts. Follow the instructions on the screen to do so and enjoy all of the content at your disposal once you complete the setup.

We'll be able to start working on the recipe once we complete all of the previous steps. It's important to note that the assets we'll be using can't be redistributed as part of an Unreal Engine project to a different user, so you won't see them within the project shipped alongside this book.

How to do it...

Free assets at our disposal? Let's start filling the coffers! I love getting my hands on 3D models and admiring the beauty of this form of art, so get ready to spend a lot of time admiring these assets if you are anything like me. Don't get carried away, though—we are here to populate a scene using Quixel Megascans assets. Let's start that journey:

1.   Open Quixel Bridge by navigating to **Window | Quixel Bridge**.

2.   Once inside, search for any assets that you'd like to use. I've chosen a plate, a bowl, and some grapes, trying to create a small still-life scene. Here is where you can find them:

     • The grapes can be found within **Home | 3D Assets | Food | Fruit**.

     • The plate is one of the ones available within **Home | 3D Assets | Interior | Decoration | Tableware**.

     • Similarly, you can download the bowl by heading over to **Home | 3D Assets | Interior | Decoration | Bowl**.

     Once you get to a particular asset that catches your attention, simply click on it and hit the **Download** button—the one located in the lower-right corner of the scene.

     Unreal will begin fetching the assets once you click on the **Download** button. This is a process that can take a bit of time to be completed, something that will vary according to the speed of your internet connection.

3.   Unreal will let you know whether it has finished the download process thanks to a visual cue that appears around the **Download** button itself. Once this operation is complete, click the **Add** button located next to **Download**. This will add the asset to the current project.

We should be looking at a screen similar to the next one throughout this entire process:

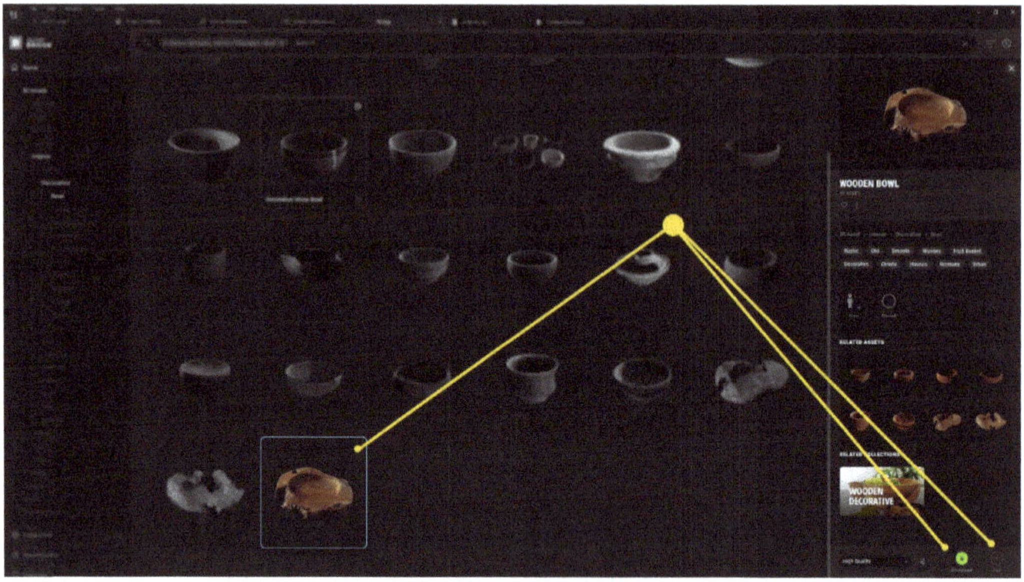

Figure 4.2 – The Quixel Bridge interface

> **Important note**
>
> Megascans assets get downloaded to a location in your computer once you hit **Download**, so you might want to remove them from there once you no longer need them, as it's not enough to delete them from your projects. To do so, head over to the **Local** category within the **Quixel Bridge** window (the bottom option in the left-hand side panel) to see what's stored on your computer. Right-click on any asset and choose the **Go To Files** option to navigate to their directory within Windows Explorer, where you'll be able to manually delete them.

It's important to note at this point that there are several options available to choose from when it comes to the quality of the assets that we download. You might have noticed a little drop-down menu to the left of the **Download** button: it is here where we can choose the quality of the assets that we get. There are several options available to us: **Low**, **Medium**, and **High** quality. These impact the size and the resolution of the assets that get downloaded onto our computers—especially when it comes to the textures.

On top of this, there's also a new option available on some models simply called **Nanite**. This feature is usually available on very high poly models that can benefit from the new virtual geometry system introduced in this new version of the engine: more on that in the *How it works...* section later in this recipe. For now, let's focus on what we can do with the models that we've downloaded.

4. Drag and drop the models that you've downloaded into the scene and arrange them in whichever way you see fit. Try to create something visually appealing but don't spend too much time on it—our focus will quickly shift to inspecting those assets. Take a look at the following screenshot in case it works as inspiration for how to composite your level:

Figure 4.3 – The composition of the scene

None of the assets we've downloaded have Nanite enabled, but they look great despite that. Well, perhaps with the exception of the grapes—they look a bit polygonal when viewed up close. We still have room to improve their appearance with just a few tweaks, though, so let's tend to that now.

5. Navigate to the model of the grapes in the Content Browser and double-click on it to open the Static Mesh Editor.

6. Look at the **Details** panel and check the option called **Enable Nanite Support** located within the **Nanite Settings** section.

7. Click on the **Apply Changes** button at the bottom of the same **Nanite Settings** section.

   Performing the previous steps will see a noticeable improvement in the quality of the asset being rendered:

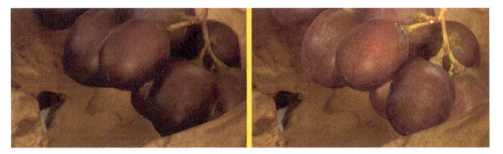

Figure 4.4 – The model of the grapes with Nanite disabled and enabled

---

Tip

You can also enable Nanite support if you right-click on the model within the Content Browser and select **Nanite | Enable**. Alternatively, you can also set it up when importing one of your own static meshes into Unreal and selecting the **Build Nanite** option within the importer options.

Enabling Nanite in the previous model means that the asset will render at its highest possible quality independently from the space that it occupies on our screens, with the added benefit of only using the necessary number of polygons required to achieve that task. The further we are from the model, the fewer polygons that are needed, and thus the fewer triangles that Nanite will need to render. This is handled dynamically by the engine, so there is no need for artists to create different levels of detail for the same model, nor do they need to specify the screen-space area that an asset needs to occupy before performing that change in detail. Doing all of that dynamically allows us to focus on more creative tasks, freeing us from all of that busy work.

Having said that, there are still cases where we still need to rely on the old tried-and-tested **Level of Detail** (**LOD**) system—especially when Nanite isn't supported by one of our target platforms. Let's look at how to work with it by looking at one of the other models located in the scene.

8. Select the wooden bowl model and open the Static Mesh Editor by double-clicking on the asset within the Content Browser.

9. Look at its **Details** panel and focus on the **LOD Picker** section. Choose between the different options available in the **LOD dropdown** panel and look at the model to see how it changes before you—the higher the LOD you pick, the fewer polygons it will have.

   By default, changes between different LODs happen automatically based on what the engine decides is right. We can change that behavior by implementing the following steps.

10. Navigate down to the **LOD Settings** section within the **Details** panel and uncheck the **Auto Compute LOD Distances** box.

11. Go back to the **LOD Picker** section and select the LOD that you want to change —we'll start with **LOD1**.

12. Once **LOD1** is selected, the section below **LOD Picker** will change its name to **LOD1**. There, we should be able to see a property called **Screen Size,** which we can tweak. Set the value there to **0.8**.

Adjusting the previous setting will effectively determine the moment when we change from one LOD to the next. The setting that controls this, **Screen Size**, is a measure of the space that the object occupies on our screens at any one time, and you can take a look at that value by focusing on the info display in the upper-left corner of the viewport in the Static Mesh Editor, as seen here:

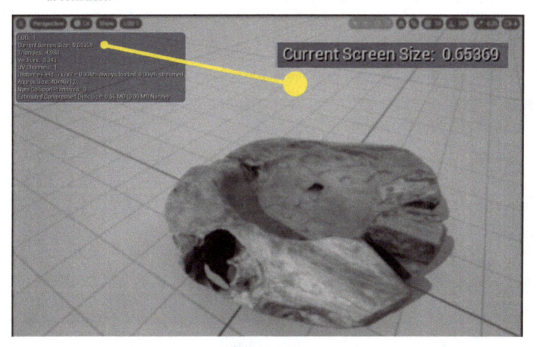

Figure 4.5 – Location of the information panel where we can check the current value of the screen size

Having multiple LODs can be great in terms of improving performance in those platforms that don't support Nanite, as we can control the number of polygons being rendered using the LOD system. The trade-off comes in the form of the extra disk space and memory needed to store those models that act as the different levels of detail, as well as the amount of time needed to create them in the first place. You can see how we would as artists prefer to use Nanite, as it frees us from quite a bit of work!

The next thing that we are going to do is to enable Nanite in the third model that we've downloaded: the concrete plate. Even though it looks like quite a simple geometry, the truth is that it is composed of 9,136 triangles—quite a big number for such a seemingly straightforward model! Nanite will help us render this asset, and we'll use the opportunity to learn more about debug visualization options through it. Let's see what this is all about.

13. Open the Static Mesh Editor for the model of the concrete planter by double-clicking on the **Static Mesh** actor within the Content Browser.

14. Check the **Enable Nanite Support** option located within the **Nanite Settings** section of the **Details** panel (just as we did when working with the grapes), then click the **Apply Changes** button.

We've said that Nanite is a virtualized geometry system that shows the appropriate amount of triangles needed to render any given model at its best at all times, but we haven't seen the triangles actually changing so far. Let's take care of that now.

15. Exit the Static Mesh Editor, head back to the main viewport, and click on the **View Modes** button. It's currently set to **Lit**, so select the **Nanite Visualization | Triangles** option instead. This is the view that we should be looking at right now:

Figure 4.6 – A look at the Triangle visualization mode for Nanite

The previous view mode will allow us to see how the "triangle soup" that makes up those models changes as the asset occupies more or less space on the screen. There are many other informative views that we can access through the same panel: **Primitives**, **Instances**, **Overdraw**, and so on. They all give us different insights into certain metrics that we might need to assess in order to understand how our scene is behaving from a performance point of view, and you can discover more about them in the links provided in the *See also* section.

Completing all of the previous steps should have left us in a much better position when it comes to working with both Nanite and content available through the Quixel Bridge library. There is still more to come with regard to both systems—and we'll be sprinkling some extra info in the next recipes—but make sure to use what we've already learned when working on your own projects!

How it works...

We've talked quite a bit about Nanite in this recipe, so it's only fair that we take a closer look at it. What is it? How does it work? It's quite important for us to understand the answer to both of those questions, as we need to know what the underlying principles are if we want to make the most of it. Let's try to provide the answers now.

At its core, Nanite is a virtualized geometry system that allows us to display as many triangles as we want. The geometry isn't always there, and that's the reason why it receives the *"virtualized"* prefix: even though a model can contain millions of polygons, the system smartly decides what needs to be rendered based on the space that the object occupies on our screens. That being the case, using models where we've enabled Nanite is great in almost every situation, especially when we compare them to standard static meshes. Provided they have the same number of polygons, a static mesh will always force the hardware to deal with all of the triangles that it contains; using Nanite, only the ones that are deemed necessary make it into the rendering pipeline.

See also

Discover more about Nanite and Quixel through the official docs:

- `https://docs.unrealengine.com/5.0/en-US/nanite-virtualized-geometry-in-unreal-engine/`
- `https://help.quixel.com/hc/en-us/sections/360005846137-Quixel-Bridge-for-Unreal-Engine-5`

Using software and hardware ray tracing

Ray tracing is a new way of rendering the contents of a scene, and one that allows us to achieve much more realistic results than were possible in the past with regard to lighting a scene. This type of technology has only been recently introduced in game engines, as it was traditionally too costly to implement in real-time environments. As a result, this technique has often been attributed to offline renderers such as the ones used in the animation industry or professional architectural visualization studios. The arrival of hardware-accelerated ray tracing graphics cards and new rendering solutions such as Lumen has enabled game studios and real-time graphics developers to tap into the potential of ray tracing, for calculating more realistic reflections, refraction, ambient occlusion, lighting, and shadows.

In this recipe, we are going to study how to implement several types of ray tracing solutions to achieve more realistic reflections, exploring both hardware and software ray tracing methods—something that will allow you to deploy this new rendering technique across many different devices.

Getting ready

Seeing as we are going to be working with ray tracing techniques in this recipe, we probably want to work with a level that can enhance those different effects we'll be studying. One of the most visually appealing ones is probably ray-traced reflections, something best viewed in levels containing highly reflective surfaces. The one provided alongside this book (04_02_Start) has you covered in that regard, but remember to place several reflective surfaces around if working on your own scenes.

Having said that, please note that a big part of this recipe is going to focus on techniques that require hardware ray tracing capabilities. As a result, you'll need access to a graphics card that contains said technology in order to complete the final part of this recipe.

How to do it...

Seeing as we'll be working with ray-traced reflections, the first thing we are going to need is a way to control the type of reflection system used in our scene. Let's take care of that first:

1. Place a **Post Process Volume** actor in the scene, something that can be done by accessing the **Quickly add to the project** menu and typing the name of that actor. If you can't type its name, simply search for the first option within the **Visual Effects** category.

2. With the new actor in the scene, access its **Details** panel and check the option called **Infinite Extent (Unbound)**. This will ensure that any effects that we apply within this actor affect the entirety of the level.

    Post Process Volume actors allow us to modify certain graphical settings on a single level, or even in parts of a map, freeing us from needing to apply those same parameters across an entire project. We want to use one such actor to be able to cycle between different rendering options, which will prove useful when comparing different reflection systems. The first one we'll be looking at is the screen-space reflections technique, a system inherited from Unreal Engine 4 that we can use to compare against the ray tracing options we'll be exploring.

3. With the **Post Process Volume** actor still selected, scroll down to the **Reflections** section of the **Details** panel and check the **Method** checkbox. This will allow us to change the default value, which we want to set to **Screen Space**. Let's take a look at the scene through one of the cameras placed on the level:

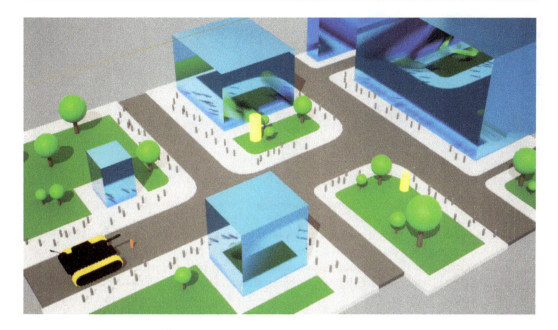

Figure 4.7 – A look at screen-space reflections

Tip

You might have noticed all the reflective surfaces present in the previous screenshot, as well as opaque objects that can be reflected on those areas. Make sure that the scene you work with has a similar layout if using one of your own, as that will be important when looking at reflections.

The screen-space reflection method uses information found within the viewport to calculate the reflections that need to be displayed back to the user. This proves to be the reason for both its strengths and also its shortcomings—using that data makes it easier for the rendering pipeline, but the accuracy of the results is limited, as the engine can't reflect areas that don't appear on our screens. That is the case with objects whose backs we can't see, or for those assets located outside the viewport. Luckily for us, Unreal Engine 5 makes use of Lumen to solve those issues.

4. Change the **Reflection Method** setting once again, this time to **Lumen**. The reflections should have now improved—let's compare them side by side with the previous screenshot:

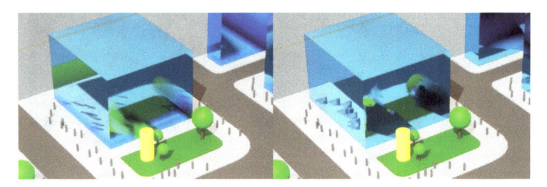

Figure 4.8 – A comparison shot between screen-space (left) and Lumen reflections (right)

Lumen reflections work quite well, especially on those areas where the screen-space technique can't solve the image because there is no available data. This is the case with some of the buildings in the scene, where the older method can't account for the back of certain objects that are in front of the reflective surfaces but whose backs aren't visible to the viewport. Lumen works quite well even in those scenarios, mostly because it is a ray-traced technique that doesn't rely on viewport data. The ray tracing phase can be software- or hardware-based in nature, the latter being more accurate at the expense of being a bit slower and requiring dedicated ray tracing hardware.

The default software ray tracing implementation works by tracing rays against the signed distance field of a given mesh, something that works on a wide range of devices. We can adjust the quality of the reflections by adjusting the quality of those signed distance fields—let's see how to do that next.

5. Open the Static Mesh Editor for the **SM_ToyTank** 3D model.

6. Focus on the **Details** panel and scroll down to the **LOD0** section. There's a setting there called **Distance Field Resolution Scale** with a default value of **1**. Change this to a higher number such as **15**.

7. Click on the **Apply Changes** button and head back to the main viewport. Let's take a look at how the reflections have changed:

Figure 4.9 – Comparison shot between a Distance Field Resolution Scale value of 1 and 15

Increasing the resolution of the signed distance field has improved the quality of the reflections by virtue of generating a more detailed asset that the software ray tracing engine can trace against. The downside to this approach is the increase in memory footprint, but the increase in the image quality is quite clear, especially when compared to the screen-space reflection method. Despite that, we need to be aware that software ray tracing can't quite produce mirror-like reflections; we need to switch to the hardware ray tracing mode if we want to get to that point. This will require adjusting certain project-wide settings, which we are about to do now.

8. Open the **Project Settings** panel (**Edit | Project Settings…**) and head over to the **Rendering** category (**Engine | Rendering**).

9. Scroll down to the **Hardware Ray Tracing** category and check the box next to the **Support Hardware Ray Tracing** option.

    You will be prompted to restart the engine at this point. Do so, but be prepared to grab a cup of your favorite drink, as this is a process that takes time given how the engine needs to recompile shaders. It's always best to undergo this operation at the start of a given project if possible, as it will take longer if we have more shaders in it.

10. Once the project restarts, head back to the same panel we were looking at before.

11. Scroll up to the **Lumen** section and check the box labeled **Use Hardware Ray Tracing when available**.

12. The **Ray Lighting** setting, located directly below the previous one, will now become available. You can choose between **Surface Cache** and **Hit Lighting for Reflections**—the latter being more accurate, the former more performant. Let's change between the two and look at the results on the screen:

Figure 4.10 – Image showing the differences between the two Ray Lighting settings

> **Tip**
>
> You can also switch between the **Surface Cache** and **Hit Lighting for Reflections** options inside the **Post Process Volume** actor. Search for **Reflections | Lumen Reflections | Ray Lighting Mode** to do so.

The results are quite decent and definitely better than any of the reflection techniques we've seen up until now. Despite that, there are still visible problems in the previous screenshot, as well as others that we'll force in just a second. Let's implement a quick change first.

13. Select the **SM_ToyCity** model and open its Static Mesh Editor.

14. Enable Nanite by checking the **Enable Nanite Support** option and clicking on the **Apply Changes** button.

You'll be able to see that the quality of the reflections has decreased as soon as you go back to the main viewport. This is happening because Lumen relies on the fallback meshes of Nanite objects to keep performance afloat when dealing with those types of objects. That being the case, we'll need to adjust the complexity of those models if we want to retain control over the quality of the reflections. This is how we can do that:

15. Still in the Static Mesh Editor for the **SM_ToyTank** model, focus on the **Fallback Relative Error** field. Change the default value of **1** to something lower such as **0.25** and click on **Apply Changes** once again.

16. You can visualize the fallback mesh by hitting *Ctrl + N* on your keyboard or by choosing the **Nanite Fallback** option within the **Show** panel.

Head back to the main viewport and look at the reflections once again. You'll see that the geometry being reflected looks more like what is supposed to and doesn't show that many compression artifacts. We can perform a quick comparison:

Figure 4.11 – Comparison shot between a fallback relative error value of 2 (left) and 0.1 (right)

Despite all our efforts, there are still some areas that Lumen can't solve in its current implementation, especially those where reflective objects show on other reflective surfaces—basically reflections on reflections. The legacy **Ray Tracing** solution is still there for us whenever we need that characteristic in place, so let's take a final look at how to know which areas Lumen can't solve, as well as how to implement that legacy system.

17. Click on the **Viewmodes** button within the main viewport (the third one from the left in the upper-left corner of the viewport) and select the **Reflection View** option within the **Lumen** category.

    Doing so will allow us to visualize the information available to Lumen when it's trying to render the reflection pass. You'll see that certain areas are left in black, something that tells us that Lumen doesn't have enough information to do its job there. Let's change the reflection method once again in order to sort out this issue.

> **Important note**
>
> For this view mode to properly work, make sure to select the **Surface Cache** option instead of the **Hit Lighting for Reflections** one we adjusted in *step 12*.

18. Change the method for the reflections once again in the **Post Process Volume** actor (**Reflections | Method**) to **Standalone Ray Traced (Deprecated)**. You'll see that the reflections have changed, but we still see the same issues where reflections on reflections should be happening.

19. Scroll down to the **Ray Tracing Reflections** subsection and expand that rollout. There's a setting there called **Max. Bounces**, set to a default value of **1**. Increase it gradually and see how the reflections start to appear; for example, let's see what a value of **1** and **6** looks like:

Figure 4.12 – Standalone Ray Traced Reflections with 1 max bounce (left) and 6 (right)

Something that we still need to be aware of is the fact that the fallback mesh is the one that shows in the reflections when dealing with Nanite objects, so be sure to keep that in mind if your results show something that you are not expecting. Having said that, we have now covered all of the available real-time ray tracing reflection systems present in Unreal, so be sure to choose the right one for you!

How it works...

Real-time reflections have always been a tricky subject, and it hasn't been until recently that things have started to improve. This technique has traditionally been quite expensive to render, so artists needed to rely on tricks whenever they wanted to depict realistic reflections: relying on scene-capture textures that got reprojected into a material, rendering the world from a different angle to use that information to drive reflections, relying on screen-space data to do the heavy lifting, and other techniques that have come and gone over the years. There were many options available to meet that challenge, but they all came with caveats that made them unfit as a general-purpose solution. Game engines have had to wait until the advent of real-time ray tracing to finally offer a method that works well for all scenarios, and Lumen is the solution put forward by the team at Epic Games.

The way it works is quite nifty: Lumen analyzes the surface of our scenes in order to know what it looks like and be able to quickly provide lighting information at different ray hit points throughout the level. The analysis of those surfaces is done offline, so the information is already there when the engine wants to query it. By default, Unreal performs this offline analysis during the import process of a model and generates 12 cards for that mesh, where a card is simply one of the views of the surface that Lumen studies. We have the ability to change that value within the Static Mesh Editor, which can sometimes increase the quality of the end results for certain assets (irregular shapes, complex geometries…).

Furthermore, we can also look at the **Surface Cache** view mode in our viewports, which will allow us to see the scene as viewed by Lumen. This visualization option can be quite useful in terms of checking which areas of the model are being covered by the system, as those that are not will be colored in pink. Adjusting the number of cards can decrease the amount of Lumen blind spots indicated by the previous view mode, as can breaking up the model into smaller pieces. Despite that, it's important to follow the guidelines set out by Epic when creating geometry that works well with Lumen: please check the documentation provided next if you want to know more about that.

See also

Do you want more information regarding Lumen and hardware ray tracing in Unreal? Make sure to check the following URLs:

- `https://docs.unrealengine.com/5.0/en-US/lumen-technical-details-in-unreal-engine/`
- `https://docs.unrealengine.com/5.1/en-US/hardware-ray-tracing-in-unreal-engine/`

Revisiting screen-space and planar reflections

Reflections are a very important visual effect that we can see in many places around us: in mirrors, puddles, glasses, and all sorts of other materials. Their importance has impacted the development of real-time rendering engines, which have tried to replicate this natural phenomenon in multiple ways.

We learned about the latest cutting-edge solution in the previous recipe when looking at software and hardware ray tracing. However, as realistic as that approach is, we have to keep in mind that not all devices support that technique. It is in those circumstances that we need to fall back to cheaper, more traditional solutions. Legacy techniques such as screen-space and planar reflections work on systems that lack the hardware needed to enable real-time ray tracing. Let's look at how to enable them next.

Getting ready

All you'll need this time around is a simple scene that contains an object where it makes sense to see clear reflections happening: a body of water, a mirror, or something like that. Try to arrange the level in a way that allows us to get some nice reflections going, maybe by placing an object aligned with the reflective surface that we can see reflected from the point of view of the camera. That's pretty much it, but remember that you can open the scene we'll be using if you load the `04_03_Start` level.

How to do it...

You might have noticed that we've had to tweak certain project settings whenever we wanted to enable specific rendering features. We did that as recently as in the last recipe when we turned on hardware ray tracing acceleration in our project. We'll start by doing something similar next, as planar reflections also require a specific setup:

1. Open **Project Settings** by heading over to **Edit | Project Settings**. Click on the first option, **All Settings**, and type `Clip` in the search box.

2. You should see a parameter named **Support global clip plane for Planar Reflections** pop up. Check the box by its side and restart the project when prompted.

3. Next, select the **Post Process Volume** actor in the scene and make sure that the **Reflection Method** setting is set to **None**.

    Doing this will enable us to use the **Planar Reflection** actor. You can add such assets to your project by visiting the **Quickly add to the project** panel and typing its name or looking inside the **Visual Effects** section if you don't want to do any typing.

4. Drag and drop the **Planar Reflection** actor into the scene.

5. With the actor selected, rotate it until it is perpendicular to the floor. We want to place it parallel to the mirror in our scene, so keep that as a reference.

Here is a screenshot of what the scene should look like at this point for reference:

Figure 4.13 – A look at the initial results after placing the Planar Reflection actor

---

Important note

It can be a bit frustrating to work with **Planar Reflection** actors, as they are only visible in the editor if you are not in **Game View**. Remember that you can enter and exit that visualization mode at will by pressing the *G* key on your keyboard while the main viewport is in focus—it won't work if you are in the **World Outliner** panel, for example!

Let's proceed to tweak the different settings that are part of the **Planar Reflection** actor to try to increase the quality of the reflections.

6. Select the **Planar Reflection** actor and look at the **Details** panel. Play with the **Normal Distortion Strength** parameter until the reflection looks right to you. The default value of 500 works well for our scene, but feel free to play with this setting if you are using your own models.

Tip

The **Normal Distortion Strength** parameter adjusts the effect that the normal for a given object has on the reflections. This allows us to have correct-looking values without having to perfectly align the **Planar Reflection** actor with our mirror.

---

7. Pay attention to the parameters called **Distance from Plane Fadeout Start** and **Distance from Plane Fadeout End**, as they control the distance at which pixels from other objects start to fade from the planar reflection. The first parameter marks the distance from the **Planar Reflection** actor at which objects start to fade from the reflection, while the second one controls the point at which they disappear altogether. Make sure to tweak these values if working with your own scene until all of the objects that you want to see reflected appear in the reflection.

8. Uncheck the **Show Preview Plane** option, just so that it doesn't bother us whenever we are in **Game View**.

9. Still within the **Planar Reflection** category, expand the **Advanced** section to access some extra settings that we might want to tweak. Focus on the **Screen Percentage** option and change it to **100**, which will increase the quality of the reflections.

Having performed the previous set of steps will give us a nice planar reflection in our mirror. Let's compare the results against the ones obtained through the screen-space reflection technique:

Figure 4.14 – A comparison shot between screen-space reflections and planar ones

Even though we now have a reliable way of achieving realistic reflections, this is a method that can be quite expensive to render, as we are basically drawing the scene twice. With that in mind, let's look at a different system that can also generate nice, detailed reflections in a cheaper way: the scene-capture technique.

This second system that we are going to explore presents some benefits as well as some potential inconveniences, and we need to be aware of them! The benefits? Well, this time we don't need to render the scene twice every so often to achieve the desired results, as this new technique for capturing reflections consists of baking our scene into a texture, which is then fed to the objects where we want those reflections to show. This can present a bit of a problem, as we'll need to manually tweak and position the baked environment in order to ensure that it blends well with our scene.

10. Delete the previous **Planar Reflection** actor so that it doesn't affect our scene anymore.

11. Head back to the **Quickly add to the project** panel and search for a **Scene Capture Cube** actor. Drag and drop it into our scene.

12. Place the previous **Scene Capture Cube** actor close to the mirror in our scene. You can take a look at the next screenshot for reference purposes:

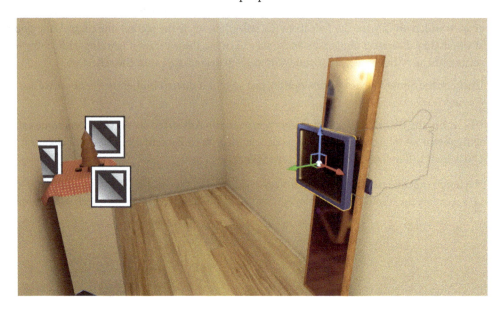

Figure 4.15 – Position of the Scene Capture Cube actor with regard to the existing mirror

The **Scene Capture Cube** actor works in tandem with **Render Target**—a texture where we'll store the view from the actor. All we are doing is placing a camera in our scene, which will create a photograph of the level for us to use as a reflection value.

13. Create a **Cube Render Target** texture by right-clicking in an empty space of the Content Browser and browsing inside the **Textures** section.

14. Give it a name you'll remember and double-click on it to access its **Details** panel. You'll find a parameter named **Size X** under the **Texture Render Target Cube** category. Assign it a value of **2048 (2K)**.

15. Head back to the main viewport and select the **Scene Capture Cube** actor we placed back in *step 12*. Look at its **Details** panel and set the previously created **Cube Render Target** texture as the default value in the **Texture Target** category.

16. There's another setting a little bit further down named **Capture Every Frame**. Disable this to increase performance.

17. Move the camera around slightly. This should update the contents of our **Render Target** texture, which you'll be able to check if you double-click on it again.

Now that we have that texture, we can use it within our mirror material. With that goal in mind, let's create a new material where we can set that in motion.

18. Create a new material and apply it to the mirror in our scene. I've named mine **M_WoodenMirror_ ReflectionCapture**. Open this new asset in the Material Editor.

19. Add a **Texture Sample** node and set the **Render Target** texture we previously created as its value.

    Even though we now have that texture inside our new material, we need to affect the way we look at it based on the position of the camera. This will give the impression that the reflection gets updated as we look into the mirror. Thankfully, Unreal has a handy node that allows us to do just that.

20. Right-click within the material graph and type `Reflection Vector WS`. This will give us what we need to make our **Render Target** texture work as intended. Create it and connect its output to the **UVs** input pin of the previous **Texture Sample** node.

21. Connect the output of the **Texture Sample** node to a **Multiply** node, which we'll create now, and set the value for the **B** parameter to something like **5**. This will brighten up the reflection we see once we apply and save the material, so make sure to play around with this value later until you are happy.

    All of the previous steps have created a reflection texture that we can see no matter where we are with regard to the mirror. We now need to blend that with the other part that makes up the mirror: the wooden frame.

22. Add three **Texture Sample** nodes and assign the following textures to them: **T_ WoodenBedroomMirror_ AORM**, **T_ WoodenBedroomMirror_ UVMask**, and **T_ WoodenBedroomMirror_ BaseColor**. All of them will be needed in the following steps.

23. Then, create a **Lerp** node: something that we'll use to mix between the texture that will act as the reflection and the color of the wood.

24. Connect the result of the **Multiply** node we created in *step 21* to pin **A** of the new **Lerp** node.

25. Connect the **T_ WoodenBedroomMirror_ BaseColor** texture to its **B** input pin.

26. Connect the **Alpha** input pin to the green channel of the **T_ WoodenBedroomMirror_ UVMask** asset. Remember to connect the **Lerp** node to the **Base Color** node of the material as well.

    Performing the previous steps has given us the color. Let's now tackle the **Roughness** value.

27. Connect the red channel of **T_ WoodenBedroomMirror_ AORM** directly to the **Ambient Occlusion** input pin of the **Main Material** node.

28. Create another **Lerp** node, which we'll use in a similar way to the one we used to drive the color values—even though it'll be the **Roughness** parameter that we'll be adjusting now.

29. Connect the green channel of the **T_ WoodenBedroomMirror_ UVMask** texture to the **Alpha** input pin, and assign a value of **1** to pin **A**. Connect pin **B** to the green channel of the **T_ WoodenBedroomMirror_ AORM** asset.

30. Plug the result of the previous **Lerp** node into the **Roughness** channel of our material.

Here's what the material graph should look like, all things considered:

Figure 4.16 – A look at the graph for our newly created material

With these last changes in place, we have effectively finished tweaking the material, and we can finally check the results if we go back to the main viewport. You'll notice that the effect is not exactly perfect, but it works quite well for specific use cases such as this one.

Remember that you can also make a dynamic effect out of this technique by resetting the property we adjusted in *step 16* back to its default value. You might want to test the effect on other surfaces to see the results, as this method tends to work best with objects larger than our mirror. However, it is a powerful technique that can add life to our scenes and limit the impact on performance that other methods can have. Be sure to test it on different types of surfaces and materials!

Figure 4.17 – A look at the reflection in the mirror using a Cube Render Target texture

How it works...

Let's take a bit of time to learn more about the way planar reflections and scene-capture cubes work.

Planar reflections

Every model in our level can be affected by a **Planar Reflection** actor, even though its effect will be most obvious in those materials that have a low roughness value. Such is the case of the mirror in our scene, as you saw in this recipe. Other surfaces, such as the walls or the rest of the props inside the level, didn't get affected as much since their roughness values didn't allow for that.

The way the **Planar Reflection** actor affects our scene is tied directly to the settings we specify in its **Details** panel. The first of those, the normal distortion strength, determines how much the normals of the affected surfaces distort the final reflection. This is especially useful when we have multiple actors that we want to affect at the same time that are not facing in the exact same direction—instead of creating multiple planar reflections that are parallel to each individual rotated object, we can adjust the normal distortion strength to allow for varying rotation angles.

Other useful settings to remember are the distance from the plane fade-out start and its twin, the end one, as they control the region where the effect is happening. However, maybe one of the most important ones is the one labeled **Screen Percentage**, as that controls the quality of the overall effect and the rendering cost of the effect.

Scene-capture cubes

The **Scene Capture Cube** actor is a very cool asset that can be used to create very different effects with it: from real-time reflections as we've done in this recipe to live camera feeds, just to cite two examples. We've already seen how to create mirror-like images in this recipe, but imagine what other things can be done using this same method. The important thing to consider is that we are rendering the scene through a camera that feeds what it sees to the selected render target, so there is a bit of a setup there that we need to be aware of. Furthermore, I'd like to make you aware of the other types of render targets beyond the cubic one we've studied: there is the standard one (simply called **Render Target**) and the more specific **Canvas** type. You can read more about them thanks to the links provided in the *See also* section, so be sure to check those out in case you are curious about the possibilities of using this system.

See also

Let me leave you with some links to the official documentation regarding planar reflections and render targets:

- `https://docs.unrealengine.com/5.1/en-US/planar-reflections-in-unreal-engine/`
- `https://docs.unrealengine.com/4.27/en-US/RenderingAndGraphics/RenderTargets/`

Creating an arch viz scene with realistic-looking glass and virtual textures

The team behind Unreal Engine 5 has gone to great lengths trying to expand the capabilities of the engine, especially when it comes to the rendering department. We've already looked at some of the new additions, such as Lumen and Nanite, but the continuous push toward allowing users to render anything they can imagine hasn't stopped with those two systems.

We'll be taking a look at two low-key newcomers in this recipe: virtual textures and the **Thin Translucent Shading Model**. The first of the two allows us to use large-resolution textures while reducing the memory footprint that they have on our machines. The second feature is none other than a new way of creating realistic-looking architectural glass.

As you can see, Unreal keeps on delivering on all fronts in order to make our lives as artists easier—so, let's take advantage of that!

Getting ready

Given how we are going to be using virtual textures and the **Thin Translucent Shading Model** in this recipe, we probably want to be working with a scene that can make those features shine. The one included in the Unreal Engine project provided by this book is a simple interior that contains several objects where we can apply different materials that take advantage of those techniques: a window where we can showcase the new glass material we'll be creating and some chairs onto which we'll apply high-resolution textures. The level in question we'll be using is 04_04_Start, so please open it if you want to continue using the same materials that you'll see printed in the next few pages. If working with your own assets, please be mindful of the type of scene that we want to depict before creating the models for it.

Something else to consider is the use of ray tracing hardware. Know that this recipe can be completed without it, but it can be a bonus to have an RTX graphics card, especially if you want to take a look at ray-traced translucency as we'll be doing at one point in the recipe.

How to do it...

The scene with which we're going to be working contains several 3D models in it, put together to give the impression that we are inside a small room looking toward the sea. Two specific assets in it are going to be the focus of this recipe: the window and the chairs. Seeing as we have to start with one of the two, let's start working on the material for the windows first:

1. Create a new material and name it something along the lines of **M_WindowGlass**.

2. Open the new asset we've created and look at its **Details** panel. We'll need to tweak a few things here—start with the **Blend Mode** setting and set it to **Translucent**.

3. The next setting that we need to tweak is **Shading Model**: change the default value to the **Thin Translucent** option.

4. Still within the **Details** panel, scroll down to the **Translucency** section and check the box for **Screen Space Reflections**. This is a must if ray tracing options aren't available.

5. Then, change **Lighting Mode** to **Surface Forward Shading**.

Tip

There's another option within the **Details** panel that you might want to revisit when working with translucent surfaces: the **Two Sided** property. Enabling this setting will sometimes work to your benefit, especially when dealing with objects where the back face should contribute to the final look.

Taking care of the aforementioned aspects is everything we need to do in order to set up the material to correctly make use of the new **Shading Model**. We now need to actually define how we want it to behave in terms of the usual properties we deal with, so let's tackle that now.

6. Create a **Vector** parameter and name it **Transmittance Color**. Set it to a blueish value or whatever works for your purposes.

7. Connect the output of the previous node to a new one that we'll need to create now, called **Thin Translucent Material**.

8. Create four scalar parameters: one to control the **Metallic** property, another for **Roughness**, a third for **Opacity**, and a final one for **Index of Refraction**.

9. Set the **Metallic** scalar parameter to **0.1** and connect it to the **Metallic** input pin of the **Main Material** node. Feel free to play around with this value until you are happy, but know that a value of 1 will prevent any reflections from showing.

10. Assign a low value to the **Roughness** property—something like **0.05**. Glasses tend to favor values close to that number, as reflections tend to be very clear across their surfaces. Remember to connect the output pin of this new node to the **Roughness** input pin of our material.

11. Set the **Opacity** property to something like **0.5** and connect it to the homonymous parameter in the main material slot. This might be a higher value than expected, but setting it to that number will make the reflections in the glass show more clearly, which works best for demonstration purposes.

12. Set the **Index of Refraction** variable to **0.8** and connect it to the **Specular** input pin in our material (this must surely feel strange, but more about it in a moment).

All in all, we should be looking at a material graph that looks something like this:

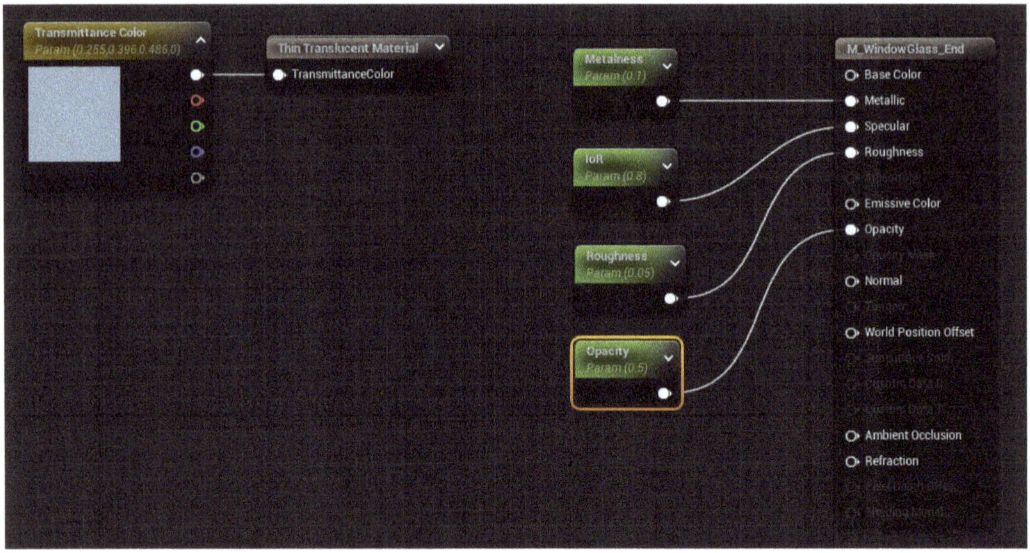

Figure 4.18 – A look at the current state of the material graph

The next thing that we need to do is to apply the material in the appropriate slot in our window models. Those present in the scene I'm working with contain two different material slots: one for the window frames and another one for the glass. Apply our new material in that slot.

Now that we have the material ready and being rendered, let's take a look at how it sits in our scene. It probably seems like a standard translucent material at first glance, but that would be a great disservice to what we've created. Even though the differences are subtle, you might be able to see a greater contribution of indoor light to the reflection on the glass, as well as faint reflections of the scene present in the material that uses the **Thin Translucent Shading Model**. Despite that, we can always make the effect clearer and more realistic by enabling ray-traced translucency. Let's take a look at how to tackle that next.

13. Select the **Post Process Volume** actor present in the scene and scroll down to the **Rendering Features** section. Look for the **Translucency** subsection and change the type from **Raster** to **Ray Tracing**.

Let's look back at the results once you enable that setting:

Figure 4.19 – Render of the same scene with ray-traced translucency disabled (left) and enabled (right)

The differences should be easier to spot this time around, as the results that ray tracing translucency enable translate into a more realistic representation of the virtual environment. For starters, the reflection of the scene present in the glass is more obvious and physically accurate than its screen-space counterpart, as that technique doesn't have all the information needed to present a realistic depiction of the interior environment. That is in itself a great feature to have when we need to work with realistic scenes. Having said that, we do need to be aware of a quirk present in the material: you might remember that we are driving the index of refraction through the **Specular** parameter instead of the appropriate **Refraction** one. As unintuitive as that might be, the refraction material input is meant to be used in a rasterized rendering pipeline—that is, when there's no ray tracing happening. Turn on that hardware-accelerated feature and you need to use the specular material input itself—more info in the *See also* section.

Completing all of the previous steps has left us with realistic-looking architectural glass. This is a very important feature present in many architectural visualization scenes. Those types of environments have a high-quality standard, and given how performance isn't usually such an important factor there, we can enable expensive rendering techniques provided they give a boost to the image quality of the scene.

Something also present in these types of environments is high-resolution textures, as they are paramount to enabling high-quality materials. We are talking about 4k+ images, whose goal is to make the scene look its best. Using those assets has often been quite prohibitive in terms of balancing their rendering cost, as they occupied a lot of space in memory and had the potential of quickly cluttering our GPUs. Their size was—and still is—a concern even today, but things have improved by quite a lot thanks to the introduction of virtual texturing methods. We'll take a further look at how they work in the *How it works...* section, but know for now that they allow the engine to only hold the relevant texture data for the images that are being drawn on the screen. We'll take a look at how to work with these types of assets next.

14. Open the **Project Settings** panel and head over to the **Rendering** section located within the **Engine** category.

15. Look under the **Virtual Textures** section and check the box next to **Enable virtual texture support**. This setting controls whether this type of asset is supported in our project, and enabling it will require us to restart the editor. While in there, check the **Enable virtual texture on texture import** option as well. Doing so will allow us to convert regular textures into virtual ones during the import process, as opposed to having to manually assign them after the fact.

    You probably know the drill by now: we'll be prompted to restart the engine after adjusting those two parameters. Feel free to grab a coffee while the restart operation completes, and let's continue where we left it after that's sorted. The first thing we'll do once the engine restarts is going to be to convert a couple of standard textures into virtual ones, so make sure that you know which assets you'll be using if working with your own images.

16. Select the textures that you want to convert into virtual ones and right-click on them. A contextual menu should have now appeared, and we need to select the option labeled **Convert to Virtual Texture**. We'll be selecting the ones called **T_Chair_BaseColor**, **T_Chair_AORM**, and **T_Chair_Normals** in this example.

17. Create a new material and name it something along the lines of **M_Chair**.

18. Open the Material Editor, create three **Texture Sample** nodes, and assign the previous textures as their values.

19. Connect the output of the **T_Chair_BaseColor** texture and connect it to the **Base Color** material input pin. Do the same for the **Normal** map, but assign it to the **Normal** input pin instead.

20. Connect the red output pin of the **AORM** texture to the **Ambient Occlusion** material input pin, then link the output of the green channel to the **Roughness** slot, and wire the blue one to the **Metallic** property. The next screenshot shows what the material graph for this shader should look like:

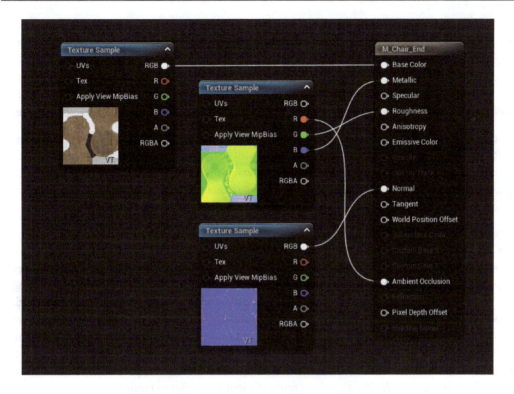

Figure 4.20 – The material graph for the chair material

Once we have that ready, remember to apply the material to the chairs in our scene in order to see it being displayed. You'll see that there's no difference when it comes to using virtual textures within a material compared to standard ones: you are free to use them in the same manner!

How it works...

We've taken a look at a couple of new concepts in this recipe, and I wanted to briefly cover them from a technical point of view before moving on.

Let's start with the **Thin Translucent Shading Model**, as that's what we worked on first. This new type of **Shading Model** allows us to realistically depict the behavior of light when it interacts with translucent objects, which is a must when we need physically accurate results in our levels. Its relevance becomes prevalent when working with real-life objects that our eyes are accustomed to seeing, such as window glass: we are very good at determining whether what we have in front of us looks fake, and previous shading models didn't do those types of objects justice.

Thin Translucent requires the use of the **Surface Forward Shading** lighting mode, as we saw in this recipe. This option allows us to get specular "*highlights from multiple lights and image-based reflections from parallax-corrected reflection captures*", as stated in Unreal's official docs, which ultimately allows for those physically accurate results we mentioned before. This lighting mode owes a lot to the forward renderer, one of the two methods that Unreal can use to realize our scenes, and one that isn't the default one—that honor falling to the deferred renderer option. More on this in the *See also* section.

The other great topic that we've covered in this recipe is the **Virtual Texture** system, which is a different way for the engine to handle images. If you think about them, textures come in all sorts of shapes and sizes—from small icons to large, high-resolution assets meant to cover big models. Dealing with those larger images can prove a challenge to the hardware that is running our apps, as the amount of memory that those assets demand can prove simply too high.

Taking that into consideration, many rendering engines have relied on creating smaller versions of those large assets (called **mips**), which can be used instead of the original ones based on the amount of space that those textures occupy on our screens and the distance from the objects onto which they are applied. This has proven to be a reliable technique but one that also has its downsides: creating those extra smaller versions requires more space on disk, and those same mips need to be stored, accessed, and loaded by the hardware responsible for dealing with them. Furthermore, objects that only show partially on our screens still load the full amount of textures that they are using for the parts that are not visible, so you can see how there was still room for improvement in that department. Virtual textures try to overcome those complications by only loading the appropriate amount of data requested by the engine, which makes them more efficient and easier to render.

See also

As always, let me leave you with some useful links regarding the different topics we've covered in this recipe:

- https://docs.unrealengine.com/4.26/en-US/RenderingAndGraphics/VirtualTexturing/

- https://docs.unrealengine.com/5.1/en-US/shading-models-in-unreal-engine/

- https://docs.unrealengine.com/5.1/en-US/lit-translucency-in-unreal-engine/

- https://www.artstation.com/blogs/sethperlstein/pdBz/ray-traced-translucency-with-index-of-refraction-in-unreal-engine

- https://docs.unrealengine.com/5.1/en-US/forward-shading-renderer-in-unreal-engine/

Varnishing wood through the Clear Coat Shading Model

We've created plenty of materials so far—woods, metals, glasses, and so on—and thanks to that, I think that we can already say that we know how to tackle many of them. Despite that, not everything that we see in real life can be described as a single material, which is what we've focused on up until this point.

There are special cases where a surface behaves like there are several materials stacked on top of each other. An example of that would be varnished wood, the topic of this recipe. Said material can be described as the combination of a wooden surface and a thin layer of varnish painted on top of it. The properties of those two materials differ, and that shows when it comes to the way light interacts with them. Consequently, the final look of a surface that has that material applied to it can't be described through the standard **Shading Model** we've been using thus far—we'll need the assistance of the **Clear Coat Shading Model**.

Getting ready

Creating varnished wood won't be too much of a problem: the **Starter Content** asset pack already provides us with some high-quality textures that we can use to describe both the underlying wood and the varnish finish of our material. We'll stick to those assets throughout this recipe, and you can open the level called `04_05_Start` if you want to follow along using the same resources.

If using your own assets, please note that you'll need textures that describe both the normals and the roughness of the base material and the clear-coat finish on top of it that you want to create. With regard to the makeup of the level we'll be working with, this is going to be a very simple one. We'll be using a copy of the very first level we worked on in this book, as we only need a source of light and a simple plane to act as a floor to which we can apply the varnished wood material. I've also thrown one of the chairs we worked on in the last recipe for good measure, just so that we have something to ground the scene a bit more.

How to do it...

Seeing as we are about to start using a new rendering technique, one of the first things that we'll need to do is to head over to **Project Settings** and enable a specific feature there as we've done in many of the previous recipes. The **Clear Coat Shading Model** is a standard feature of Unreal, but using two normal maps alongside it isn't. Let's make sure to enable this functionality first:

1. Open the **Project Settings** panel and head over to the **Rendering** section located within the **Engine** category.

2. Look for the option named **Clear Coat Enable Second Normal** inside the **Materials** section and enable it. You'll need to restart the engine once you do this, so be ready to wait until the shaders finish recompiling.

3. Create a new material and give it a name that represents what we'll be creating—I've gone with **M_VarnishedWoodFloor_ClearCoat** this time. Open it in the Material Editor so that we can start working on it.

    There are many things that we need to pay attention to when creating a clear-coat material. If we try to order the materials from the bottom layer to the upper one, we'll see that we need to provide the base color information for the underlying material, its roughness and normal values, as well as potentially the **Metallic** property. We'll also need to define the roughness and the normals of the clear-coat layer situated on top of the base shader. Let's start from the beginning.

4. Create a **Texture Coordinate** node. You can do this by the usual method of right-clicking and typing its name, but you can also hold the *U* key on your keyboard and click anywhere within the Material Editor graph.

5. Add a **Scalar Parameter** node and name it **Base Texture Scale**. We'll use this node to control the tiling of all of the textures used to drive the appearance of the base layer of our new material.

6. Include a **Multiply** node right after the previous two we created, and connect its inputs to the outputs of the previous ones. No particular order is needed here thanks to the commutative property!

    We'll use those three nodes we just created to drive the tiling of the base layer textures. Next up: adding those images! Let's start setting up the **Base Color** and **Roughness** properties.

7. Create a **Texture Sample** node and assign the **T_Wood_Floor_Walnut_D** texture as its default value. Connect its **UVs** input pin to the output of the **Multiply** node we created in the previous step.

8. Drag a wire out of the **RGB** output pin of the previous **Texture Sample** node and connect it to the **Base Color** property of our material. This will have taken care of the **Base Color** property.

9. Moving onto the **Roughness** parameter, create another **Texture Sample** node and set the **T_Wood_Floor_Walnut_M** texture as its value.

10. Just as we did when working with the **Base Color** texture, connect the **UVs** input pin of the new **Texture Sample** node to the output of the **Multiply** node we created in *step 6*.

11. The texture we've chosen to work as the roughness value isn't really suited to be used as one since it was designed as a mask. We can still adjust it to fit our purposes: create a **Cheap Contrast** node and place it immediately after the previous **Texture Sample** node.

12. Connect the output of the **Roughness** texture to the **In (S)** input pin of the new **Cheap Contrast** node.

13. Add a **Constant** node and set its value to **-0.2**. Connect its output to the **Contrast (S)** input pin present in the **Cheap Contrast** node.

14. Create a **Multiply** node and connect one of its input pins to the output of the **Cheap Contrast** function. Set the value of the other input to 2 and connect the output to the **Roughness** input pin of the **Main Material** node. This is the graph that we should be seeing:

Figure 4.21 – A look at the current state of the material graph

Having done that has taken care of both the **Base Color** and **Roughness** properties for the base layer of our new clear-coat material. Let's focus now on the normals.

15. Create another **Texture Sample** node and assign the **T_Wood_Floor_Walnut_N** texture as its default value. Connect its **UVs** input pin to the output of the **Multiply** node we created in *step 6* once again.

16. Seeing as we'll want to modify the strength of the normals so that they are more apparent (as they can get a bit blurred once we define the clear-coat properties), create a **Flatten Normal** node and connect its **Normal (V3)** input pin to the output of the previous **Texture Sample** node. The **Flatten Normal** function might make us think that its purpose is that of decreasing the intensity of the normals we provide it, but we can hack our way around it to achieve the opposite effect.

17. Create a **Constant** node and assign a value of **10** to it. Connect its output to the **Flatness (S)** input pin of the previous **Flatten Normal** node.

18. Add a **Clear Coat Normal Custom Output** node and connect its input pin to the output of the previous **Flatten Normal** function. Even though you should search for the name Clear Coat Normal Custom Output, the name displayed by the node is that of **Clear Coat Bottom Normal**.

This is what that part of the graph should look like:

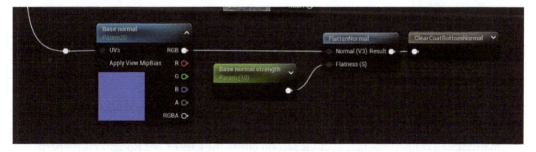

Figure 4.22 – A closer look at the nodes affecting the normals of the bottom layer of the material

The last node we used, **Clear Coat Normal Custom Output**, is the one responsible for computing the normals for the bottom layer of the clear-coat material. The standard material input called

Normals will be used to drive that setting for the clear-coat layer, as we'll see next!

19. Create a second **Texture Coordinate** node. We'll use this one to control the tiling of the normals we are about to define for the clear-coat layer.

20. Add another **Scalar Parameter** node to control the tiling. Set its default value to **10**.

21. Multiply the previous two nodes to completely control the tiling.

22. Create a new **Texture Sample** node and set the **T_Ground_Moss_N** asset as its default value, a texture that is part of the **Starter Content** asset pack. We'll use it to drive the normals of the clear-coat layer.

23. We want to be able to control the intensity of the normals, so we are going to approach this operation in the same way we did when dealing with the base layer. Start by including a new **Flatten Normal** function and connecting its **Normal (V3)** input pin to the output of the previous **Texture Sample** node.

24. Add a new **Scalar Parameter** node to control the strength of the normals and set its value to **0.9**. We are actually using this node as intended this time, as we'll be decreasing the intensity of the normals!

25. Connect the output of the **Flatten Normal** node to the **Normal** input pin of our material. This is the graph that we should be looking at now:

Figure 4.23 – The nodes affecting the normals for the clear-coat layer of the material

Now that we've defined the normals for the clear-coat layer, it's time to enable that part of the material and define its roughness settings. Let's take care of that now, which will see us completing this material.

26. Create two **Scalar Parameter** nodes and set their default value to **0.2** and **0.1** respectively.

27. Connect the output of the first of those two new values to the **Clear Coat** input pin of our material, and connect the output of the second one to the **Clear Coat Roughness** property.

The two previous values control the spread of the clear-coat layer and the roughness that the engine should be applying to the material. We've chosen a low number in both places so that the spread ends up being very subtle and the reflections become more apparent. With that done, proceed to apply the material to the place in the scene and review the results:

Figure 4.24 – A final look at the clear-coat material being applied

A final look at the scene reveals the interaction of both material layers with the light that reaches them—we can see the normals of the clear-coat layer modulating the light that touches the surface of the floor at the same time that we can see the seams separating the wood boards. All of that without the need to create two different pieces of geometry—quite nifty!

How it works...

Clear-coat materials are a nice addition to the engine, especially since the arrival of the second normal channel toward the end of the development cycle of Unreal Engine 4. They serve a very specific purpose, but even though their potential usage is confined to a narrow subset of real-world materials, achieving the same effect would require a more complex setup using some of the other means available to us. We could always rely on a second translucent object situated right on top of the first one, or fall back on Unreal's material layer system, but the clear-coat option makes it easier for us by default.

Focusing now on the way this **Shading Model** works, we had the chance to observe that it is almost like defining the material attributes of two materials in a single graph. We specified the **Roughness** and **Normal** values for the base material and the clear-coat layer independently from each other, just as we would if we were creating two separate shaders. In spite of this, the material graph reminds us that we are dealing with just a single material, as there are certain settings that only affect the base pass. An example of that would be the **Base Color** property, which only affects the underlying layer.

The clear coat is treated like a small thin film placed on top of the base object, and that is something that we need to keep in mind when working with these types of assets.

See also

As usual, let me leave you with a few useful links that you can check if you want to know more about the **Clear Coat Shading Model**:

* https://docs.unrealengine.com/5.1/en-US/shading-models-in-unreal-engine/
* https://docs.unrealengine.com/5.0/en-US/using-dual-normals-with-clear-coat-in-unreal-engine/

5

Working with Advanced Material Techniques

Welcome to the fifth chapter of this book! So far, we've spent most of our time discovering how to create different kinds of shaders: wooden surfaces, metals, cloth, rubber, wax, glass, and more. We have seen plenty of combinations that focused on defining the different ways in which light behaves when it reaches the objects that we are trying to describe. That is one way to use the Material Editor!

But there are more. The material graph can also be used to describe extra geometrical detail that can be added to the models that we work with, or to build decals that can then be applied on top of other models. We'll also see how artists can blend between different textures in a single model thanks to the vertex color technique, or light a level using emissive materials. These are just examples of the versatility that the Material Editor empowers, something that we are about to study in the following recipes:

- Using vertex colors to adjust the appearance of a material
- Adding decals to our scenes
- Creating a brick wall with Parallax Occlusion Mapping
- Taking advantage of mesh distance fields in our materials
- Lighting the scene with emissive materials
- Orienting ourselves with a logic-driven compass
- Driving the appearance of a minimap through Blueprint logic

As always, here is a little teaser just to whet your appetite:

Figure 5.1 – A little tease of the things we'll be working on

Technical requirements

As always, let's start with a link to the Unreal Engine project we have provided with this book: `https://packt.link/20u7B`.

You'll be able to get a hold of all of the assets I've used to create the recipes in this chapter: the models, the textures, the levels, and the materials. Be sure to use them if you don't have access to other resources!

Using vertex colors to adjust the appearance of a material

As we said in the introduction to this chapter, we are going to be working inside the material graph to unlock certain effects that go beyond just describing different material attributes. We'll start by looking at an important and useful technique called **Vertex Painting**, which allows us to assign specific color values to the vertices that make up our models. We'll be able to reuse that information within the material graph to create masks that can drive the appearance of a material.

This can be a very useful technique when texturing large props as it allows us to manually paint certain material effects right within the viewport – almost as if we were hand painting a 3D model. I'm sure you'll find this technique very useful once we take a proper look at it, so let's not delay any further!

Getting ready

Vertex Painting is a technique that is not very complicated to demonstrate: all we'll need is a 3D model and a material. These are things that you can get a hold of thanks to the Starter Content, but remember that you can always download the Unreal Engine project that was used to create this book through the link provided in the *Technical requirements* section. You'll find the same assets that we'll use in the following pages there, so make sure you grab a copy if you want to follow along using those resources. Otherwise, feel free to create your own 3D models and bring them along for the ride: you'll be able to work with them, so long as they are UV mapped correctly.

If you're working with the Unreal Engine project provided alongside this book, open the level named `05_01_Start` to begin our journey.

How to do it...

Vertex Painting requires at least a couple of different elements that we'll need to set up – a material that can be used with this technique and the vertex values to be painted on the meshes on which we are going to be operating. We'll take care of those two parts in that order, so let's start things off by creating a new material:

1. Create a new material anywhere you fancy within the Content Browser and give it an appropriate name. I've gone with **M_VertexPaintingExample_End** for this example.
2. Next, assign the previous material to the model that you want to work on. If you've opened the map provided with this book, that will be the lamp in the center of the level.

3. Finally, open the Material Editor by double-clicking on the newly created material.

    The next steps will probably make more sense if we briefly talk about what we want to achieve using the Vertex Painting technique. That is simple: assigning different textures and effects to certain parts of our model by interactively painting their vertices. Taking the lamp in the center of the level as an example, we are going to make the base of the object look different from the rest of the body. As we want to apply different textures, the first order of business is going to be bringing all of the necessary assets into the material itself so that we can choose which ones to use when we paint the vertices. Let's do that now.

4. Start by creating two texture samples within the Material Graph Editor.

5. Set the first of them to use the **T_Lamp_Color** texture. Apart from that, select that node and look at its **Details** panel, and change **Sampler Type** from **Color** to **Linear Color**.

6. The second one should make use of the **T_Lamp_Base** asset. Just like before, make sure that the **Sampler Type** property for this **Texture Sample** is set to **Linear Color**.

7. Next, include a **Lerp** node and place it after the previous two texture samples, and connect the first of them to its **A** input pin. The second one should be connected to input pin **B**.

8. With the new **Lerp** node in place, connect its output to the **Base Color** input pin in the **Main Material** node.

    Doing this will see us interpolating between both textures, something that we'll use to define the look of the base of the lamp and its main body. Having said that, we still need to assign something that can act as the interpolation value: the key element of this recipe, the **Vertex Color** node.

9. Create a **Vertex Color** node by right-clicking and typing that same name anywhere within the material graph.

10. Proceed to drag a wire from its **Red** output pin and connect it to the **Alpha** input pin of the **Lerp** node we created back in *step 7*. The material graph should look something like this:

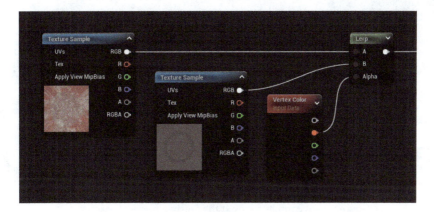

Figure 5.2 – A look at the current state of the material graph

The **Vertex Color** node gives us access to the color values of the vertices of the mesh on which the material is being applied. You can think of painting vertices as assigning specific values to them, which we can then use for other purposes, such as driving the **Alpha** channel of a **Lerp** node, as we did in the previous step. In it, we used the **Red** channel of the **Vertex Color** node to drive the **Alpha** parameter of the **Lerp** node. The reason why we are using the output from that channel is because that is where we are going to be painting the vertex values later, but we could use any other channel if we wanted to. With that done, let's make the material a little bit more interesting by setting up the roughness and metallic properties of the material.

11. Add four constants to the graph — we'll use two to drive the **Metallic** property of the material and another two to adjust the **Roughness** attribute.

12. With the constants in place, assign values of **0** and **1** to two of them – we'll use those to drive the **Metallic** property of the material. As you may recall, a value of 0 equals a non-metallic object, while a value of 1 has the opposite effect. This is just what we want to see in the body and the base of our model.

13. As for the second set of constants, give them whichever value you fancy – I've gone with **0.5** and **0.25**, values that I intend to use for the body and the base, respectively.

14. Next, create a couple of **Lerp** nodes to make it possible to interpolate between the previous constants.

15. Connect the **Red** channel's output pin of the **Vertex Color** node to the **Alpha** input pins of the two new **Lerp** nodes.

16. After that, wire each of the previous constants to the **A** and **B** input pins of the new **Lerp** nodes, just like we did for the **Base Color** input pin.

17. Finally, connect the output of the **Lerp** nodes to the **Metallic** and **Roughness** input pins present in the **Main Material** node. This is what we should be seeing at this point:

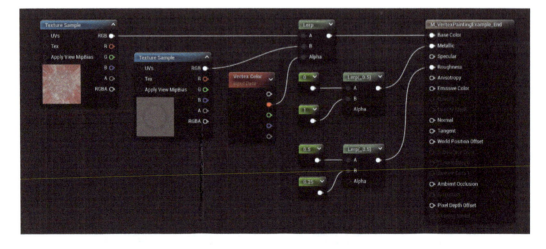

Figure 5.3 – A complete look at the finished material graph

As you can see, this is quite a simple material, but one that will allow us to see the Vertex Painting technique in action. The previous steps have left us with a material that we can apply to the model in our scene, but doing that won't be enough to show the technique we are demonstrating in this recipe. We'll also need to specify which areas of our model should be painted with the textures we've included before: let's take care of that now!

18. Select the model that you want to work with – in our case, it's going to be the spherical object in the middle of the level, the one called **SM_Lamp_Simplified**.

19. Assign the material we just created to the selected sphere.

20. Next, head over to the **Modes** panel (either by navigating to the top left of the viewport or using the *Shift + 4* keyboard shortcut) and select the **Mesh Paint** option. The following screenshot shows the location of this new area, as well as some of the values we'll be using in later steps:

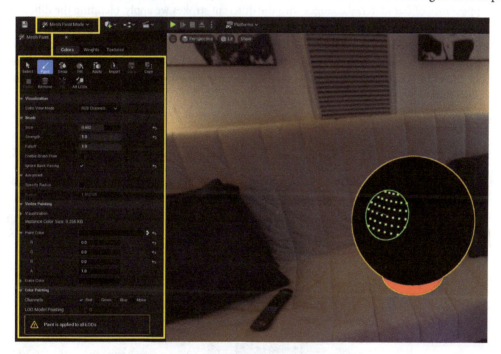

Figure 5.4 – The Mesh Paint mode panel and the settings I'm using to paint the vertices of the model

21. While in that panel, stay within the **Colors** section and click on the **Paint** option. This is the mode that will allow us to interactively assign colors to the vertices of our model.

22. There are certain settings that we can tweak to help us with the painting task that we have in front of us. The first thing we can do is look at the **Visualization** area and choose a different **Color View Mode** than the one selected by default. Choose the **Red Channel** option, as that will allow us to see the values that we are about to paint in that texture channel.

23. Next, feel free to play around with the different settings that control the behavior of the brush that we'll be using to do the painting. Settings such as **Size**, **Strength**, and **Falloff** influence the way we work with this tool, so adjust them to your liking.

24. Focusing on the **Color Painting** area now, select the **Red** channel while making sure that the others are not selected. You might remember that we chose to use the information output by the **Red** channel back in *step 10*, so that's the one we need to be painting on now.

25. Make sure that **Paint Color** is set to white and that **Erase Color** is set to black. Those settings can be found in the **Vertex Painting** section, the one above the previous one we visited. You can take a look at the settings I've used in *Figure 5.4*, the screenshot you saw on the previous page.

With that done, it's finally time to paint the vertices of our model. This can be done with your mouse or with a pen device, provided you have one. Feel free to experiment and see how the appearance of our material changes with the different strokes we apply, blending the different textures we set up in our material. The idea here is to assign different colors to the vertices situated at the base of the model, so go ahead and paint those white while leaving the upper part of the sphere black. At the end of this process, you should hopefully be looking at something like the following screenshot, once you revert to **Off Color View Mode** or once you exit the **Mesh Paint** mode:

Figure 5.5 – The final look of our material

Finally, know that you can exit the **Mesh Paint** mode by selecting one of the other ones available in the **Modes** panel, such as the default **Select** one we usually work with. You can expand this technique by painting in multiple channels and using that information to drive the appearance of the material, and even play with the strength of the tool to create interesting blends between several assets. Something you could find very useful is the ability to slightly tweak the appearance of a large surface by using this method, just by creating different textures and blending between them. There are multiple possibilities – all in all, it's a handy feature!

How it works...

Let's take a little bit of time to go over the **Mesh Paint** panel and the different options that we have at our disposal there. Even though we've used ones that we needed to tweak in the recipe, it's good to get more familiar with some of the settings that we didn't touch on. The first few options available to us were those contained in the **Brush** section, which allowed us to tweak the brush we were using when we painted the vertex colors. Let's cover what they do in further detail:

- One of the most crucial settings there was the **Size** parameter, which affects the area affected by the brush, letting us adjust how much of the model we cover at once.

- The **Strength** setting signals the intensity with which we paint, while the **Falloff** option controls the size of the region where the intensity of the tool fades between the chosen strength and zero – something that allows us to seamlessly blend between adjacent areas. As an example, setting **Intensity** to a value of **1** will fully assign the selected color to the affected vertices, whereas a value of **0.5** will assign half of the selected shade.

- The last two settings in this section are called **Enable Brush Flow** and **Ignore Back-Facing**: the first allows us to continuously paint across the surface (updating our strokes every tick), while the second controls whether we can paint over back-facing triangles. This can help us in situations where we don't want to accidentally paint areas that we are not currently looking at.

Moving beyond the **Brush** section, we can find the **Vertex Painting**, **Color Painting**, and **Visualization** areas. The first one allows us to select which color we want to paint with, while the second one determines which of the RGB channels we affect, as well as letting us paint multiple LODs at the same time. Finally, the third section allows us to choose which channel is shown on the viewport so that we can see what we are painting instead of going at it blindly.

Adjusting the previous parameters to our advantage will make our lives easier when assigning colors to vertices. This data can then be accessed within the Material Editor thanks to the **Vertex Color** node we used in the material we created in this recipe. This node can read the colors assigned to the vertices of the model where the material is being applied, allowing us to drive different effects, as we've done in this recipe.

See also

Even though we've looked at the Vertex Painting technique in this recipe, there are some more things that we can say about that topic. There are at least two things that I'd like to talk about before we move on to a different recipe: the different scenarios where we might benefit from using this technique and the possibility of not only painting vertices but textures as well.

The first of these scenarios is the one where we use this technique to remove the visible tiling pattern that sometimes shows up on large-scale surfaces. We can do that by blending between two very similar but different textures. Think of this as a way of adding variation to your levels in an interactive way, without having to blend between multiple different noise patterns and images within your material graph.

The second one relates to another tool that we can find within the **Mesh Paint** tab that works similarly to **Vertex Color Painting** – that is, **Texture Painting**. This tool allows you to modify any of the textures that are already applied to the mesh you have selected in your level. The only thing that you need to do is select the texture that you want to be operating on from the **Paint Texture** drop-down menu, then start painting on top of the model however you like! You can only choose to paint with a solid color, which might not work that well on the **Base Color** property of a realistic textured asset – but be sure to try it out, as it can work wonders when you want to modify certain values such as the **Roughness** and **Metallic** properties!

Adding decals to our scenes

Decals are a great thing, as they allow us to project certain materials into the world. This opens up multiple possibilities, such as having greater control over where we place certain textures or the ability to add variety to multiple surfaces without complicating our material graphs. They are also useful in a visual way, as they allow us to see certain changes without the need to move back and forth between the Material Editor and the main scene.

Furthermore, Unreal has recently introduced a new type of decal that allows not only for planar projections but also mesh-based ones, which can be useful whenever we want to project something into a non-planar surface (as we are about to see).

So, without further ado, let's jump right into this recipe and see what decals are all about!

Getting ready

Just like we mentioned at the beginning of this recipe, we are about to explore both the standard decals that Unreal has included since it launched and the most recent mesh-based ones. This means that we'll need to set up a scene that allows for both of those techniques to show, ideally including a planar surface, such as a floor or a wall, and a more complicated one, such as a sphere where a planar projection just wouldn't work. Keep those conditions in mind if you want to set up your own level!

If you just want to jump straight into the action, know that we've included a default scene for you, just like we always do. Its name is `05_02_Start`, and you can find it in the Unreal Engine project included in this book.

How to do it...

As we mentioned in the introduction to this recipe, there are two techniques that we'll be exploring here: **Deferred Decal** and **Mesh Decal**. We'll start by focusing our attention on creating the materials that we'll be using to power those assets. You can think about this technique as a two-part process – the first one being defining the shader that we want to project and the second one being the method that we want to use to project them. Let's start creating the materials first:

1. Start by creating a new material and giving it whatever name you think is appropriate. I've gone with **M_DeferredDecal_HockeyLines** as we are going to create the boundaries of a hockey field through it.

2. Open the Material Editor for the new asset and select the **Main Material** node.

3. Focusing on the **Details** panel, set **Material Domain** to **Deferred Decal** and **Blend Mode** to **Translucent**.

    The previous steps tell the engine how our new material should be used, which, as we said earlier, will be as a decal. The next part we need to tackle is creating the material functionality that we want to include – displaying the hockey lines in our level.

    The first order of business is going to be creating a couple of **Texture Sample** nodes. Let's do that now.

4. Continue by assigning the **T_HockeyLines_Color** and **T_HockeyLines_Normal** textures to the previous two nodes. These are the textures that we'll eventually use to drive the **Base Color** and **Normal** properties of our new material.

5. Now that we have a Normal map ready, wire the Texture Sample node containing it to the Normal input pin of our material.

6. Next, connect the **Alpha** channel of the **T_HockeyLines_Color** asset to the **Opacity** input pin in the **Main Material** node. This will drive the opacity of the decal.

7. Continue by adding a **Multiply** node and connect the **RGB** output of the previous **Texture Sample** node (the one that contains the **T_HockeyLines_Color** image) to its **A** input pin. Set the **B** input to use a value of **500** (remember that you can do that by adjusting the value in the **Details** panel with the **Multiply** node selected). We are performing this multiplication to brighten the final result of the image.

8. Then, connect the output of the previous **Multiply** node to the **Base Color** input of the **Main Material** node.

This is the state of the material graph we should be seeing at this point:

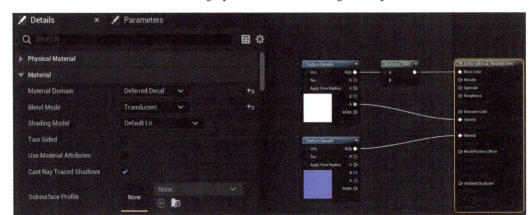

Figure 5.6 – The current state of the material graph

That's everything we need to do for the first decal. We are now going to create a second material that we'll also use as decal material, but one that we'll use according to the **Mesh Decal** technique we'll see later. Let's tend to that now.

9. Create another material and give it a name. I've gone with **M_MeshDecal_HockeyGoal** this time around, as we are going to use it in the hockey goal model present in the level.

10. Next, double-click on the new asset to open the Material Editor and select its **Main Material** node. Then, head to the **Details** panel, as we need to change some of the properties there – just like we did with the previous material.

11. Continue by setting **Material Domain** to **Deferred Decal** and **Blend Mode** to **Translucent**. These are the basic parameters that need to be set up whenever we want a material to act as a decal.

12. With that out of the way, create three **Texture Sample** nodes – one for **T_ HockeyGoal_ MeshDecal_ Color**, another one for **T_ HockeyGoal_ MeshDecal_ AORM**, and a final one for **T_ HockeyGoal_MeshDecal_Normal**. Assign those assets to each **Texture Sample** node.

13. We want to wire the output of the Normal texture to the **Normal** input pin of the **Main Material** node, just like we did in the previous material. After that, connect the **Green** output pin of the **T_ HockeyGoal_ MeshDecal_ AORM** texture to the **Roughness** input pin of our material, and hook the output of the color texture to the **Base Color** node.

14. Next, drag a wire out of the **Green** output channel of the color texture and create a **Cheap Contrast** node at its end. We are going to use that channel to drive the opacity of the decal.

15. Continue by creating a **Constant** and connect it to the **Contrast** input pin of the previous **Cheap Contrast** node. Give it a value of **2**, which will work well in this example, though be sure to play with that value once we finish creating this decal, as that will directly impact its opacity (as we are about to see!)

16. Connect the output of the **Cheap Contrast** node to the **Opacity** pin of our material.

   We are almost done creating this second material, but something that we need to adjust at this point is the **World Position Offset** pin, which we can see is part of the **Main Material** node. This parameter, if enabled, controls the offset that gets applied to the model in which the material is being implemented.

   None of the shaders we have created in previous recipes made use of this setting, but it is one that we'll need to explore at this point. The reason for this is that mesh-based decals get projected from 3D models, and those models used for casting can sometimes overlap with the ones that are supposed to receive the effects of the decal. That being the case, we need to enable **World Position Offset** to ensure that the effects always show and that they are not being obscured because of the locations of the different geometries involved in this technique: we need to make sure that the geometry that projects the material is closer to the camera than the object where the effect gets projected. To do so, we'll need to get the position and the direction of the camera through a couple of new nodes, just so that we can offset the projection toward the camera to make sure that the decal is always visible. We'll learn how to do that next.

17. Create a **Camera Direction Vector** node by right-clicking and typing that name.

18. After that, add a **Multiply** node and place it to the right of the previous **Camera Direction** one, and connect that to its **A** input pin.

19. Include a **Constant** and give it a negative value, something like **-0.5**. This will ensure that we offset the model toward the direction of the camera. Connect it to the **B** input pin of the previous **Multiply** node.

20. Connect the output of the **Multiply** node to the **World Position Offset** pin of the **Main Material** node.

   You can take a look at all of the nodes that make up the material graph in the following screenshot:

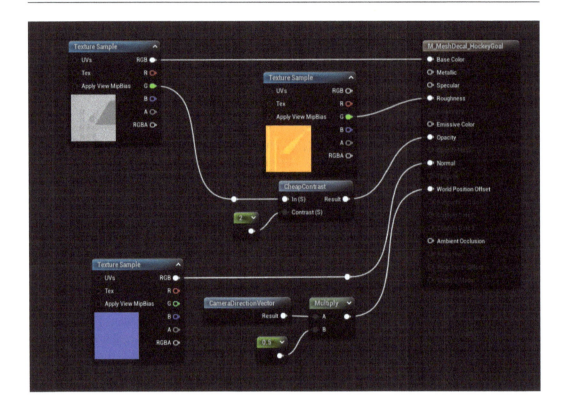

Figure 5.7 – The material graph of the mesh-based decal material

We'll have a working material that is ready to be tested once we click on the **Apply** and **Save** buttons. That being the case, let's head back to the main editor and start creating the decals themselves.

21. Create a **Decal Actor** by heading over to the **Quickly add to the project** tab and searching for that name in the search box. If you don't want to type, you can also find it in the **Visual Effects** subcategory. Drag and drop it into the main viewport.

22. With the new actor selected, focus on the **Details** panel and assign the **M_ DeferredDecal_ HockeyLines** material as the asset to use in the **Decal Material** setting.

23. Play with the decal's scale and position until the lines show and align themselves with the world (you can take a look at *Figure 5.8* to see what the expected result should be).

We've already taken care of the first decal projection, so let's set up the mesh decal next. The steps to do this are even simpler than before, as you are about to see.

24. Duplicate the model that is acting as the hockey goal. You can do this by pressing *Control + D* once you have the asset selected.

25. With the new actor selected, look at its **Details** panel and change the material that is being applied there. Set it to the **Mesh Decal** material we have just created to see it in action and check out the results in the following screenshot!

Figure 5.8 – A look at the decals we have just created

As you saw in the previous couple of examples, decals are a great way to add details to our scene. They open up the door to using multiple different effects: from breaking up the apparent repetition present in large-scale surfaces to allowing dynamic real-time effects, such as bullet holes in video games. Be sure to remember them when you are faced with those types of challenges!

How it works...

Mesh decals are one of my favorite features of Unreal Engine, however strange that might sound. This is because they offer a type of functionality that I wanted the engine to include for a long time, and I couldn't have been happier when they finally added support for them in a later version of the previous Unreal Engine 4. They give us the freedom to project textures based on any mesh that we select, not just planes, opening the door to using decals on curved geometry – something that looks better than the previous planar solution.

Given how we rely on a 3D model to perform the projection of the decal, it's important to include the **Camera Direction Vector** node to ensure that the textures are always positioned between the camera and the surface onto which they are being projected. As its name implies, the **Camera Direction Vector** node gives us the angle at which the camera is facing, a valuable piece of information that we can use to offset the position of the material. We did that through a **Multiply** node and a single **Constant** in this recipe. This is something that we didn't need to do when working with the standard **Decal Actor**, as we have the option to control its size and position to make sure that we always see the decal material we select.

See also

As always, check out Epic Games' documentation regarding decals: https://docs.unrealengine.com/5.0/en-US/decal-actors-in-unreal-engine/.

Creating a brick wall using Parallax Occlusion Mapping

Creating a more detailed level usually involves adding more geometrical detail to the objects used to populate it, something that can be achieved in multiple ways. We've already used Normal maps, a special type of texture that modifies the way light behaves when it reaches the surface of a given object. We also used Nanite meshes in the previous chapter, a novel method of bringing more triangles into our scenes. It is now time to explore **Parallax Occlusion Mapping** (**POM**), a texture-based approach to adding more detail to 3D objects.

This technique relies on textures to create more geometrical detail, in a process that might feel similar to using **Normal** maps. Unlike those textures, POM goes a bit further by actually pushing the pixels that represent our models outwards or inwards, creating a real 3D effect that we just can't get with **Normal** maps. Relying on images to do this kind of job frees the CPU from having to deal with more complicated meshes, something that can alleviate the burden on that specific piece of hardware. We'll learn how to implement this technique in the next few pages.

Getting ready

Something that you'll need in this recipe is a texture that contains depth information as we'll need that type of data to drive the POM technique we are about to see. There is one provided alongside the Unreal Engine project bundled with this book, so feel free to open the `05_03_Start` level if you want to use the same assets we'll be seeing in the next few pages.

As always, feel free to bring your own textures if you want to apply this technique to your custom models. Alternatively, you can use some of the available resources that you'll find in the Starter Content, such as the one called **T_ CobbleStone_ Pebble_ M**, which comes bundled with it.

How to do it...

The first step we'll take in this recipe is going to be the same one we've taken many times before when starting a new lesson: we need to create a material where we can apply the technique we are about to demonstrate. Let's not dwell on this for too long and get it out of the way quickly:

1. Create a new material and give it a name – I've gone with **M_ Walls_ Parallax**.

2. Apply the material to whichever model you want, if you're working on your own level, or the **Material Element 1** slot of the walls model if you've opened the same scene I'll be using.

    With that done, let's open the Material Editor for our new asset and start populating its graph with nodes that will bring our parallax effect to life. The first node we'll need to create is the **Parallax Occlusion Mapping** node, which will act as a hub for many of the nodes that we'll create after it. It is a function that expects several different inputs to work properly, as you can see in the following screenshot:

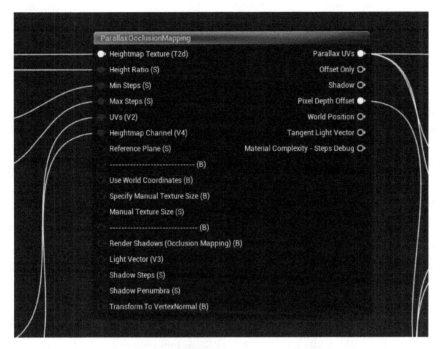

Figure 5.9 – A look at the POM function

3. Let's start by right-clicking anywhere within the material graph and typing `Parallax Occlusion Mapping` – select that option and add the node to our material. We will refer to this node as "POM" for future reference.

4. The first parameter expected by the **POM** function is a **Heightmap Texture**. That being the case, let's create a **Texture Object** and connect it to the **Heightmap Texture (T2d)** input pin of the **POM** node.

5. Select the **Texture Object** node and set the depth texture you want to use in the **Details** panel. We'll use **T_ Walls_ Depth** in our example.

    The next set of nodes we'll need to add are all going to be constants, which we can turn into scalar parameters if we so choose. This will allow us to interactively tweak them once we create a Material Instance. Even though we'll cover them briefly, you can visit the *How it works…* section of this recipe to learn more about them and the role they play when it comes to adjusting the different properties of the POM node.

6. The first node that we'll create is going to be a **Scalar Parameter**. Give it a name after creating it, something like **Height Ratio**. We'll use this to affect how much our texture extrudes and intrudes concerning the model onto which we are applying the material. Give it a value of **0.02**, but make sure to play around with it later to see how this affects the overall effect. Connect it to the **Height Ratio (S)** input pin of the **POM** node.

7. Next, continue by creating another new **Scalar Parameter**, one that we will connect to the **Min Steps (S)** input pin of the **POM** node. Give it a similar name to that input pin and set it to something like **8**. Both this and the parameter we'll create in the next step will contribute to the quality of the final effect we are tackling in this recipe, so make sure you adjust these settings according to your needs after completing the material setup to see how the result changes.

8. Include another **Scalar Parameter**, and name it something like **Max Steps**, but don't connect it to the **Parallax Occlusion Mapping** node just yet – we'll adjust this parameter in the next few steps before doing so.

9. The fourth **Scalar Parameter** should be named something like **Temporal AA Multiplier**, and you can give it a value of **2**. We'll use it to control the smoothness of the anti-aliasing effect alongside the next few nodes we'll be creating.

10. Continue by creating a **Lerp** node and hook its **B** input pin to the previous parameter we created, leaving the **A** input pin with a value of **0**.

11. Right-click within the material graph and look for a node called **Dither Temporal AA**. Connect that to the **Alpha** channel of our **Lerp** node.

12. After that, create a **Multiply** node, then connect both the **Max Steps** scalar parameter we created in *step 8* and the output of the previous **Lerp** node to it.

13. Finally, connect the **Multiply** node to the **Max Steps** input pin of our **POM** node.

As mentioned previously, make sure you head over to the *How it works...* section of this recipe to learn more about how these last steps work from a theoretical point of view. Just so you don't have to wait until then, what we've done so far affects the amount by which our material will seem to be extruded, as well as the quality of that extrusion, which is controlled through the **Minimum Steps** and **Maximum Steps** scalar parameters we created. The reason why we didn't just plug the **Max Steps** scalar parameter directly into the input pin of the **POM** node is that we want to smooth out the effect using the **Dither Temporal AA** node; the result might look a bit blocky otherwise.

With that out of the way, the next settings we are going to take care of are going to be the **Heightmap** channel and the tiling of the texture itself. Tiling is something that we've already dealt with in the past, so that should be quite straightforward to handle. However, you might find it strange to have to specify which texture channel we use for the height map. The reason for that is merely because of the way the **POM** node has been set up to work. The **POM** function expects us to provide the depth data through the **Texture Object** parameter we wired into it, but it doesn't know if that depth information is provided through the combined **RGBA** output or if it is stored in one of its specific channels. The way we tell the function which channel to use from that **Texture Object** is by manually specifying the channel, as we are about to do.

14. Let's continue our material-creation process by creating a **Vector Parameter** and giving it a name – such as **Channel**. Give it a value of **Red**, or **(1,0,0,0)** in RGBA format.

15. Next, include an **Append** node and hook the output of the **Red** and **Alpha** channels of the previous node to this one's input pins.

16. Wire that to the **Heightmap Channel (V4)** input pin of the **POM** node.

Note

The reason we are appending the **Alpha** pin to the **RGB** output of the vector parameter is that the input pin of the **POM** node expects a **Constant4Vector**. The **RGB** output by itself only contains three channels, so appending the **Alpha** pin to that grants us the 4-channel vector we need.

---

17. Create a **Texture Coordinate** node and leave the default value of **1**.

18. Then, add a **Scalar Parameter** and set it to **7**. We'll use it to control the tiling of the material in case we create a Material Instance.

19. Seeing as we want to control the tiling of the textures, place a **Multiply** node after the last two and wire them up. Connect its output to the **UVs (V2)** input pin of the **POM** node.

    Completing the previous steps has left us with all of the parameters needed to feed the different pins of the **POM** function that we want to tweak. Seeing as we've dealt with quite a few nodes so far, let's take a moment and look at the state of the material graph:

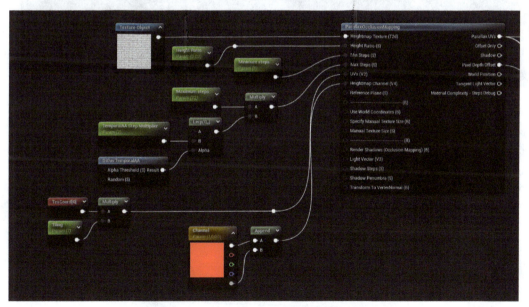

Figure 5.10 – A look at the nodes we have created thus far

Now that we have the **POM** function all wired up, we can use the different information that it provides us with as the base for the next set of textures that we'll use. Let's start creating them!

20. Let's continue by creating three **Texture Sample** nodes and assigning the following textures to them: **T_ Walls_ Depth**, **T_ Walls_ AORM**, and **T_ Walls_ Normal**. Connect their **UV** input pins to the **Parallax UV** output pin of our **POM** node.

21. Wire the output of the first texture to the **Base Color** input pin of the material. Next, connect the **Red** channel of the second texture to the **Ambient Occlusion** input pin and the **Green** channel to the **Roughness** slot. Wire the output of the **Normal** map to the **Normal** input pin of our material.

22. Finally, hook the **Pixel Depth Offset** output node of the **POM** function to the **Pixel Depth Offset** input pin of our main material node. We are all done!

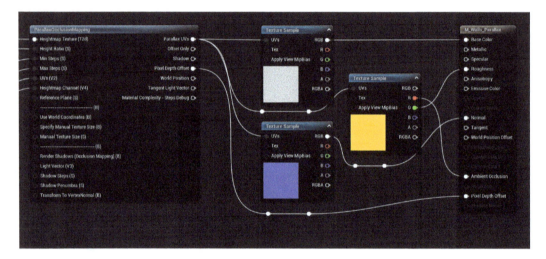

Figure 5.11 – A look at the last few nodes we've created

Now, all we need to do is save our work and apply the material to the appropriate model in our scene. You should be looking at a white brick wall that extrudes out of the plane in which it was originally applied, and that effect remains visible no matter the angle you are viewing it. Remember, all of this was achieved without us using any extra geometry!

Figure 5.12 – A final look at the brick wall material we have just created

How it works...

As you've seen, the **POM** function is quite a complex one, as it includes multiple different inputs that we need to set up properly for the effect to work well. However daunting that may seem, things will start to make more sense once we go into a little bit more detail on what they do and how each setting we tweaked contributes toward building the final effect.

The first thing that we need to be aware of is how the effect works. POM operates on top of Unreal's standard rendering pipeline, which consists of rasterizing the meshes that make up our level and, in turn, assigning a world position value to each pixel in our scene. In other words, each pixel that we see on screen is the result of a calculation made by the engine, which takes into account the location of our models and any other effects or materials that are being applied.

POM takes the world position of a given pixel, as calculated by the engine, and modifies its transform value according to the different settings that we specify in the **POM** function. The first setting that we needed to tweak is **Heightmap Texture**, an asset that determines the difference in the position of a given pixel concerning the underlying model. In our example of a brick wall material, the white parts of the depth texture tell the renderer to extrude those areas outwards, while the darker tones push pixels inwards.

The second setting that we tweaked was **Height Ratio**, which serves as a modifier to increase or decrease the values of our previous height map – thus making the effect more or less notorious.

On top of that, we added a couple of extra constants that affected the **Min Steps** and **Max Steps** scalar parameters of the effect, which, in turn, decide on the quality that we end up seeing. More samples mean that the effect will appear to be more realistic, while less means the opposite. The reason why we call them steps is that this effect is similar to what we see in splines or curved surfaces in a 3D package – the more subdivided they are, the more rounded they appear.

It's also for that very reason that we used the **Dither Temporal AA** node, which increases the smoothness between the different steps we created. This makes them blend better and do so in a more seamless way.

Finally, the last two settings we needed to tweak were the **UV** and **Heightmap** channels. The former is quite a straightforward operation as we need to specify how much we want our material to tile. The latter is quite curious – instead of specifying which texture channel we want to use in the original **Heightmap** texture we created, we need to do so here. This is a quirk of the **POM** node we are using, which expects a **Texture Object** as texture input rather than a **Texture Sample**. The type we are forced to use cannot specify the channel we want to use, thus making these last settings necessary.

See also

As we mentioned previously, POM is a GPU-driven effect. This means that the CPU doesn't have access to what's being created there, which is an important thing to keep in mind. As with any game engine, many different processes are happening at the same time, and some of those get taken care of by the CPU while others rely on the graphics unit. Such is the case of something that affects us in this case – the collision system. Whenever we use the POM technique, we usually apply it with the intent of adding detail and transforming the surface onto which we are projecting the effect. As a result, we end up with a different surface or model – one where its visible bounds differ from those of the original model. This is an important thing to consider when we need to have collisions enabled in that same model, as those are computed by the CPU and don't know anything about what's happening GPU side. This could lead to perceptible differences between the location where a collision between two objects should occur versus where it happens.

A possible workaround to this problem is to create a custom collision mesh that takes into account the result of the POM technique – maybe by adding a collision volume that respects the new visual bounds we see after we apply the aforementioned technique. Whatever we decide to do in the end, make sure you keep the nature of this effect in mind to prevent any possible issues when working with other systems that are driven by the CPU.

Finally, let me leave you with some extra documentation on the POM technique. POM can be a very interesting topic from a technical point of view, and learning more about it can give you a good insight into how computer graphics work. I'd like to leave you with a good link to a site that explains it well, and one that can give you a foot up into a bigger world if you want to read even more: `http://online.ts2009.com/mediaWiki/index.php/Parallax_Occlusion_Map`.

Taking advantage of mesh distance fields in our materials

You can say that each different recipe we've tackled in this book has explored something new within Unreal's material creation pipeline. We've seen new material features, useful nodes, and smart shader-creation techniques, and this recipe is not going to be any different in that regard. Having said that, we are now going to look outside the Material Editor to tackle the next technique.

We are going to be focusing on a specific feature called **Mesh Distance Fields**, an attribute of the 3D models that we work with that will allow us to know how close or far we are from them – enabling us to change the appearance of a material based on that information. You can think of them as the visual representation of the distance from a volume placed around the model to the surface of the 3D mesh. This data gets stored as a volume texture, something that can be very useful for us as it allows us to create dynamic effects such as distance-based masks or ambient occlusion-driven effects.

With that information at our disposal, we are going to create a simple cartoonish ocean material that shows foam around the objects that intersect it. Let's see how to do that!

Getting ready

The level we'll be working on is quite a simple one – in fact, it only contains a couple of 3D models: a plane and a sphere. Feel free to load the `05_04_Start` level to start working on it, or create your own one using similar assets to the ones I mentioned.

Something critical to completing this recipe is making sure that the **Mesh Distance Fields** setting is enabled in the project where we are working. You can do so by heading over to **Edit | Project Settings** and looking under **Engine | Rendering | Software Ray Tracing**. There is a property there called **Generate Mesh Distance Fields** that we need to make sure is enabled, so please ensure that you do that before jumping into the next section.

How to do it...

Let's start this recipe by looking at the scene that we have in front of us. If you're working with the one provided with this book, we should be able to find a couple of 3D models: a plane and a sphere, currently intersecting each other. To use the Mesh Distance Field technique appropriately, it's useful to know which models are going to be affected by it, as we might need to treat those differently from the rest. In our case, we'll want to apply a material to the plane, which will take the distance fields of the other objects into account. Let's start tweaking that model first:

1. Select the plane in the middle of our level and open its Static Mesh Editor. Remember that you can do this by double-clicking on the thumbnail inside the **Details** panel.

2. Inside the new editor, look at the **Details** panel and scroll down to the **General Settings** category. Look for the option named **Generate Mesh Distance Fields** and make sure it's turned off.

3. Go back to the main viewport and, with the plane still selected, look at the **Details** panel. Scroll down to the **Lighting** section and turn off **Affect Distance Field Lighting**.

4. After that, select the sphere and open the Static Mesh Editor, just like we did with the plane.

5. Once inside that editor, look once again for the **Generate Mesh Distance Fields** setting and turn it on.

6. Scroll a little bit toward the top of the **Details** panel and find the setting called **Distance Field Resolution Scale**, which is located toward the bottom of the **LOD0** section. Set it to **10** and click on the **Apply** button located immediately below it.

Here is a screenshot indicating the location of that setting:

Figure 5.13 – Location of the Distance Field Resolution Scale setting

The previous few steps have ensured that the sphere in our scene has a **Mesh Distance Field** representation, while the plane doesn't. It's important to disable this feature on the model on which we are going to operate as its distance fields would otherwise conflict with those of the other objects around it. We want to apply a material that is aware of the Mesh Distance Fields of other objects, but not of the object on which it is being applied, as that would lead to confusion as to which field it has to read – that's why we need to remove it!

With that out of the way, we can now concentrate on creating the material that will make use of this feature of the engine.

7. Create a new material and give it a name. I've gone with **M_ SimpleOcean_ End** as that's what we'll be creating in this recipe.

8. Open the material graph for the new asset and add the first node to our graph — one named **Distance To Nearest Surface**. This node gives us the distance between the pixels affected by the material that makes use of this function and the nearest surface to them, something that gets determined thanks to the distance fields present in the static mesh models with which we are working. The values thrown by this function are in centimeters, Unreal's standard units. Given how we'll want to use the results as masking material in a **Lerp** node later on, we'll need to bring them closer to the 0 to 1 range expected in that function.

9. Add a **Scalar Parameter** and name it something like **Distance Offset**, while setting its value to **0.015**. We'll use this node to adjust the values thrown by the previous **Distance To Nearest Surface** function.

10. Create a **Multiply** node and hook the output of the previous two nodes to the input pins of this new one.

The value we chose for our newly created **Scalar Parameter** will control the extent of the area that we'll use for masking purposes. As mentioned previously, the **Distance To Nearest Surface** function gives us the distance from each pixel of the affected object to the nearest model that has distance fields enabled. Seeing as that value is in centimeters, we need to adjust it to bring it closer to the 0 to 1 range expected by the **Lerp** node. The value that we choose to modify that node will directly impact the distance covered by that 0 to 1 range, with lower values making the area wider. As an example, if a pixel is 50 units away from a given model and we apply a 0.02 factor to it, we'll get a value of 1, marking that as the limit of the effect. If we choose a lower value as the factor, something such as 0.01, the new limit would be moved to the 100-pixel mark, thus effectively extending the area that we cover.

Now that we've got that out of the way, and seeing as this is going to be a simple material, something we can do is introduce a break in the previous gradient we created and make it look a little bit more like a toon shader. We'll do that by segmenting the gradient into smaller steps, which we will do now.

11. Create a **Scalar Parameter** and give it a name similar to **Number of Steps**. We'll use this value to fragment the gradient we have so far – the bigger the number, the higher the fragmentation.

12. Include a **Multiply** node and use it to connect the previous **Scalar Parameter**, as well as the output of the **Multiply** node from *step 10*.

13. Add a **Floor** node. This will give you the bottom value of a float: for example, if you have a value of **0.6**, the result after creating the node will be **0**. Connect it to the output of the previous **Multiply** node.

14. Place a **Divide** node after the previous one and connect the output of **Floor** to its **A** input pin. Then, connect the **B** input pin to the output of the **Scalar Parameter** node we created in *step 11*. Refer to the following screenshot for more information:

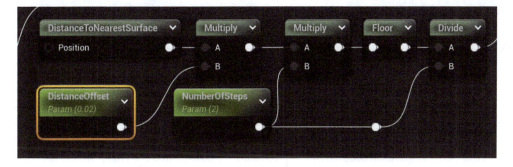

Figure 5.14 – A look at the previous nodes we created

The previous set of nodes has given us a banded gradient, something that works well with toon-like shading techniques. The main reason why the **Floor** node can create gradients of this type is that it takes a float value and discards the decimal part. Because we have a gradient, each pixel has a different value, so the **Floor** node gets rid of the seamless variation in favor of a more stepped one since it gets rid of intermediate values – the ones between integers.

Knowing this, let's get back on track with creating our material by putting the previous sequence to work as a mask.

15. Create a couple of different colors by placing two **Constant4Vector** nodes in the material graph and assigning them different values. I've gone with blue and white.

16. Add a **Lerp** node and connect the previous **Constant4Vector** nodes to its **A** and **B** input pins. The **Alpha** pin should be connected to the output of the **Divide** node we created in *step 14*.

17. Connect the output of the **Lerp** node to the **Base Color** property of our material.

18. Include a simple **Constant** to modify the value of the **Roughness** parameter. Something like **0.6** works well in this case!

19. Finally, save the changes you made to the material and assign it to the plane in our level.

Once those steps have been completed, this is the material we should be looking at:

Figure 5.15 – A final look at the material being rendered

Ta-da! Look at what we now have! This simple material, which is used in tandem with the **Mesh Distance Field** property of our models, can work very well for any kind of distance-based effect. Things such as ambient occlusion-based dirt, or water effects like the one we've just seen, are great examples of that. Furthermore, this technique is dynamic, meaning that the effect will update itself in real time as things move around your level. There are many possibilities, so make sure you play around with it for a while to see how far you can take it!

How it works...

Mesh Distance Fields can feel a bit daunting at first, especially since they are something that not many artists encounter very often. We'll try to introduce the basis on which they operate and, more importantly for us, how to increase their quality whenever we need to.

Trying to simplify the concept as much as possible without going into too many technical details, Mesh Distance Fields can be thought of as the way that the engine stores the distance at which the different parts of a model are from a volume texture. Let's take a look at the following screenshot:

Figure 5.16 – A visualization of the Mesh Distance Field for the sphere in our level

The previous screenshot shows the visual representation of the **Mesh Distance Field** property for the sphere located in the middle of our level. You can enable this view mode by clicking on **Show | Visualize | Mesh Distance Fields**. Distance fields are calculated offline when we import a model, or when we adjust its distance field resolution, as we did in *step 6*. We can tap into this pre-computed data to drive different effects, from the one we saw in this recipe to dynamic flow maps, just to name a couple.

Something that we need to be aware of is how we can affect the final quality of the fields we are creating. We can do so through the setting we saw at the beginning of this recipe: the **Distance Field Resolution** multiplier. Going beyond the default value of 1 will make the volume texture weigh more, but we sometimes need that extra quality in very detailed objects or models that contain thin elements. Remember to tweak that whenever you need to!

See also

We've covered the way Mesh Distance Fields work from a practical point of view, going a bit into the technical details of how they are created as well. If you are still curious about the theory behind this important technique, make sure you check out the following link for more information: https://docs.unrealengine.com/5.1/en-US/mesh-distance-fields-in-unreal-engine/.

Lighting the scene with emissive materials

We used lights in previous recipes whenever we needed to illuminate a scene. You might remember what we did in *Chapter 1* when we explored the **Image Based Lighting** (**IBL**) technique. Lights are one of the key types of actors within Unreal, and their importance is critical within the rendering pipeline. We wouldn't be able to see much if it wasn't for them!

Having said that, the engine puts another interesting option at our disposal whenever we need to light a level: emissive materials. As their name implies, this type of shader allows objects to emit light, helping us shade the levels we are creating differently.

Of course, they come with their own set of pros and cons: on the one hand, they allow us to use custom 3D models to light different areas, enabling interesting-looking effects such as neon lights or light coming from TV screens – something that would prove quite difficult to tackle using standard lighting techniques. On the other hand, emissive materials can cause noise artifacts to appear, so we need to be aware of how and when it is safe to use them.

We are going to be taking a look at those scenarios in this recipe through the use of a salt lamp.

Getting ready

Even though we are going to be using some custom 3D models and textures in this recipe, this is just so that we have something pretty to look at. If you want to use your own assets, know that there aren't many requirements in terms of what you'll need to produce. You'll want to have a simple mesh where you can apply the emissive material we'll be creating and a small scene where you can test the results of the light-baking process we'll be tackling. Even using the assets provided as part of the engine can work here.

The one thing that you'll probably want to have is a somewhat dark scene. This is because of the nature of this recipe: since we are going to be creating materials that emit light, the results are going to be more obvious if the scene is dark. We'll work on two different scenarios, with and without dynamic lighting, but don't worry about this yet as we'll tackle that in the recipe itself.

With that out of the way, feel free to open the `05_05_Start` level if you want to use the project provided alongside this book. I'll open that one now, as we are ready to start the recipe!

How to do it...

The first step we'll want to take when starting this recipe is going to involve adjusting the intensity of the lights currently placed within the level on which we are going to be working. Regardless of whether you've opened the same level I'll be using or your own, we want to make sure that we have quite a dark scene in front of us just so that we can see the effects of what we are about to create. If you've opened the scene we have provided, make sure to do the following:

1. Select the **BP_LightStudio** Blueprint asset within the World Outliner and look at the **Details** panel. The third visible category, named **HDRI**, has an option called **Use HDRI**. Make sure that box is unchecked so that we can better appreciate the results of the next steps we'll be enabling.

    Once that's out of the way, let's proceed to create the material on which we'll be working.

2. Create a new material within the Content Browser, the one we'll use as the light caster. Name it whatever you like – I've gone with **M_LightBulb**.

3. Select the **Main Material** node and change **Shading Model** from **Default Lit** to **Unlit**.

    We'll be ready to start creating the material logic once we've performed the previous couple of steps. Creating an emissive light is simple, so we'll try to spice things up along the way. The first bit that I want to make you aware of is that we'll try to stay close to the original look of the lamp, which you can check by looking at the lamp present in the scene when you first open it. We'll use the same textures to achieve that.

4. Create a **Texture Sample** node and assign it to the **T_Lamp_Color** texture. We've included this asset in the **Content | Assets | Chapter05 | 05_05** folder of the Unreal Engine project that accompanies this book.

5. Next, add a **Cheap Contrast** node after the previous **Texture Sample** node and connect its **In (S)** input pin to the output of the previous **Texture Sample** node.

6. After that, include a **Scalar Parameter** and name it something like **Contrast Intensity** – since that's what it will control. Connect it to the **Contrast (S)** input pin of the **Cheap Contrast** node, but not before assigning it a value (**2** works well for our purposes).

    The previous set of nodes has taken a regular texture and created a black-and-white version of it. This will let us mask certain areas of the lamp so that some of them emit more light than others. Take a look at the following screenshot:

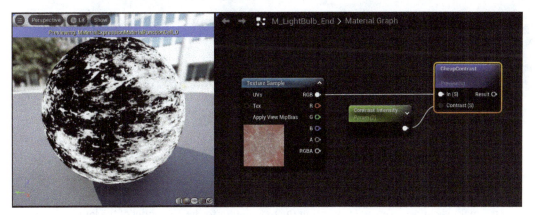

Figure 5.17 – The masked version of the color texture for the lamp

7. Include a **Multiply** node and connect the output of the **Texture Sample** node to its **A** input pin.

8. Next, create a **Scalar Parameter** to feed into pin **B** of the previous **Multiply** node. As this will be controlling the brightness of the lighter areas, name it accordingly – something like **Light Area Brightness** will do. Remember to also assign a value, something like **500**.

    We'll use the previous couple of nodes to control the intensity of the lighter areas of the mask we created thanks to the **Cheap Contrast** node. Next up – taking care of the darker parts!

9. Drag a cable out of the original **Texture Sample** node and create a new **Multiply** node.

10. After that, add a new **Scalar Parameter** and name it something similar to **Dark Area Brightness** – we'll use this to control the emissive intensity in those areas. Set its default value to **2**.

11. With that done, add a **Lerp** node somewhere to the right of the existing graph. Connect the output of the **Multiply** node we created in *step 7* to input pin **B** of the new **Lerp** node and connect the other **Multiply** node we created in *step 9* to input pin **A** of that same **Lerp**.

12. Next, connect the **Result** pin of the **Cheap Contrast** node to the **Alpha** pin of our new **Lerp**.

13. Finally, connect the output of the **Lerp** node to the **Emissive Color** input pin of the material and save your creation. The following screenshot shows the material graph we should be seeing at this point:

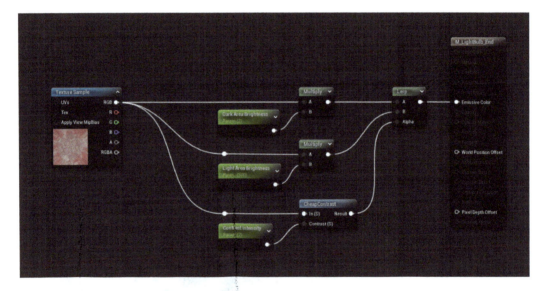

Figure 5.18 – The graph of our material

With that last step in place, we can almost say that we are done with the recipe… at least in terms of setting up the material! Despite that, you might notice that nothing changed back in the main viewport, even after assigning the new shader to the 3D model of the lamp. Chances are your screen will still look as dark as when we turned off the **Use HDRI** option back in *step 1*. The material emitting visible light or not is something that depends on its distance from the camera to the mesh – there is currently a cutoff point where emissive models stop contributing to the lighting of our scene. Move closer and farther from the lamp to see this effect for yourself. This is something that we probably don't want to happen, so let's take care of that next.

14. Select the 3D model of the lamp and look at its **Details** panel. Scroll down to the **Lighting** section and expand the **Advanced** panel, where you'll find an option named **Emissive Light Source**. Check that box to correct the previous behavior.

Finally, you'll see our emissive material work alongside Lumen:

Figure 5.19 – The emissive material lighting the scene

Quite nice, isn't it? The results we get are interesting, but we need to keep in mind that the previous figure shows a static representation of the scene: you might notice that the quality of the effect isn't the same on your screen. We need to be aware that even though Lumen can pick up the lighting produced by emissive materials, the results tend to be a bit noisy, so it's best not to rely entirely on this technique. As a result, the previous scene can look much nicer when paired with the original HDRi we were using at the beginning – try to re-enable it and adjust its intensity for a noise-free effect.

Having said that, we can now say that we know how to work with emissive materials and dynamic lighting using Lumen. But what about static lighting? These types of shaders can work well in that scenario, even by themselves, as the light-baking process can take care of removing the artifacts we tend to see with Lumen. We'll look at how to enable that next.

15. Select the 3D model of the lamp and focus on the **Details** panel once again. Head down to the **Lighting** section and expand the **Lightmass Settings** panel. You'll find an option that we need to check there called **Use Emissive for Static Lighting**.

16. Next, select the **Post Process Volume** node present in the scene and set the property called **Global Illumination Method** to **None**. This will prevent Lumen from taking care of the global illumination, something that we already explored in the *Using static lighting in our project* recipe in *Chapter 1*.

17. For Unreal Engine 5 to support static lighting, we need to make sure that the **Allow Static Lighting** option located within the **Engine / Rendering** section of the **Project Settings** area is turned on. It should be by default, but there's no harm in checking and knowing about this parameter in the process!

18. Finally, head back to the main viewport and click on the **Build Lighting Only** option located within the **Build** panel. Wait for the process to complete, and compare the results to what we got earlier.

The resulting image should be quite similar to the one we saw earlier, with the added benefit that this time, we have more control over the quality of the lighting. To achieve that, all we need to do is increase the resolution of the lightmaps used by our static models and/or adjust the **Lightmass** settings – Unreal's static lighting solver. Remember that you can adjust each model's resolution by heading over to their static mesh editor and tweaking the **Min Lightmap Resolution** property. As for the **Lightmass** settings, they can be tweaked within the **World Settings** panel by going to **Window** | **World Settings** | **Lightmass settings**. If you want to know more about that, make sure you revisit the *Using static lighting in our project* recipe.

How it works...

So, how does it all work? What are the principles that govern the light-emitting properties of emissive materials? Those might seem like big questions, but they are not that difficult to answer. However, the response does change, depending on whether we are dealing with static or dynamic lighting, so let's tackle each type separately.

When it comes to static lighting, you can think of using emissive materials as using static lights. The engine needs to calculate the lighting pass for the results to show, and you need to pay attention to certain **Lightmass** settings to achieve your desired result.

Using this method is quite cheap on our machines. The material is lighter to render than most of the other ones we've seen so far, as it uses the **Unlit Shading Model**. The light itself is also very simple and won't drain resources from our computers as it's of the static type. This means no dynamic shadows, and it also means that we have to build the lighting of our level if we want to use it. The most demanding aspect we need to pay attention to is the **Lightmass** settings we use when building the light since that process can take a long time, depending on the size of our level and its complexity. Other than that, feel free to use this method as much as you want as it can be a cheap way to add lighting complexity to your scenes. Many artists use this technique to fake detail lighting, where instead of having real lights that define the overall look of the scene, they use this baked method to highlight certain areas.

If using Lumen, the behavior of the emissive materials will be different than explained here. This is due to the nature of this dynamic technique, which turns all emissive surfaces into dynamic emitters at no performance cost. The downside, as we saw, is the possible introduction of noise artifacts, depending on the size and the brightness of the material itself. Unlike static lighting, controlling the quality is more difficult in this case, but a good place to look for settings that can control this behavior is the **Lumen Global Illumination** section within the **Post Process Volume** nodes that we placed.

Lumen Scene Lighting Quality, **Lumen Scene Detail**, and **Final Gather Quality** are all settings that control the quality provided by Lumen, so make sure you tweak those if you want to increase the rendering quality.

See also

Emissive materials and lighting are tied to the global illumination system used in Unreal – regardless of whether you're opting for the static (pre-computed) or the dynamic (Lumen) way. Let me leave you with more documentation on the topic in case you want to learn more: `https://docs.unrealengine.com/5.0/en-US/global-illumination-in-unreal-engine/`.

Orienting ourselves with a logic-driven compass

We are used to applying materials to 3D models – after all, that's what we've done in most of the recipes in this book. Having said that, there are other realms where materials and shaders can leave their mark, some of which we've already seen. As an example, you might remember the decals that we worked with back in the *Adding decals to our scenes* recipe in this very same chapter: decals are not 3D models, and as such one of the first things we needed to adjust was the **Material Domain** property of the new material we created in that recipe.

Another field where we can use materials is in the user interface of our game, and that's where we are going to focus our attention during the next few pages. We are going to learn how to create a compass that we can use to orient ourselves within the game world, and such an element is usually part of the user interface. As a result, we are going to dip our toes into two new topics: user interfaces and materials that can work on those UIs. Exciting times are ahead!

Getting ready

Seeing as we'll be working with a compass in this recipe, we don't need anything else beyond the textures that we'll use to drive the appearance of our new material. There will be two: an image that we'll use to display the marker that is going to signal which cardinal direction we are facing and another texture that displays the directions themselves. We need to keep them separate as we want both to move independently from each other so that the marker can change its orientation. On top of that, we need the images to have transparent backgrounds so that we can see the images that are behind them once we stack them on top of each other. For reference, this is what the textures I'll be using look like:

Figure 5.20 – A look at the two textures used to create the compass

Apart from that, there's nothing else that we'll require to tackle the next few pages. Even though that's technically true, orienting ourselves in an empty space would probably be a bit difficult. Make sure you have a scene with some props scattered around so that it's easier to orientate yourself in it. Alternatively, feel free to open the 05_06_Start level provided alongside this book, where you'll find a scene ready for you to use.

How to do it...

As I mentioned in the introduction to this recipe, the material we are about to create is going to work as a UI element. That being the case, the first step is going to be the creation of our first-ever UI! If you haven't dealt with UIs before, make sure to check the *See also* section at the end of this recipe, where we'll explore the setup of the UI editor. Having said that, let's create one straight away:

1. Start by heading over to the Content Browser and creating a new **Widget Blueprint**. This is one of the types of assets that we can create whenever we want to work with UIs in Unreal. To create one, right-click anywhere within the Content Browser and search for the **Widget Blueprint** option available within the **User Interface** section.

2. Remember to name the new asset by selecting it in the Content Browser and hitting *F2* on your keyboard. I've gone ahead and named my version **Game Compass**.

3. Open the new widget and select a **Canvas Panel** from the **Palette** tab. You can find it by looking inside the **Panel** subsection or by typing its name into the **Search Palette** search box. Once you find it, drag it into the main viewport.

4. Just like we've done before, select a **Scale Box** this time and drag it into the main viewport.

5. Next, select an **Image** from the **Palette** tab of the UI editor and drag it into the main viewport. Name it something such as **T_Compass** if you want to remember this variable more easily.

6. After that, focus your attention on the **Hierarchy** panel, which will show you all of the elements that currently make up our UI. Make your **Scale Box** the parent of the **Image** property by dragging and dropping the latter into the former.

7. With that done, select your **Scale Box** and look at the **Details** panel. Select the upper-middle anchor in the **Anchors** drop-down menu.

8. Type in the following settings for your **Scale Box**: **Position X** left as its default of **0.0**, **Position Y** changed to **75.0**, **Size to Content** turned on, and **Alignment** set to **0.5** for **X** and **0** for **Y**. Our goal here is to place our **Scale Box** in the upper middle of the screen in a centered position. Refer to the following screenshot if you have any doubts regarding the values to use:

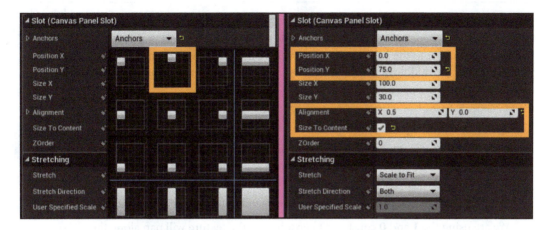

Figure 5.21 – The settings that should be used to correctly position the Panel Slot

Again, be sure to check out the *See also* section if you've never dealt with the UI editor before. Completing all of the previous steps will take care of almost every setting we need to tweak within the UI itself, but we'll still need to assign a material to the image to work as a compass. Let's do that next.

9. Create a new material within the Content Browser and give it an appropriate name, something such as **M_Compass**. Open the Material Editor so that we can start tweaking it!

10. Start by selecting the **Main Material** node and looking at its **Details** panel. In there, change the **Material Domain** property from the default **Surface** option to the **User Interface** one. That's all we need to do to have our materials work on UI elements!

11. Still in the **Details** panel, change **Blend Mode** from **Opaque** to **Masked**. This will let us use the alpha channel of the images in our UI material to control what is visible.

Having already configured the material to work as a UI element, let's continue with creating the actual logic within the shader. Something we need to take into consideration is how the compass will look in the end – we'll have an animated texture that will show us the cardinal directions as the player rotates, and another fixed image working as a maker. We saw both of those textures in the *Getting ready* section, so let's implement them now.

12. To start this process, create a **Texture Coordinate** node and add it to the graph.

13. Next, create a **Panner** node and feed the previous **Texture Coordinate** node to the **Coordinate** input pin of this new node.

14. After that, add a **Scalar Parameter** and give it a name similar to **Player Rotation**, as we'll use it to get the current rotation of the player.

15. Include an **Add** node right after the previous **Scalar Parameter** and connect the output of the latter to the **A** input of the former. Choose **0.25** as the value for the **B** input pin.

16. Connect the result of the **Add** node to the **Time** input pin of the **Panner** node.

> **Important note**
>
> The reason we assigned a value of **0.25** to the **B** input pin of the **Add** node is that we need an initial offset due to the way the image we will be using later was created. In it, the North indicator is in the middle of the image, and we need to offset it a little bit to match Unreal's real North position.

17. Now, let's add a couple of **Constant** nodes and assign them values of **1** and **0**.

18. After that, include a **Make Float 2** node and connect the previous two **Constant** nodes to it. Make sure that the **1** node is connected to the **X** input pin and the **0** node is connected to the **Y** input pin.

19. Wire the result of that previous node into the **Speed** node of the **Panner** node.

    We are using the **1** and **0** constants to ensure that the texture will pan along the *X*-axis at a constant speed, and not on the *Y*-axis. The actual amount by which we'll perform the panning operation will be driven by the **Player Rotation** scalar parameter, which we will dynamically link to the player later on.

20. Continue by creating a **Texture Sample** and connecting its **UVs** input pin to the output of the previous **Panner** node. Assign the **T_Compass_Directions** texture to it.

21. Create a **Vector Parameter** and give it a name similar to **Directions Color**, as we'll use it to affect the tint of the movable part of this material.

22. Multiply the previous two nodes by creating a **Multiply** node and connecting its input pins to the output of the previous **Texture Sample** and **Vector Parameter**.

    Even though we have something that we could already use as an animated compass, we still need to add the static marker to help indicate the direction the player is facing. We'll do that next, but not without first taking a look at the state of the material graph – just to make sure that we are all on the same page!

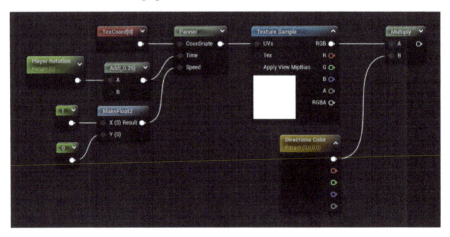

Figure 5.22 – A look at the current state of the material graph

23. Next, add a **Texture Sample** node and the **T_Compass_Fixed** texture to it.

24. After that, include another **Vector Parameter** and name it something such as **Marker Color** since it will control the color of the marker that indicates the North position in the compass.

25. Proceed to create a **Multiply** node and connect the output of the previous **Texture Sample** and **Vector Parameter** to both of its input pins.

    The previous steps will let us easily tint the fixed part of the compass material should we choose to create a Material Instance. Now, let's add both sections together and define the **Opacity Mask** property of the material.

26. Continue by adding a **Lerp** node after the previous **Multiply** node and connect the output of the latter to the former's **A** input pin.

27. Next, connect pin **B** of the **Lerp** node to the output of the **Multiply** node we created in *step 22*.

28. As for the **Alpha** pin of the new **Lerp** node, connect it to the **Alpha** channel of the **Texture Sample** node we created back in *step 20*.

29. With the **Lerp** node sorted, connect its output pin to the **Final Color** input pin of our **Main Material** node.

30. After that, create an **Add** node and place it just before the **Opacity Mask** input pin of our material. Connect them both.

31. Wire the **Alpha** channel of both **Texture Sample** nodes present in the material into the **A** and **B** input pins of the previous **Add** node. And that's it – we now have the material ready!

    After all of the previous steps, let's take a look at the material graph before tackling the final part of this recipe:

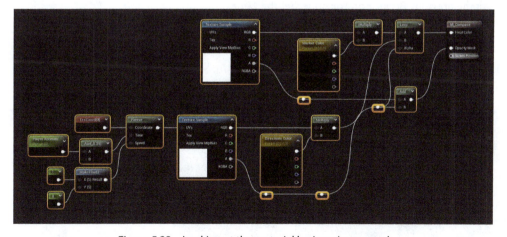

Figure 5.23 – Looking at the material logic we've created

Now that we have a working material, let's create a **Material Instance** out of it and apply it to our UI!

32. Right-click on the material we created within the Content Browser and select the **Create Material Instance** option.

33. Double-click on the new instance to tweak the different editable settings however you like, especially **Directions** and **Marker Color**, so that they look good on your end.

34. Head back to the UI we created and select the image in the **Hierarchy** panel. Looking at the **Details** panel, expand the **Brush** option and assign the Material Instance as the image's value.

35. Set the **Image Size** property to **1,532** in its **X** component and **24** in the Y one, or something that clearly shows the compass at the top of your screen. This can vary depending on your screen's resolution, so make sure you choose something that looks right to you.

We are now in a position where we can feed the player's rotation to the scalar parameter we set in our material and drive the appearance of the compass through it. Let's jump to the Event Graph to do that!

36. While still in the UI editor, switch over to the **Graph** tab and drag and drop the image variable into the main Event Graph.

37. Next, create a **Get Dynamic Material** function and connect it to the output of the **Event Tick** node.

38. After that, create a **Get Player Controller** function.

39. Drag a wire from the output of the previous node and search for the **Get Controller Rotation** function.

40. With that done, pull another wire out of the **Get Controller Rotation** function and look for the **Break Rotator** node.

41. Now, create a **Divide** node and connect its **A** input pin to the output of the **Z (Yaw)** output pin of the previous **Break Rotator** node. Set the value of the divisor to **360**.

42. Next, drag a wire out of the **Return Value** output pin of the **Get Dynamic Material** function and start typing `Set Scalar Parameter Value` to add that function to the graph.

43. Type the name of the scalar parameter created in *step 14*. Make sure it is the same name; otherwise, the whole thing won't work! Let's take a moment to review the code before moving forward:

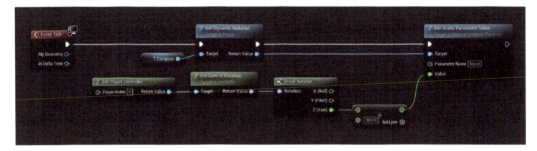

Figure 5.24 – The setup of the Event Graph of our Widget Blueprint

With that done, let's show the UI on our screens by adding some logic to our **Level Blueprint**.

44. Open **Level Blueprint** (**Toolbar tab | Blueprints | Open Level Blueprint**) and drag a wire out of **Event Begin Play** to create a **Create Widget** node.

45. Next, create a **Get Player Controller** node and wire it to the **Owning Player** input pin of the previous **Create Widget** node.

46. Choose the right widget in the **Create Widget** drop-down menu, the one we created in *step 1*.

47. Pull a wire out of the **Return Value** output pin of the **Create Widget** node and add an **Add to Viewport** one.

With that done, you should be able to see the results for yourself once you hit the **Play** button! Having the ability to know where the North is can be very helpful, especially in certain types of games and apps. Think, for example, about recent hits in the video game industry, such as Fortnite, where the user is presented with a world where certain navigational tools, such as a map and a compass, are almost essential. With this new tool at your disposal, you already know how to tackle these problems, so make sure you put them to good use! Let's finish this recipe by taking a final look at what you should see on your screen once you hit **Play**:

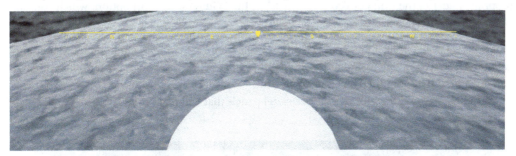

Figure 5.25 – A final look at the compass at the top of the viewport

How it works...

As usual, let's recap what we did in this recipe just to make sure that everything makes sense. The core of the functionality of this recipe lies in the logic we created within the material graph for the compass, so let's make sure we review that carefully.

The first and most crucial component of our material graph is probably the **Panner** node and all of the nodes that feed into it. This is the area where we specify how much to pan the texture, which will impact what appears to be the North once the material is properly set up. We are relying on a **Scalar Parameter** named **Player Rotation** to know how much to pan the texture, but the actual amount provided by this node won't be known until we implement the functionality within the Event Graph of the UI. The other nodes that feed into the **Panner** node, such as the **Texture Sample** node and the nodes that affect **Speed**, are there to provide a custom value – we set everything to a value of **1** in all of those fields except for the speed in the **Y** direction, which needs to be **0** as we are only planning on panning along the **X** axis.

The rest of the material graph should be more familiar to us as we are doing things we've already done in the past – such as setting the colors of certain textures by multiplying them times a color we set as a variable or adding the results of our calculations together thanks to the **Add** and **Lerp** nodes. The next bit of interesting code comes in the shape of the Event Graph we set up as part of the UI, where we modify the value of our **Scalar Parameter**, which we defined within the material. We do that by getting a reference to the dynamic material that is being applied to the image, which is our previously created **Material Instance**, and passing a reference to **Player Rotation Scalar Parameter**. The values we send are calculated according to the **Z** value of the player's rotation, which tells us where it is looking in the world. It's as simple as that!

See also

In case you haven't dealt with UIs before, let's take a quick look at how to work with them now. This is going to be especially useful in this recipe; if you are not familiar with how they operate, make sure you read the following helpful tips!

The first bit we need to know is how to create them – to do that, just right-click anywhere within the appropriate folder of the Content Browser and select the **Widget Blueprint** option, which you can find in the **User Interface** section.

The next bit we need to talk about is the UI editor itself. There are two parts to it: the **Designer** and **Graph** tabs. The second one is similar to **Blueprint Graph**, which we are already familiar with, but things are a bit different in the **Designer** panel. This is an editor where you can create any UI you want within Unreal, and therefore includes several panels that will make your life easier and which we'll explore next:

Figure 5.26 – Breakdown of the UI editor

Let's check out these features:

1. **Palette**: This is the place where you can select from the different elements you have at your disposal to create your UI. Buttons, images, vertical boxes, or safe zones are just a few examples of the available elements, and each has some unique features that make them more suitable for certain situations than others. Getting familiar with this section is very useful since this is where we'll select all of the different assets we want to add to our widget.

2. **Hierarchy**: The hierarchy panel is a very important one as the different elements that we can create tend to expect a certain structure. For instance, different assets from the **Palette** tab can accept a different amount of children, which is important to know to properly organize our UI. A button, for example, only accepts a single child, whereas a **Vertical Box** accepts many. This is the place where we can parent and child different components to each other so that we get the functionality that we want out of them.

3. **Animations**: If you want to create little animations on your UI, this is the place to be. Imagine you want to create a nice fade-in effect when you load the app, just so that the UI doesn't load abruptly — this is the place where you would create and edit the different animations.

4. **Designer**: The designer is where you can arrange and check the different elements that you create. All of them will appear here, and you'll be able to visualize what you are doing in real time.

5. **Details**: If you have something selected, chances are you are going to look at the **Details** panel to see or change how that specific element looks or works.

Something to note is how the panels you see here can change positions according to the resolution of your screen or the ones you have selected. If you want to add or remove certain ones, feel free to look under the **Window** setting and select all of the ones that you need!

Now that you know where to locate each element within the UI editor, you should be able to complete this recipe. Play around within the editor itself for a little bit, get comfortable with the different panels, and you should be ready to go in no time.

Driving the appearance of a minimap through Blueprint logic

Now that we are familiar with dynamic UI elements, let's continue exploring this topic by creating a minimap! This can be a nice extension to the compass we've just created, as both elements often appear side by side in games and apps. Furthermore, it will add a new layer of interactivity between what's happening in the game and its visual representation on the UI. Let's see what this is all about!

Getting ready

You'll need a couple of things to tackle this recipe: a texture that can act as a map for our level and an icon to represent the player's location. Both of those elements are included alongside the Unreal Engine project distributed with this book, but I'll also show you how to create them using the content bundled with the engine. Make sure you check the *See also* section if you are interested in that!

If you want to work with the same assets you'll see me employing in the next few pages, go ahead and open 05_07_Start level contained within the **Content | Levels | Chapter 05** folder. See you in the next section!

How to do it...

As usual, the first few steps in our journey involve creating the assets with which we'll be working in the next few pages. The first of those will be the UI, the element tasked with hosting the minimap we wish to create in this recipe. Let's start doing that:

1. To begin with, create a new UI element by right-clicking anywhere within the Content Browser and searching for the **Widget Blueprint** option, which is available within the **User Interface** section of the contextual menu. Remember to give the new widget a name upon creating it – I've chosen **UI_Minimap** this time.

2. Open the new widget by double-clicking on it and proceed to add a **Canvas** and a **Scale Box** to it. Set the anchor point of the last element to be in the middle of the screen.

3. As for the alignment, set it to **0.5** in both the **X** and **Y** fields, and tick the **Size To Content** checkbox option.

4. Set **Position X** and **Position Y** to **0**, which should place our **Scale Box** in the center of the **Canvas** area. The following screenshot shows the current setup:

Figure 5.27 – The settings chosen for the Scale Box

So far, we've taken care of creating a new UI and adding a **Scale Box** that will constrain the position of the minimap within our widget. The next steps are going to deal with creating the textures that will drive the appearance of the minimap itself.

5.   Start by creating an **Overlay** panel and child it to the previous **Scale Box**. This will allow us to stack multiple widgets, which is perfect since we want to have multiple layers in our minimap: one for the base map, and a second one that indicates the player's location.

6.   Next, and with the **Overlay** panel selected, tick the checkbox next to the **Is Variable** text located at the top of the **Details** panel. We'll need this to access its properties at runtime.

7.   Create two images and make them both children of the previous **Overlay** panel.

8.   Rename the first of them to something such as **I_Minimap** and assign the minimap texture to it. If you are using the assets provided in the Unreal project that comes with this book, said texture is called **T_Minimap**.

You might want to resize the previous texture a little bit so that it looks good within the UI. A value of 800 x 800 pixels works well in our case, instead of the default of 1,024 x 1,024.

9.   The second image will be the icon for the player location, so rename it accordingly (**I_PlayerLocation** will work!) and select the icon that you like best. I'll be using the texture called **T_Minimap_Locator**.

> Tip
>
> Even though we are using the **T_Minimap_Locator** asset in our UI, know that you can find another one provided as part of the Engine Content, named **Target Icon**, that will also do the job.

Seeing how we've included a few new elements, such as the **Overlay** panel, in the previous steps, let's take a quick look at their setup:

Figure 5.28 – The hierarchy panel showing both the new Overlay panel and its two children, the images

With all of the previous elements added to our UI, we can now start implementing the logic that will drive the behavior of our minimap. Unlike in previous recipes, where we built the logic within a material graph, this time, most of that logic is going to stay within the Graph Editor of the UI itself. This is because we don't need to work with a material this time around, as we'll only need to animate the image we are using as the locator on top of the texture representing the minimap. With that in mind, let's open up the Graph Editor of the UI and take it from there.

10. Let's continue by creating a couple of new variables of the **Vector** type. You can do so by clicking the + icon in the **Variables** section of the **My Blueprint** panel.

11. The first of those two new assets should be called something such as **Lower Right World Position**, and the second one **Upper Left World Position**. These are going to be the physical coordinates of the corners of the texture that is visible on the minimap – so we'll need to take note of that data next.

12. Next, we'll need to write down the location of the lower right and the upper left map bounds, just so that we can feed that data to the vector variables we created in the previous step. To do so, we can simply create a temporary actor, place it in those locations in the world that we mentioned, and write down the coordinates. If you are working on the same level that I am, the values should be **X= 350, Y= 180, Z= 0** for the **Lower Right World Position** vector and **X= -15, Y= -180, Z= 0** for the **Upper Left World Position** one. Make sure you assign those values to the variables we created in *step 10*.

    With those coordinates written down and assigned to the previous vector variables, we can now continue to expand the logic in the Event Graph.

13. Continue by dragging both of the previous vector variables describing the two world positions into the UI graph.

14. After that, create a **Vector – Vector** node after them, and set the **Lower Right World Position** vector as the minuend and the other vector as the subtrahend.

15. Next, right-click on the vector output pin of the **Vector - Vector** node and select the **Split Struct Pin** option, as we'll need access to the **X** and **Y** coordinates separately.

16. With that done, create a couple of **Float/Float** nodes and place them to the right of the previous nodes. Position one above the other.

17. Connect the **X** and **Y** float output from the previous **Vector - Vector** node to the divisor input pin of each of the **Float/Float** node.

18. Provided we checked the **Is Variable** checkbox back in *step 6* for the **Overlay** panel, we should now have access to it via the **Variables** category inside the **My Blueprint** tab. Get a reference to it on the UI graph.

19. Drag a wire out of the output pin of the **Overlay** node, then select the **Get Desired Size** node. Right-click on the **Return Value** output pin and select the **Split Struct Pin** option, just like we did for the **Vector - Vector** node.

20. After that, connect the **Result Value X** output pin from the previous **Get Desired Size** node to the dividend of the **Float/Float** node, which is using the **X** output of the **Vector – Vector** node as the divisor. Do the same with **Result Value Y**, wiring it to the other **Float/Float** node.

With all of those nodes in place, this is the graph we should be seeing at this stage:

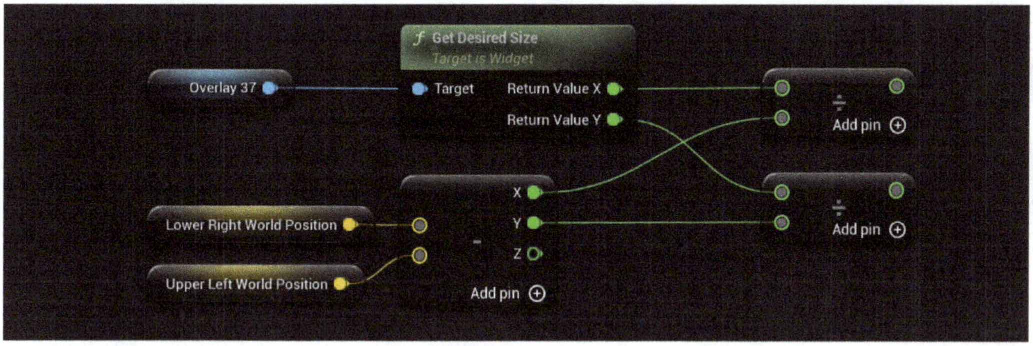

Figure 5.29 – A look at the previous nodes we have created

I'd like to say something before things start to get messy: every float operation we will perform in this recipe will only involve one axis. If you look at the previous example, you'll be able to see that we are dividing the output of **Return Value X** between the X coordinate of the **Vector - Vector** node. Likewise, we are operating on the Y values on the other **Float/Float** node. Just as in that example, all of the following nodes we'll create will follow that rule. I think this is pertinent to say at this stage as we are about to create a large number of nodes and we could get confused as to what goes where. When in doubt, make sure that you are not mixing **X** and **Y** values in your graph.

21. Continue by duplicating the **Upper Left World Position** vector and placing it a little bit further down the graph.

22. Next, right-click somewhere below the **Upper Left World Position** vector and look for the **Get Player Camera Manager** node.

23. After that, drag a wire out of that last node and create a **Get Actor Location** node at the end.

24. Subtract the **Return Value** property of the **Get Actor Location** node from the **Upper Left World Position** vector by creating a **Vector - Vector** node and wiring things appropriately.

25. Just like we did with the **Vector - Vector** node in *step 14*, right-click on the output pin and select the **Split Struct Pin** option.

26. Then, create a couple of **Float - Float** nodes and add them after that last node. Leave **0** as the minuend of both nodes and wire the subtrahend to the **X** and **Y** outputs of the previous **Vector - Vector** node.

Let's quickly do a bit of catch-up to ensure that we are all on the same page at this point:

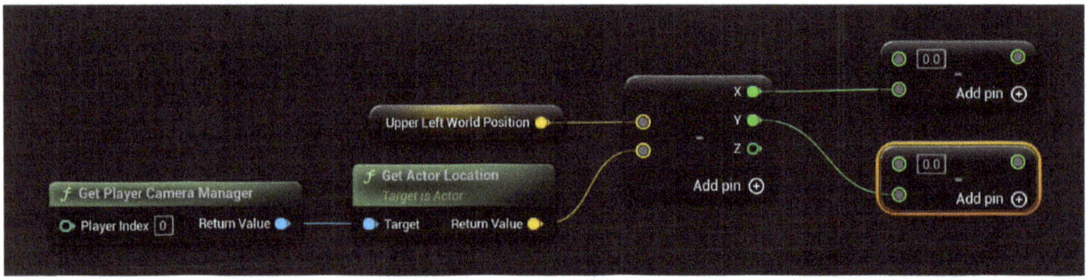

Figure 5.30 – A look at the last few nodes we have created

This set of nodes will give us the position of the player, which will allow us to calculate where the player position texture icon should be placed on top of the minimap.

27. Continue by creating a **Multiply** node, and connect the first of its input pins to the result of the **Float/Float** node that is dividing **Return Value X** of the **Get Desired Size** node between the **X** float value of the **Vector - Vector** operation from *step 16*. As for the second input pin, that should be connected to the output of one of the **Float – Float** nodes we created in *step 26*, the one where we are subtracting the result of **X** from **0**.

28. Now, create a second **Multiply** node, and in a similar way to what we did before, connect the first of its input pins to the result of the **Float/Float** node that is dividing **Return Value Y** of the **Get Desired Size** node between the **Y** float value of the **Vector - Vector** operation. The second input pin should be connected to the output of the other **Float - Float** node we created in *step 26*, the one where we are subtracting the result of **Y** from **0**.

29. Next, create a couple of **Float - Float** nodes and place them after the previous two **Multiply** ones.

30. Connect the result of each of the previous **Multiply** nodes to the minuend of the last **Float - Float** nodes, following the rule of not mixing the **X** and **Y** paths. Set the subtrahend to be half the resolution of our player icon's resolution, which should be **64** if you are using the same assets I am.

31. After that, create a couple of **Clamp (float)** nodes. Set the **Value** input pin to be the result of the previous **Multiply** nodes – one for the wire connected to **X** and the other for **Y**.

32. Don't connect the **Min** input pin. Instead, use the number -**64** in that field. The actual figure you should use is half the resolution of the texture you are using as the icon, with a negative sign in front. The one we are using is 128 pixels, hence the -64 number.

Let's take a moment to review the nodes we have created in the last few steps:

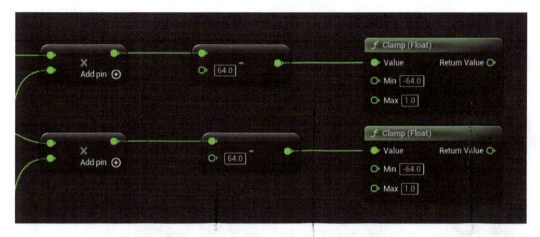

Figure 5.31 – The last few nodes we have created

Everything we've done so far is a comparison between the real-world bounds of our level and the widget's dimensions. We've done this so that we know where we need the player icon to be in widget space to match the position of our player. Now, we need to complete this logic by feeding the last values into the previous **Clamp** nodes.

33. Continue by including another set of two **Float - Float** nodes and place them somewhere after the **Get Desired Size** node we created back in *step 19*.

34. After that, connect the **Return Value X** and **Return Value Y** output pins of the **Get Desired Size** node to the minuend of the previous two **Float - Float** nodes.

35. Set the value of the subtrahend of the **Float - Float** nodes to be half the resolution of our player icon one. If you are using the assets we are providing with this project, that number should be 64.

36. Next, connect the output of the previous **Float – Float** nodes to the **Max** input pin of the **Clamp** nodes we created back in *step 31*.

37. After that, we are going to play a little bit with one of the images that make up our UI. Get a reference to the **Player Location** image into the Event Graph and drag a wire out of that node. Start typing `Set Render Transform` to create one such node.

38. Right-click over the **In Transform** input pin of that last node and select the **Split Struct Pin** option, and wire the results of the previous **Clamp** nodes into the **In Transform X** and **In Transform Y** input pins.

39. After that, connect the execution pin out of the **Event Tick** node to the input execution pin of **Set Render Transform**.

Just like we've done before, let's take another look at these last nodes we have created:

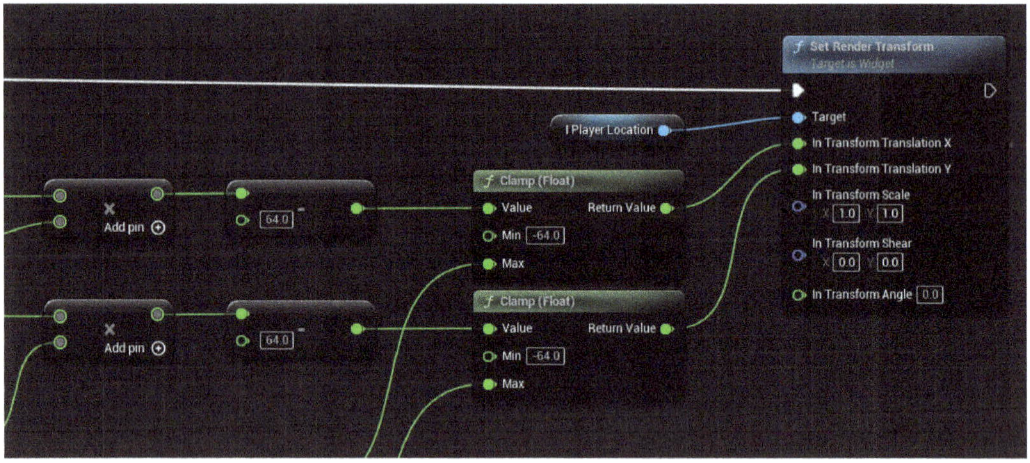

Figure 5.32 – A look at the final part of the Event Graph

Now that our widget has been set up, we need to add it to the screen and make it visible. This is something we already did in the previous recipe when we created the compass, so let's do it again.

40. Go back to the level we are working on and open **Level Blueprint**.

41. Drag a wire out of the execution pin for **Event Begin Play** and add a **Create Widget** node. With that last node selected, assign the widget we created in this recipe for the minimap to the **Class** drop-down box.

42. Next, connect a **Get Player Controller** node to the **Owning Player** input pin of the **Create Widget** node.

43. Finally, create an **Add to Viewport** node and wire both its execution pin and its **Target** to the appropriate output pins of the previous **Create Widget** node.

Now, if you hit **Play**, you should be able to see our widget pop up before showing where the player is. Move around the level to see how the position of the icon changes, all thanks to the logic that we have created! Before we go, let's take a final look at what we should be seeing on our screens:

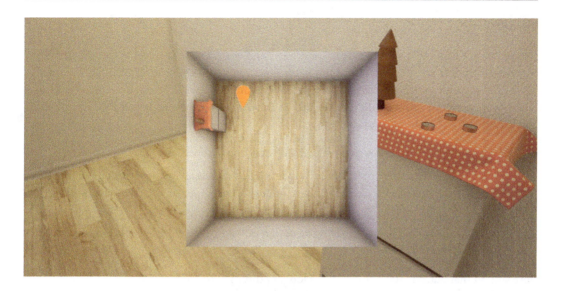

Figure 5.33 – The minimap being shown on the screen

How it works...

Even though this recipe included multiple different steps, the logic behind it all was quite straightforward. We focused on creating the minimap and the functionality that made it work – even though setting things in motion took a little bit of time, the underlying principles were simple.

The first step was, as always, to create the assets we would be working with. We are talking about two different elements at this stage: the image of the minimap and the icon for the player's location. They work together thanks to the **Overlay** panel within the Widget Editor, which puts one on top of the other.

The second and last part of the process was to code that functionality into the Event Graph; we did that by comparing the real-world coordinates of our play area to that of the widget we were using. Having a way of translating what was happening in world space coordinates to those used in the UI made it possible to update the player's location and show where it was within the minimap. And that enabled the whole system to work!

See also

As we said at the beginning of this recipe, here's a method you can follow to get your own textures for the minimap. We are going to talk about a top-down render of the scene, much like the one I captured for our level, which you can use later on as the minimap itself. You can also take that to an image-editing program and tweak it there to make it more stylish or simply give it a different look. In any case, here's how you can start doing this:

1. Drag a camera into the scene and make it look at it from above.

2. Ensure that the camera rotation is right – that is, make sure that the top part of the image the camera is producing is effectively the North of our scene.

3. Once you've done that, head over to the little menu icon located to the left of the viewport drop-down menu. You'll be able to find an option in there called **High Resolution Screenshot** if you look toward the bottom.

Before you hit that button, make sure you are looking through the camera's view. Your final image should look something like a top-down projection of the level, so make sure that's the case. You can also print screen an image, as seen from the position of the camera that covers the area that is going to be used as a minimap, which is a quick and dirty option. The point is that it shouldn't be difficult to get hold of this type of texture! One thing to keep in mind before we move on is to try to keep the edges of the image very close to the boundaries of our level, as those should match the playable area.

6

Optimizing Materials for Mobile Platforms

Moving away from the most powerful gaming and computing platforms means that we, as developers, have to adjust to less capable devices, and demanding elements, such as shaders, need to be tweaked in order to maintain performance.

In this chapter, we will take a look at how to bring our materials into a mobile environment and how to optimize them, using many of the built-in tools provided by Unreal.

In particular, we will learn about the following topics in this chapter:

- Increasing performance through customized UVs
- Creating materials for mobile platforms
- Using the forward shading renderer for VR
- Optimizing materials through texture atlases
- Baking a complex material into a simpler texture
- Combining multiple meshes with the HLOD tool
- Applying general material optimization techniques

Here's a sneak peek at what we will be creating:

Figure 6.1 – A look at some of the effects we'll be looking at in the following recipes

Technical requirements

As usual, you can find a link to the Unreal Engine project accompanying this book here:

`https://packt.link/20u7B`

There you'll find all of the assets that we will work on in this recipe, letting you replicate the same scenes you'll see later.

Increasing performance through customized UVs

Games are generally computationally expensive, so it makes sense to improve performance whenever we can. With a computer's GPU, a vertex shader is run for every vertex on a model, and pixel shaders are run for every pixel on the screen. For those cases when there are fewer vertices than pixels, Unreal has a feature called **Customized UVs**, which can give you a performance boost if you run it on just the vertex shader, instead of also using a pixel one. In this recipe, we'll take a look at how to take advantage of that feature by tiling a texture more efficiently.

Getting ready

To easily see the differences in UVs, you should have a texture where you can easily tell where the edges of it are. In this case, I will be using the **UE4_Logo** texture, which is included in the **Engine Content | VREditor | Devices | Vive** folder in the Content Browser.

> Tip
>
> The **UE4_Logo** texture is an asset that is part of the engine – there is no need to even include the Starter Content in your project if you want to use it. Make sure to enable the **Show Engine Content** option in the Content Browser, which you can find through the **Settings** button located in the top-right corner of the Content Browser itself.

How to do it...

Before we can modify the UVs of a material, we need to create a material to use. Let's get started:

1. Create a material and name it **M_CustomizedUVs**. Double-click on it to enter the Material Editor.

2. From the editor, create a texture sample by holding down the *T* key and then clicking to the left of the **M_CustomizedUVs** main material node. Connect the top pin to the **Base Color** input in the main material node.

3. Then, with the **Texture Sample** node selected, set its **Texture** property to use an image that is clearly recognizable, one where we can clearly see whether it repeats multiple times across a surface – so nothing like grass or concrete, where we wouldn't be able to easily identify its

boundaries. As mentioned before, I've used the **UE4_Logo** texture, which you can see in the following screenshot:

Figure 6.2 – A look at the selected texture

4.   Afterward, deselect the node by clicking elsewhere on the screen. The **Details** panel will fill with information about the material in general. This information will also show up if you select the **M_CustomizedUVs** result node.

5.   From the **Details** tab, click on the search bar and type in the word Custom. You should see the **Num Customized UVs** property. Set that to **1** and press *Enter* to commit the change:

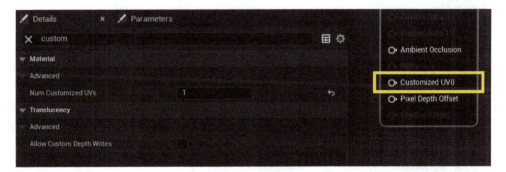

Figure 6.3 – The Num Customized UVs property and the new input pin in the main material node

If all went well, you should see from the result node that a new property was added to the bottom of it, **Customized UV0**.

Now that we have the ability to modify the UVs on the material, we need to create some UVs that we can modify. This is done through a node called **Texture Coordinate**.

6. To the left of the **Customized UV0** property, right-click and type in Tex; from there, select the **Texture Coordinate** option.

7. For the sake of demonstration purposes, connect the output pin of the newly added **Texture Coordinate** node to the **Customized UV0** input pin in our material. Once connected, select the **Texture Coordinate** node, and from the **Details** tab, set the **UTiling** and **Vtiling** properties to **4**. Let's take a quick look at how the material has changed after that last tweak:

Figure 6.4 – The preview window showing the before and after states of the material

As you can see, this causes the image to tile four times along the X and Y axes. This is useful in those circumstances where we want a texture to repeat multiple times across a given surface, allowing us to increase the perceived detail without needing to use larger images.

To show this concept being taken further, we can also modify the UVs in other ways.

8. Delete the connection between the texture coordinate and the **Customized UV0** channel by holding down *Alt* and clicking on the connection.

9. Create a **Panner** node between the previous two ones. Afterward, connect the output of the texture coordinate to the **Coordinate** pin of the **Panner** node. Then, connect the output pin from the **Panner** node to the **Customized UV0** channel.

10. Select the **Panner** node and set the **Speed X** property to **1** in the **Details** panel. Afterward, set the **Speed Y** property to **2**.

If all went well, you should see the material moving around the object without any modification having to be made to the other channels!

How it works...

UV mapping is the process of taking a 2D texture and drawing it onto a 3D object. Generally, models have their own default UVs that can be created in external modeling programs. In this recipe, we saw how we can affect them within Unreal Engine to increase the number of times a given texture repeats over the surface of our 3D objects. This technique is usually referred to as "tiling a texture," something that can be also used to tackle moving fluids, such as a waterfall or lava. In those cases, we can pan the UVs of a model instead of adjusting their size, giving the impression of a flowing image. With that in mind, you can refer to the *Displaying holograms* recipe in *Chapter 3, Making Translucent Objects*, where we used the **Panner** node to also adjust the UVs of a model – but without using the Customized UVs technique we've seen in this recipe. Challenge yourself by going back to that recipe and using a **Panner** node alongside the **Customized UVs** property we've just studied!

See also

For more information on Customized UVs, check out the following link: `https://docs.unrealengine.com/5.1/en-US/customized-uvs-in-unreal-engine-materials/`.

Creating materials for mobile platforms

When developing materials for mobile platforms, there are some things that we need to be aware of related to performance. Due to a number of limitations with the hardware, the materials that we create for it need to be more humble, in terms of the effects that we can enable and the fidelity of the resources that we can use within it. As you spend more time building materials for mobile devices, you will discover that there are often trade-offs that are required, such as reducing complexity for the sake of your application size or frame rate. In this recipe, we will explore some of the limitations inherent to these platforms by creating a simple material that could be used on those devices, while also discussing different settings that we can take advantage of.

Getting ready

Any textures that you want to use on mobile platforms need to have a resolution of 2,048 x 2,048 or lower, preferably a square texture to the power of 2 (64, 128, 256, 512, 1,028, or 2,048), as that is the most efficient use of memory.

> **Important note**
> For more information on creating textures for mobile platforms in UE5, check out `https://docs.unrealengine.com/4.27/en-US/SharingAndReleasing/Mobile/Textures/`.

How to do it...

This recipe will go over two very specific parts – creating a standard material and then seeing how we can tweak it for mobile platforms. Let's start with the first one:

1. Create a material and name it **M_MobileExample**. Double-click on it to enter the Material Editor.
2. After that, create a texture sample by holding down the *T* key and then clicking to the left of the main material node. Connect the top output pin, **RGB**, to the **Base Color** input in our material.

3. With the **Texture Sample** node selected, go to the **Details** tab and set the **Texture** property to something that also has a normal map. I used the **T_Brick_Cut_Stone_D** texture from the Starter Content.

4. Next, create another texture sample and assign the normal map texture to it (in my case, **T_Brick_Cut_Stone_N**). Connect the top pin of the newly created texture sample to the **Normal** input pin of our material. You can take a look at the state of the material graph in the next picture:

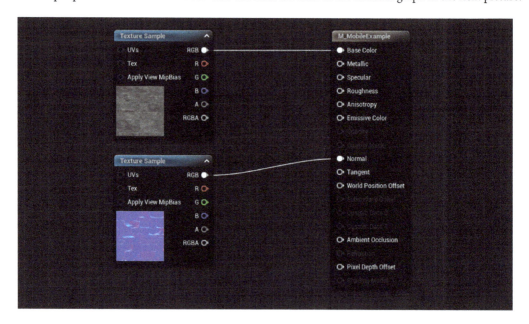

Figure 6.5 – A look at the state of the material graph

Seeing as we are working with a shader designed for mobile platforms, we want to pay special attention to the rendering cost of the new material we are creating. We can do this by inspecting the **Stats** or **Platform Stats** panels, two information displays that can be accessed by clicking the corresponding buttons located at the top of the material graph editor. Make sure that the **Platform Stats** panel is at least open before proceeding.

Now that we are done creating a basic material, let's take a look at one specific setting that we can use to make our material less computationally expensive:

5. Select the **M_MobileExample** main material node. From the **Details** tab, go to the **Material** section and click on the downward-facing arrow to open the advanced options. There, enable the **Fully Rough** property.

As you can see in the following screenshot, the **Platform Stats** window now displays fewer instructions in the **Pixel Shader** section!

Figure 6.6 – A comparison shot between the instruction count
for the same material after the last adjustment

The **Fully Rough** setting, when enabled, will force the material to ignore the **Roughness** channel, making it completely rough. This will save a number of instructions, as seen in *Figure 6.6*, and reduce one texture sampler, allowing us to use an extra texture for a different purpose. Optimizations such as this one are very interesting when working with mobile devices, where each of these wins works toward alleviating the burden on the hardware that runs our apps and games, making those experiences smoother and more performant.

Beyond that, we can also take a look at some extra settings present in the **Details** panel that will make our shaders more efficient. As those are more theoretical in nature, we'll review them in the next section, *How it works...*. Be sure to stick around for that!

How it works...

As we've seen before, mobile platforms are more limited than stationary ones – be they desktop computers, game consoles, or servers. As a result, mobile GPUs have certain limitations that we need to be aware of. For example, we can use fewer texture samples than on desktop platforms, and we need to be mindful of all of the math nodes we apply within the material graph. Having said that, there are certain adjustments that we can implement that will make things less computationally expensive.

To begin with, the Material Editor contains a section called **Mobile** that has two settings in it – **Use Full Precision** and **Use Lightmap Directionality**. These are intended to reduce the rendering cost of our materials, something especially interesting when working with mobile devices. Enabling the first setting will see Unreal using less precise math in order to save on memory and computation time. If enabled, the **Use Full Precision** property will use the highest precision available on the mobile device. This will solve certain rendering issues at the cost of being more expensive to use. Generally, you'll leave this property disabled unless you notice issues in how the material looks.

Use Lightmap Directionality has an effect that is very apparent when using a normal map. Basically, it will use the lightmap to show the light at the top of the normal map and shadows at the bottom. You'll need to use static lighting and build the project in order to see the lightmap information being used. If disabled, lighting from lightmaps will be flat but cheaper.

Another interesting section is the **Forward Shading** one, one that contains the **High Quality Reflections** property, which, when disabled, can increase the number of texture samples you can use by two. Turning this off will remove these more precise reflection calculations from the pipeline, substituting them for cubemaps to display the reflections.

While some of these properties will have less quality than the traditional fully-formed material, the settings can reduce the number of instructions on the vertex shader and the texture samplers used, which you can see from the **Stats** toolbar in the Material Editor.

It's also worth noting that, since there are so many mobile devices out there, there are a number of feature levels that can potentially be supported. To ensure maximum compatibility, the following channels can be used without any changes:

- **Base Color**
- **Roughness**
- **Metallic**
- **Specular**
- **Normal**
- **Emissive**
- **Refraction**

One area where we are limited is in the available shading models – only **Default** and **Unlit** are supported, and we should limit the number of materials with transparency or masks, as they're very computationally expensive.

> **Tip**
> You can find more information on creating materials for mobile devices at https://docs.unrealengine.com/4.26/en-US/SharingAndReleasing/Mobile/Materials/.

See also

As well as the techniques and properties that we learned about in this recipe, make sure to keep in mind the Customized UVs technique we discussed in the previous recipe, *Increasing performance through customized UVs*. Beyond that, we'll also take a look at some other more general approaches to keeping things light rendering-wise in an upcoming recipe, *Applying general material optimization techniques*. That recipe should prove quite helpful to ensure that your materials can work on many different types of mobile devices.

Using the forward shading renderer for VR

Working with VR gives artists and designers a lot of interesting challenges. Even though VR experiences typically involve high-powered computers, we can still encounter performance issues due to the large screen size required. Unreal Engine 5 contains a different rendering system, called **Forward Rendering**, which will give us a good performance boost at the expense of removing certain graphical features. In this recipe, we'll explore how to enable this alternative rendering method in our projects.

Getting ready

In order to play your game in VR, you will need to have a VR headset that is plugged in and ready to be used. Unreal gives us the option to check out our levels in VR mode once it detects a connected headset, which occurs when the **VR preview** button is clicked, located within the **Change Play Mode and Settings** area of the editor – we'll see exactly where once we start the recipe. If the option is still grayed out after connecting a headset, close the UE5 editor and restart it.

Something else that you should have available is any kind of scene that is ready to be played, preferably one which uses static lighting. That type of illumination is cheaper to render and, thus, more suited to VR experiences – and it is a type that we already discussed in the first chapter, in the *Using static lighting in our projects* recipe.

How to do it...

The **Forward Shading** property is disabled by default, but we can enable it fairly easily. Let's see how to do it:

1. From the Unreal Editor, go to **Edit | Project Settings**.
2. Once inside the **Project Settings** panel, go to the menu on the left and scroll down until you reach the **Rendering** category, and then select it.
3. Now, scroll down to the **Forward Renderer** section and enable the **Forward Shading** property.

4.   At this point, you will be prompted to restart the editor. Click on **Restart Now**.

Figure 6.7 – The location of the Forward Shading option

Once the editor restarts, you'll be able to use the other **Forward Renderer** options and features. Note that it may take a considerable amount of time to restart, due to the fact that Unreal will need to rebuild all of the shaders in your project, including the sample content.

5.   Open the **Project Settings** menu again by going to **Edit | Project Settings**, and then select the **Rendering** option.

6.   Scroll down the menu until you get to the **Default Settings** section. From there, change the anti-aliasing method to **Multisample Anti-Aliasing (MSAA)**:

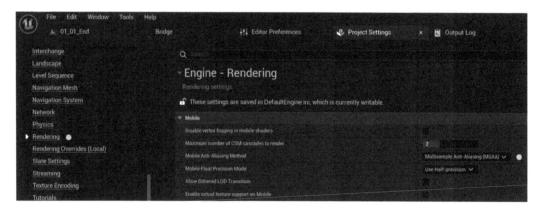

Figure 6.8 – The location of the MSAA setting

7. You may be asked to rebuild your lighting if you are working with a level that makes use of static lighting. Do so if you need to, and then, from the **Change Play Mode and Settings** menu, click on the dropdown and select **VR Preview**:

Figure 6.9 – The location of the VR Preview button

With that, you'll be able to play your game in VR making use of the forward renderer!

How it works...

Unreal Engine 5 makes use of a deferred renderer by default, as it gives artists and designers access to many interesting rendering features. However, those features are computationally expensive and may slow down the experience provided by VR games. The forward renderer provides a faster experience on average, leading to better performance at the cost of losing some features, and has additional anti-aliasing options that help visuals greatly in VR projects.

> **Important note**
>
> The **Forward Shading** renderer is still being worked on, and the number of available features will increase over time. Make sure to check out the following site for a comprehensive list of all of the available features at the time of writing, as well as any that appear in the future: `https://docs.unrealengine.com/5.1/en-US/forward-shading-renderer-in-unreal-engine/`.

See also

As VR experiences can be very computationally demanding, make sure to implement any of the features we discussed in the previous recipe (*Creating materials for mobile platforms*) if you need to improve performance on your VR projects.

> **Important note**
>
> If you are interested in learning even more about creating art with VR in Unreal Engine, check out Jessica Plowman's *Unreal Engine Virtual Reality Quick Start Guide*, also available from Packt.

Optimizing materials through texture atlases

Often referred to as a sprite sheet in the game industry, **texture atlases** are a great way to optimize game projects. The general concept is to have one big image that contains a collection of smaller ones. This is often used when there are smaller textures that are used frequently, in order to reduce the overhead on the graphics card being used, which has to switch between different texture memory locations. In this recipe, we'll explore how to take advantage of this technique, how to use texture atlases within a material, and how to access the different textures contained within it.

Getting ready

To complete this recipe, you will need to have a single texture that contains multiple smaller ones inside of it. If you do not have one, you can make use of the flipbook texture from the Engine content, located within the following folder – **Engine Content | Functions | Engine_MaterialFunctions02 | ExampleContent | Textures**.

As we said in the first recipe of this chapter, make sure to enable the **Show Engine Content** option in the Content Browser to be able to find the flipbook texture. You can find that option through the **Settings** button, located in the top-right corner of the Content Browser itself.

How to do it...

One way we can use texture atlases is by modifying the UVs on an object. Let's do that now:

1. Create a material called **M_TextureAtlas**, and double-click on it to enter the Material Editor.

2. Once inside the editor, create a texture sample by holding down the *T* key and then clicking to the left of the main material node. Connect the top **RGB** output pin to the **Base Color** input pin on the main material node.

3. With the **Texture Sample** node selected, go to the **Details** tab and set the **Texture** property to the one you chose (if using your own assets), or go for the flipbook texture I mentioned, located in the aforementioned folder in the *Getting ready* section.

To make it easier to see this particular image, we can change the preview display to show a flat plane. This will allow us to work with the texture in a flat environment, making the whole texture atlas concept easier to understand.

4. From the preview window on the left side of the Material Editor, click on the button displaying a plane. You can see which one I'm talking about in the following figure:

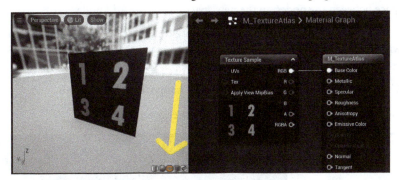

Figure 6.10 – The location of the plane preview option within the material preview viewport

5. Move the camera so that you can see the plane. You can click and hold the left mouse button to rotate the camera and then use the mouse wheel to zoom in and out.

6. Next, create a **Texture Coordinate** node to the left of the previous texture sample.

7. After that, connect the output pin of the **Texture Coordinate** node to the **UVs** input of the texture sample.

8. Select the **Texture Coordinate** node. From the **Details** tab, change the **UTiling** and **VTiling** properties to **0.5**. We should now be looking at an image similar to the following one:

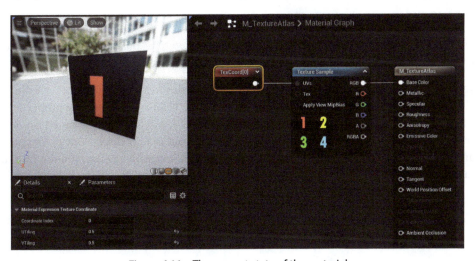

Figure 6.11 – The current state of the material

You should see after the shader compiles that the material will only display the top-left half of the image (**1** on the flipbook texture). To display a different image, we can offset the UVs being used by using an **Add** node.

9. Remove the connection between the **Texture Coordinate** node and **UVs** input pin by holding down *Alt* and clicking on the connection. Move the texture coordinate a little to the left to add some space between the nodes; in the new space, create an **Add** node.

10. Next, create a **Constant2Vector** node by holding down the *2* key and clicking on an empty space within the material graph. Connect the output pin of the new node to the **B** input pin of the **Add** node.

11. Select **Constant2Vector**. From the **Details** tab, change the **R** value to **0.5**.

Once the shader compiles, you should see **2** on the screen!

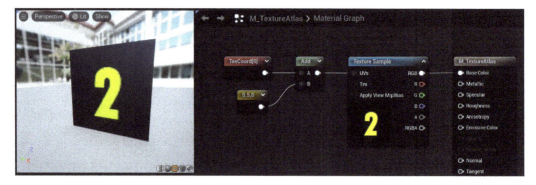

Figure 6.12 – A final adjustment to Constant2Vector

How it works...

To access the different aspects of the texture atlas, there are two steps that need to be performed – scaling the image and then offsetting it to the section you want to see.

First, you have to scale the image to display the portion you wish to see. The **Texture Coordinates** node allows us to zoom in and out on the image by using the **UTiling** and **VTiling** properties. A value of **1** means it will display 100% of the image, **0.5** means it will display 50%, and **2** would be 200%, effectively duplicating the image. In this example, we want to only display one of the images at a time, so **0.5** is the value we want to use. In the case of each image being even, you can find this value mathematically by taking the number of parts you want to display divided by the total number of parts (in our case, 1/2 or 0.5).

The second aspect is to decide which tile to use. In this case, by default, we see the number one being displayed. To see the others, we will need to offset the UVs. We can do this thanks to the **Add** node. In this case, we add **Constant2Vector** to offset the image in the *X* and *Y* axes, with the **R** property being used for the *X* offset and **G** for the *Y* one. The value we need to use is a percentage, just like the U and V tiling properties. Using **0.5** in the **R** property reduces the image to 50% of the size of the image to the right. Using **0.5** in the **G** value would move us down 50%. Following that logic, using a value of **(0.5, 0.5)** would display the number **4**.

You can see this concept being expanded on and used in greater detail with the **FlipBook** node, which includes variables to display the images gradually. You can see an example of that node in action in the **M_Flipbook** material included in the Unreal Engine project provided alongside this book, in the folder named **Content | Assets | Chapter06 | 06_04**. You can see a preview of that node in the following screenshot:

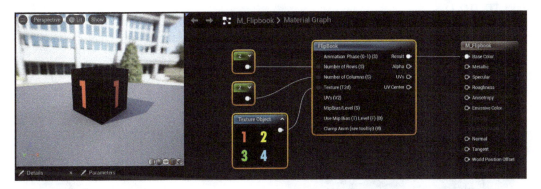

Figure 6.13 – A look at the FlipBook material function

If you double-click on the **FlipBook** node, you'll be able to see each of the steps used to create it. This type of behavior is often used in games, such as for sprite sheets in 2D games to create waterfalls by panning the UVs of an object. It's a great way to animate an object with a lower performance cost.

See also

For more info on texture atlases, check out `https://en.wikipedia.org/wiki/Texture_atlas`.

Baking a complex material into a simpler texture

Studying different material techniques throughout this book has left us with a wide variety of tools at our disposal when it comes to creating shaders. Some of those instruments are more expensive to render than others, so we also looked at several optimization techniques in the previous recipes that should help us keep performance in check. Having said so, there are times when the best optimization approach is to simply bake down all of the complex node graphs we have created into simple textures, removing all of the computationally expensive functions in favor of standard, static textures.

In this recipe, we'll learn how to decrease the rendering cost of a model by reducing both the complexity and number of shaders applied to it, all thanks to the **Merge** tool panel.

Getting ready

To get ready, you will want to have a mesh that contains two or more material slots placed within a scene you would like to use. If you don't feel like making one yourself, you can open the 06_05_Start level located in the **Levels | Chapter 06** folder of the Unreal Engine project accompanying this book. This map contains the engine model named **SM_MatPreviewMesh_01**, which is part of the engine content.

How to do it...

One of the easiest ways to combine materials together is through the **Merge Actors** tool:

1. From the Unreal Editor, select the object that you want to bake. In this recipe's case, we'll select the **SM_MatPreviewMesh_01** model.

2. Next, navigate to **Tools | Merge Actors**:

Figure 6.14 – The location of the Merge Actors tool

3. With the **Merge Actors** menu now open, make sure that the mesh you want to bake is selected.

4. After that, check the option currently selected in the **Merge Method** box; the default value should be that of **Merge**, which is what we want to use.

5. The next step revolves around the **Merge Settings** section – expand the **Settings** drop-down arrow to show all of the options available to us.

6. Once that's done, the first parameter we will focus on is the **LODSelection Type** setting located halfway through the **Settings** panel. For the merge material operation to work, we need to move away from the default **Use all LOD Levels** option and choose **Use specific LOD level** instead. Make sure that the **Specific LOD** field located underneath the previous setting continues to say **0**.

7. With that previous step in place, navigate toward the top of the list of options, and check the box next to **Merge Materials**. This will allow us to simplify the setup of the **SM_MatPreviewMesh_01** model.

8. Next, expand the **Material Settings** section and change the **Texture Size** field to read **512** in both boxes (instead of the default **1024**). This will ensure that the textures that get created are of a lower resolution than the default preset.

9. Click on the **Merge Actors** button. You will see a menu asking you where you'd like to save the new mesh and materials – the location I selected is the **Assets | Chapter06 | 06_05** folder, and I've chosen the **SM_MERGED_MatPreviewMesh** name for the new asset. Click on the **Save** button and wait for Unreal to do all the heavy lifting.

    You can take a look at the **Merge Actors** tool and some of the chosen settings in the following screenshot:

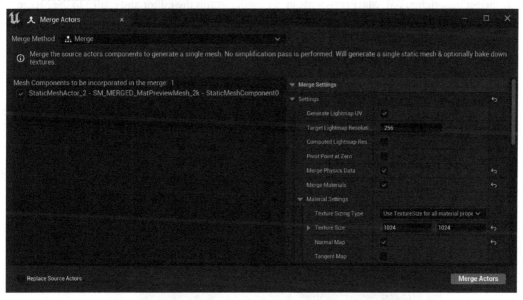

Figure 6.15 – A look at the Merge Actors tool and some of the selected settings

10. Unreal will proceed to show you the content that has been generated once it finishes the creation process by moving you to the appropriate folder within the Content Browser. From there, drag and drop the newly created static mesh into your level, and let's review the results:

Figure 6.16 – The original actor and the simplified version, side by side (the original is on the left)

As you can see, we now have a new mesh with a single material instead of the two that made up the original model, but we seemed to have lost a bit of quality as well. This is due to the default parameters used when creating the simplified asset, which heavily decreased the number of texture samples and properties used for the sake of optimization. To get an example of something closer to the original, we can customize some additional options.

11. Select the original **SM_MatPreviewMesh_01** model within the Editor and return to the **Merge Actors** menu.

12. Next, click on the arrow to the left of the **Material Settings** option and change the **Texture Size** property to read **2048** in both boxes.

13. Continue scrolling down a bit, and make sure to tick the checkboxes next to the **Roughness Map**, **Specular Map**, and **Ambient Occlusion Map** options.

14. Feel free to uncheck the **Allow Two Sided Material** setting, as we are not going to see the back faces of the model we are merging. This will save a bit on performance, which is not a bad thing!

15. After that, click on the **Merge Actors** button and go to the folder where want to place the files. I chose the same folder as before and used the name **SM_ MERGED_ MatPreviewMesh_2K**. Once you've picked out your options, click on the **Save** button and wait for Unreal to complete the merge process.

16. Just like before, Unreal will focus on the folder where you've saved the new asset. Drag and drop the new static mesh into your level, and feel free to compare it against the other two versions, as we'll see in the next figure:

Figure 6.17 – From the left, the 2K merged asset, the original, and the highly simplified one

As you can see, this looks a lot more like the original material but at the cost of performance. Tweak the **Material Settings** properties until you are content with the quality of your material.

> **Important note**
>
> You might see that the depth effect calculations present in the cobblestone material have disappeared when creating the simplified version. Note that we are limited to the effects present in the **Materials Settings** section, so keep that in mind when simplifying your models.

How it works...

The **Merge Actors** tool is traditionally used to combine multiple static meshes into a single new actor. This is typically done later in development, after playtesting a level and ensuring that you are not going to move objects or make impactful changes, in order to improve the performance of a game or app. We're using this tool in this recipe because, in addition to combining meshes, it can also combine multiple materials together into a single shader with the UVs set up correctly.

The tool is non-destructive, so any of the changes you make won't actually change the original assets. Once a merge occurs, it will automatically generate textures and materials based on the parameters chosen in the **Material Settings** area of the **Merge Actors** tool. In our first attempt, the tool made two textures (one for the diffuse color and another for the **Normal** map), a material that uses those two images, and a new static mesh. The second time we created a mesh, we added three additional textures and quadrupled the resolution size, reaching a value of 2,048 pixels, which is the highest resolution we can use in current mobile devices.

See also

For more information regarding the **Merge Actors** tool, check out Epic Games' documentation on the topic: `https://docs.unrealengine.com/5.0/en-US/merging-actors-in-unreal-engine/`.

Combining multiple meshes with the HLOD tool

The amount of actors contained in a level affects the performance of the real-time experiences we create. Each unique model added to the scene represents another entity that the engine needs to track, and even though innovations such as Nanite have decreased the burden that having more and more polygons represents, we still need to tackle the problem we face when dealing with thousands of different, unique meshes.

The **Hierarchival Level of Detail (HLOD) System** takes center stage when trying to solve that problem. In essence, its function is to group different models based on how far we are from them and merge them into individual entities, thus reducing the burden that these objects have on the hardware that displays them. We'll learn about this in the next few pages, so buckle up for the ride!

Getting ready

You should have a map that contains multiple static meshes within it in order to apply the techniques we are going to be studying. If you don't have one, you can also open the 06_06_Start map included in the UE project accompanying this book, which looks like the following figure:

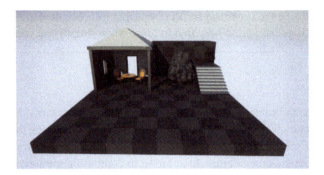

Figure 6.18 – A look at the scene we are going to be working with

How to do it...

Before we can use the HLOD tool, we need to know where to find it. There are two main areas that we need to pay attention to, as we will see:

1. Make sure that the **World Settings** menu is visible by checking that option inside the **Window** menu.

2. From the **World Settings** tab, scroll down to the **HLOD System** section – this is the area where you'll find many of the settings used to play with this tool.

    The **HLOD** section of the **World Settings** panel contains the instructions needed for the HLOD tool to operate under the **Hierarchical LOD Setup** parameter. That setting is itself an array, as it can contain as many entries as you choose. Each different one you add contains the different settings that the tool will use to complete the grouping and merging of the different objects it finds in the scene. Having multiple entries means that the grouping and merging passes will happen as many times as the number of entries that are contained in the array, something useful when you want to perform that operation several times. This is useful when working on big areas, as it gives you the option to keep on grouping actors as you move farther and farther away from them.

    There are some settings that we can tweak there, in the **HLOD** section of the **World Settings** panel, but some others can only be adjusted from a different panel, **Window | Hierarchical LOD Outliner**. Make sure to have both panels open when tackling the next steps!

3. Click the **Hierarchical LOD Outliner** button, which will open the window that allows us to generate the different groups (named **clusters**), generate the proxy meshes, and apply the different settings we selected in the **HLOD System** section of **World Settings**. Keep both sections in focus, as we are going to be working on both.

    You can take a look at the two areas where we are going to be working in this screenshot:

Figure 6.19 – The Hierarchical LOD outliner (left) and the HLOD System section (right)

With those two areas as our focus, let's continue by setting up the different configurations that we will use to group the different static mesh actors together.

4. Go back to the **HLOD System** section of the **World Settings** panel and expand the area called **Hierarchical LOD Setup**. You should see a default section named **HLOD Level 0** – expand it to reveal the two groups that it contains, **Cluster generation settings** and **Mesh generation settings**, and focus on the first one. The first of them contains the options that define how static meshes are going to be grouped together in that HLOD level, while the second one is used to determine what the output mesh will look like.

5. After that, expand the **Cluster generation settings** area to reveal the parameters contained within. Focus on **Desired Bound Radius** and set it to a value of **250.0**, as seen here:

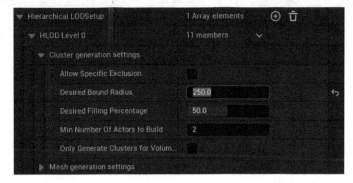

Figure 6.20 – The settings used for HLOD Level 0

> **Important note**
>
> The **Desired Bound Radius** setting determines the distance at which the tool will try to merge objects, so static meshes that can be grouped within that range will be grouped together and marked for merging. Remember that Unreal's default units are centimeters, so a value of 2,000 translates to 20 meters, or (roughly) 65 and a half feet.

6. As we want to demonstrate things further, let's create another HLOD level by clicking on the + icon to the right of the **Hierarchical LOD Setup** area. If you look at *Figure 6.20*, it's the icon located to the right of the top of the image, next to the bin icon. If you open up the newly added **HLOD Level 1**, you should see that the engine has automatically filled the **Desired Bound Radius** parameter with a larger value than the one we had for **HLOD Level 0** – this makes sense, as each new level we add should try to contain actors that are larger and larger each time.

7. With the previous step implemented, head back to the **Hierarchical LOD** outliner and click on the **Regenerate Clusters** button. Doing so will proceed to create temporary clusters of the objects that the engine thinks it should combine; you can check which ones those are by simply selecting their name on the **Hierarchical LOD** outliner and looking at the scene, where a red wireframe sphere should become visible. Expanding the names on the outliner itself will reveal which objects have been considered for merging.

You can take a look at this behavior in the following screenshot:

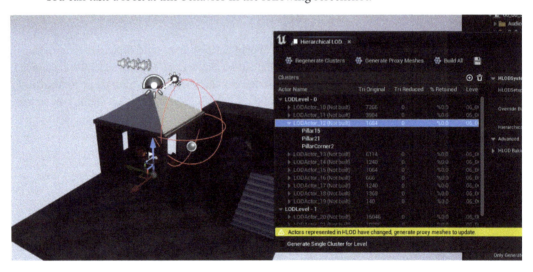

Figure 6.21 – The HLOD System in action, with one of the LOD actors selected

As mentioned previously, selecting any of the LOD actors will show a sphere placed around the objects that are being used. The higher the LOD level, the more the actors will be combined, and the larger the sphere will be. This will reduce the number of draw calls that will be used the farther away the user is.

8. Next, click on the **Generate Proxy Meshes** button to start building the LOD actors. This will typically take a while to complete, so you may want to take a break while your computer is working.

9. When finished, make sure to click on the **Save** icon to save your work. This is located to the right of the **Build All** button.

    Upon saving, you'll see that a new folder called **HLOD** will have appeared next to the location where your level is currently saved. This directory contains all of the models, materials, and textures that have been generated for your assets. We had no say in how those were created, but we can tweak the generation parameters by exploring another section of the **HLOD System** menu in the **World Settings** panel.

10. Go back to the **HLOD System** section of the **World Settings** panel, and look for the **Mesh generation settings** subsection contained within the **HLOD Level 0** area.

11. The first option there, called **Transition Screen Size**, allows us to control the point at which the original models get swapped for their optimized grouped counterparts. The point at which this happens is represented by the amount of screen space that those objects occupy, with a value of **1** representing the bounding box occupying the entirety of the screen. The current value is **0.315**, or 31.5% of the screen – let's change that to **0.6** to make the substitute mesh appear sooner.

12. The next option, **Simplify Mesh**, allows us to create a decimated version of the original assets. This means that the model resulting from combining the original ones will contain fewer triangles than the sum of those originals. Check that box.

13. Upon doing so, you'll see that the next section we can explore is the **Proxy Settings** one. This will allow us to tweak the parameters controlling the generation of the mesh that will replace the original objects we are merging, with the added benefit that the new model will also have fewer triangles than the original. The parameter that controls the fidelity of the resulting mesh is the very first one, simply called **Screen Size**. The engine will try to create a simplified mesh that looks identical to the original material when the content occupies the designated amount of pixels indicated by the **Screen Size** field. Seeing as we've bumped the **Transition Screen Size** parameter, let's bump the default **300** value to a more generous **600** instead.

14. The next area we can expand is **Material Settings**. The parameters found here are very similar to the ones we encountered in the previous recipe, *Baking a complex material into a simpler texture*. Make sure to check the boxes next to **Roughness** and **Metallic** maps this time, as some of the models we are combining display those properties.

15. Once that's sorted, let's move on to **HLOD Level 1** to tweak some of the settings found there. Open the **Mesh generation settings** area once more, and increase **Transition Screen Size** to **0.33**. Leave the rest as it is.

16. Upon completing the previous steps, go back to the **Hierarchical LOD** outliner and click on **Regenerate Clusters**. When complete, click on **Generate Proxy Meshes** and save your results once the tool finishes its task.

Once the engine computes the models, you'll be able to check the results by heading over to the **View Modes** panel and selecting **Hierarchical LOD Coloration** from within the **Level of Detail Coloration** category. This will allow you to see the points at which the transition between the original models and their merged versions happens, as shown here:

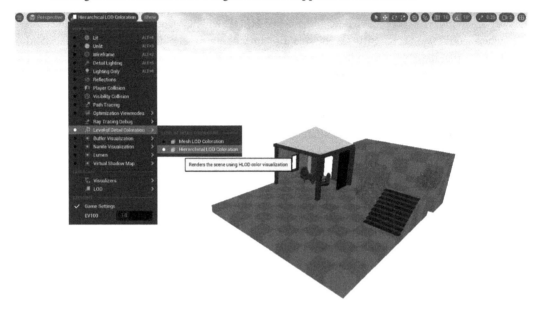

Figure 6.22 – A final look at the scene and the Hierarchical LOD Coloration view mode

Remember that you can also inspect the models by navigating to the **HLOD** folder next to your level in the Content Browser, where you'll be able to adjust things such as their collision complexity.

How it works...

Before we finish this recipe, let's take a look at some of the settings that are part of the HLOD tool. One of the first we saw was **Desired Bound Radius**, a parameter that notes how far to look for actors to combine together. The larger the radius, the more objects that can be combined. In levels that have sparse static meshes, such as deserts, the radius may need to be larger, but for complex and detailed areas, it may need to be smaller. The more levels that you have, the more possibilities to blend there are.

By default, all of the meshes will be exactly the same, but the objects will share a single material that contains the same properties and works exactly as in our last recipe. However, if you have large levels, it makes sense to simplify meshes that are further away to improve performance, but you'll often need to spend time tweaking the properties to get something that feels right for your level.

It's important to note that HLOD objects can only be used with static meshes, so moving objects will not be able to use this tool. We are required to generate proxy meshes every time we make changes to our properties because the system works in a very similar way to how baking lighting works. We're doing work ahead of time in order to improve performance at runtime.

See also

If you'd like more information on the topic, check out Epic's official docs regarding this tool: `https://docs.unrealengine.com/5.0/en-US/using-the-proxy-geometry-tool-with-hlods-in-unreal-engine/`.

Applying general material optimization techniques

One of the ways that we can develop materials that work on both low- and high-end devices is through the material quality-level system. This recipe will discuss the creation of materials with quality settings in mind, allowing us to create one material that can be used on a variety of devices.

Getting ready

In order to complete this recipe, you will need to have a material that you would like to look different, depending on what quality level your game is running on. We'll create one in the following section, but feel free to use one of your own if you think that it would benefit from having different quality settings.

How to do it

To start, let's create a simple material that we can use to demonstrate the material-optimization techniques we'll be studying later on:

1. Create a material and name it something along the lines of **M_QualitySettings**. Double-click on it to enter the Material Editor.

2. From the editor, create a texture sample by holding down the *T* key and then clicking to the left of the main material node.

3. With the **Texture Sample** node selected, go to the **Details** tab and set the **Texture** property to something that also has a normal map. I used the **T_Brick_Cut_Stone_D** texture from the Sample Content.

4. Next, proceed to create a **Quality Switch** node, and connect the top **RGB** pin from the texture sample to the **Default** pin on the **Quality Switch** node.

5. After that, connect the output pin of the **Quality Switch** node to the **Base Color** input pin of the main material node.

   To make it very clear that our material is being modified, let's use a color when we switch to the low-quality level.

6. Below the texture sample, create **Constant4Vector** by holding down the *4* key, and then set it to a color of your choice; I used red, which is **(1,0,0,0).** Connect the pin from the right side of **Constant4Vector** to the **Low** pin of the **Quality Switch** node:

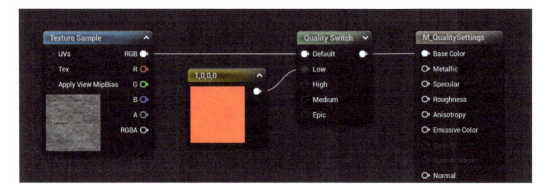

Figure 6.23 – A look at all of the nodes we've created so far

7. Once that's done, create another texture sample and assign the normal map texture to it (in my case, **T_Brick_Cut_Stone_N**).

8. Connect the top pin of the newly created texture sample to the **Normal** pin of the main material node.

9. After that, apply the material to any model that you want – it doesn't matter if it looks good or bad; we only want to demonstrate the effects of the **Quality Switch** node in the viewport. Feel free to apply it to any model that you have in mind, or simply create a new level and drag a cube into it where we can apply the new material.

As things stand, the material that you should see in front of you probably uses the high-quality level preset, provided we haven't changed this beforehand. This is what we should be looking at:

Figure 6.24 – The default look of the material being applied, using the high-quality preset

To see the effects of the **Quality Switch** node in action, we will need to actually set the material quality level of our scene.

10. From the Unreal Editor, go to **Settings | Material Quality Level** and choose **Low**.

> **Tip**
>
> You can also adjust the quality level of the game during play through the console, by pressing the ` key and then typing `r.MaterialQualityLevel`, followed by a number, with **0** for **Low**, **1** for **High**, **2** for **Medium**, and **3** for the **Epic** quality setting.

At this point, you'll have to wait for Unreal to compile all of the shaders for the selected quality level. Once finished, you will see that the material now uses the **Low** channel from the **Quality Switch** node, as shown in the following figure:

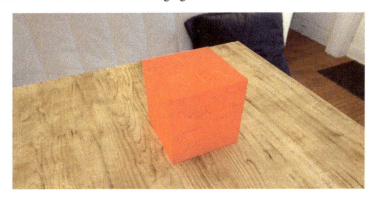

Figure 6.25 – The same material being rendered under the low-quality preset

How it works...

The **Material Quality Level** property allows us to use less intensive mathematical operations on our shaders when we target devices with less powerful graphics cards.

The **Quality Switch** node works much as a switch statement does in programming – it runs the relevant code, based on the **Material Quality Level** value. If nothing is provided to a pin, the default will be used. It is important to note that you are required to use the **Default** channel, as not doing so will cause an error.

You can add as many **Quality Switch** nodes as you want. In fact, you will need to have one for each of the channels that you want to act differently, depending on what is being done. With this in mind, you now have the knowledge to create materials that will support almost any kind of device and get the optimization that you're looking for!

See also

If you want to learn more about performance, make sure to check out one of Epic's dedicated pages on that topic: `https://docs.unrealengine.com/4.26/en-US/TestingAndOptimization/PerformanceAndProfiling/Guidelines/`.

7
Exploring Some More Useful Nodes

Unreal contains many different and useful nodes – some of which we've already seen, while others are still left to be explored. It would probably be too ambitious to try and cover them all, given the huge number of functions available at our fingertips, however, it's also true that the more examples we see, the better prepared we'll be when we need to create a new shader. That being the case, we'll continue to look at some of those useful nodes that we haven't had the chance to explore so far in the book.

So, we will cover the following recipes in this chapter:

- Adding randomness to identical models
- Adding dirt to occluded areas
- Matching texture coordinates across different models
- Using interior cubemaps to texture the interior of a building
- Using fully procedural noise patterns
- Adding detail with Detail Texturing

As always, here is a little snippet of what we'll be seeing next:

Figure 7.1 – Some of the materials we'll be creating in this chapter

Technical requirements

The materials we'll be creating in the next few pages can be tackled with the usual tools we've used up to this point: models, textures, and standard Unreal Engine assets that you will be able to use as long as you have access to the engine. As a result, the only things that you'll need are the computer that you've been using to work in Unreal and your passion for learning something new.

Let me also leave you with a link you can use to download the Unreal Engine project that I've used to create this book: `https://packt.link/20u7B`

Feel free to download the project and load it up in Unreal if you want to work on the same scenes you'll see me adjusting!

Adding randomness to identical models

The first recipe we'll tackle in this chapter will deal with instances of a 3D model. You might have heard about this type of asset, as they are a common feature in many different 3D content-creation packages. We use the name instance to refer to the identical copies of an asset that get scattered throughout a scene, taking advantage of a feature of our graphics cards that allows us to render these copies much faster than if we were dealing with unique individual objects. This can be a very powerful technique for increasing performance, especially when we work with assets that appear multiple times in a level: vegetation, props, modular pieces, or other similar assets. Having said that, dealing with identical models doesn't necessarily mean that we want them to look the same; we sometimes want a certain variation to occur, if only to break up any apparent visual repetition across the level.

In this recipe, we will explore a simple little node that will let us tweak the look of each individual instance, allowing us to assign a random value to each copy so that we can alter their look within our materials. Let's see how that's done!

Getting ready

You can follow along using your own assets, or by opening the scene we provide alongside the Unreal Engine project accompanying this book. The only thing you'll need to complete this recipe is a simple 3D model and a material that you can apply to it, so feel free to import any that you have or use one from the Starter Content if you don't want to use the ones we'll be providing.

I'll be using the toy tank model and the **M_ToyTank** and **M_ToyTank_Textured** materials that we created in *Chapter 2* in the recipes called *Instancing a Material* and *Texturing a small prop*. In any case, the stars of the show will be a couple of new nodes that you can have access to through the Material Editor, so there are no special requirements beyond that.

If you want to use the same scene you'll be looking at in the next few pages, feel free to open the level named 07_01_Start located within the **Content | Levels | Chapter07** folder.

How to do it...

As we want to add visual variety to multiple instances of the same model, the first few steps in this recipe will deal with creating those assets. Let's see how we can do that by taking the following steps:

1. Start by creating a Blueprint somewhere within your Content Browser. To do so, right-click anywhere within the Content Browser and select the **Blueprint Class** option from the **Create Basic Asset** category.

2. Next, pick **Actor** as the parent class and assign a name to the new Blueprint – I've chosen **BP_ToyTankInstances**. Double-click on the asset to open the Blueprint Editor.

3. Once inside the editor, click on the **Add** button at the top left corner of the **Components** section of the viewport. This will allow us to add a new component to our Blueprint.

4. Upon clicking the **Add** button, start typing Instance to reveal the **Instanced Static Mesh** component option. Select it to add it to the stack.

5. After that, select the new component and perform a drag and drop operation, releasing the new **Instanced Static Mesh** component on top of the existing **Default Scene Root** component. This will make it the default component of the new Blueprint.

6. With our new component still selected, focus on the **Details** panel and set the **Static Mesh** property to use whatever model you want in this recipe. I've gone with the trustworthy **SM_Tank** model on this occasion.

7. Next, assign a material to the **Element 0** drop-down menu of the **Materials** section. It doesn't matter which one you choose, as we'll be creating a new one later. For now, I've chosen the **M_ToyTank** material we created in the *Instancing a material* recipe back in *Chapter 2*.

8. Next, head down to the **Instances** section and add different entries by clicking on the + button. Create a few entries here, as we need several models to really appreciate the differences between them. I've gone ahead and placed nine instances.

9. Navigate to the **Transform** section of each instance and adjust the location value, as they will spawn on top of each other if we don't tweak that. You can see the location of that panel in the following screenshot:

Figure 7.2 – Adjusting the transform of the different instances we've spawned

10. Click on the **Compile** and **Save** buttons to apply the changes we've made, and drag the Blueprint into our level. It should look something like the following screenshot:

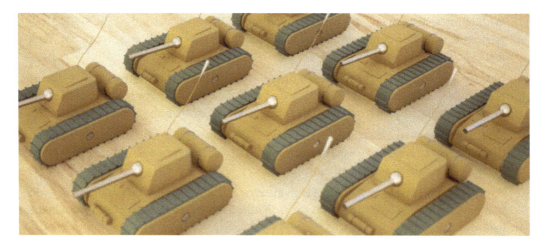

Figure 7.3 – The result of dragging the Blueprint into the level

The arrangement of the toy tanks you see in the preceding screenshot results from the different transform values I've chosen, which will probably look different for yours based on the values you use. All of the previous steps should have left us with a Blueprint that we can use to test the effect of the material we are about to create. So, with that in mind, let's proceed to create the new shader.

11. To start with, duplicate one of the materials we had previously created for the toy tank asset – there is one called **M_ToyTank_Textured** available in the **Content | Assets | Chapter02 | 02_03** folder within the Unreal project that accompanies this book that will work well for our purposes. We created that material in the *Texturing a small prop* recipe, also in *Chapter 2*. Working from that one will serve our purposes, as we have already set up the right textures for it. If you don't have that one ready, create a similar one by revisiting that recipe!

> **Important note**
>
> It doesn't matter whether you duplicate the previous material or create a new one — what we'll do inside it will be very similar. Of course, using an existing material will save us time when we're setting up certain parameters, but you can ignore those and just focus on the new nodes we'll be creating later on.

We'll focus on a specific part of the material — the part that deals with the color of the tank's main body. So, no matter whether you choose to use the same shader or create a new one, the steps we'll introduce here can be applied to both scenarios.

12. Create a **PerInstanceRandom** node and place it somewhere within the material graph.

13. Add a **Multiply** node and connect the previous **Per Instance Random** node to its **A** input pin.

14. As for the **B** input pin of the **Multiply** node, connect the output of the wood **Texture Sample** node that is currently driving the main color of the body of the toy tank to it. The original material which we've used as a base for this one is the **Texture Sample** node, which was connected to the **B** input pin of a **Lerp** node.

15. Reconnect the output of the **Multiply** node to the **B** input pin of the **Lerp** node that was previously driven by the wood texture. The graph should now look something like the following screenshot:

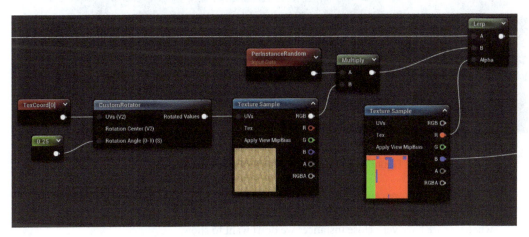

Figure 7.4 – The setup of the Per Instance Random node within the graph

> **Important note**
>
> The preview window of the material might show a large dark area on the part of the shader that was previously affected by the wood **Texture Sample** node. This is due to the nature of the new node we used, the **PerInstanceRandom** node, which can only be displayed when working with instances of a 3D model – otherwise, it simply shows as a black value, as demonstrated by the preview window.

The **PerInstanceRandom** node can be directly connected to the **Base Color** input pin of our material instead of being used like it was previously. Feel free to do that if you use a new material instead of the duplicate we've already shown in the previous steps. The new node will assign a random grayscale color to each item of the instanced static mesh array. This isn't really a color as much as a floating-point value, which we can use for other purposes: for example, to modify the **UV** coordinates of the textures that we apply within the material, as we'll see next:

16. Head over to the location of the **Texture Coordinate** node that is driving the **T_ Wood_ Pine_ D** texture and create a new **PerInstanceRandom** node above it. For reference, this is the same **Texture Coordinate** node you saw on the far left of *Figure 7.4*.

17. Once in there, add a **Multiply** node after the new **PerInstanceRandom** node, and connect both by plugging the output of the **PerInstanceRandom** node to the **A** input pin of the new **Multiply** node we've just created.

18. Unhook the **Texture Coordinate** wire from the **UVs (V2)** input pin of the **CustomRotator** function and hook it to the **B** input pin of the previous **Multiply** node.

19. Reconnect the **UVs (V2)** input pin of the **CustomRotator** function to the output of our new **Multiply** node. The graph should now look something like the following screenshot:

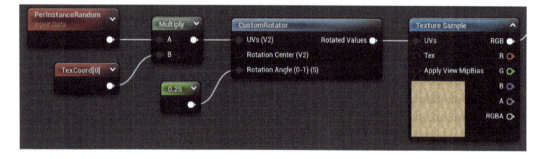

Figure 7.5 – The revised section of the material graph

Another addition we can make to our material is to introduce the **Camera Depth Fade** node, which we can use to hide instances based on how far we are from them. This can be useful whenever we have very close instances we don't want to show.

20. Select the main material node and look at its **Details** panel. In there, change **Blend Mode** from the default **Opaque** option to the **Masked** option.

21. Next, create a **Constant** node and give it a value of **100**.

22. Then, add a **Camera Depth Fade** node and connect the previous **Constant** node to its **Fade Length (S)** input pin.

23. Finally, connect the **Result** output pin of the **Camera Depth Fade** node to the **Opacity Mask** input of our material node.

    Now that we've implemented all of those changes, the last thing we need to take care of is applying our new material to the **Instanced Static Meshes**.

24. Head back to the Blueprint we created at the start of this recipe and assign the new material to the **Instanced Static Mesh** component. This is the same operation we performed in *step 7*, so refer to that instruction if you have any doubts!

    That's all we need to do in this recipe, so let's check out the results of our work!

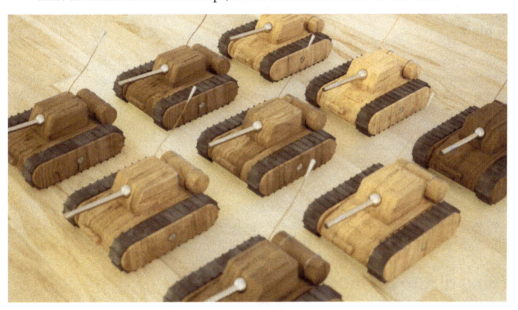

Figure 7.6 – The final look of the scene

As you can see, this technique can be very useful whenever we want to differentiate between the different instances of the same model present across our levels, making it ideal for level design work. It combines the performance boost that we get by using this component with the variability of the material node. Make sure to keep this in mind whenever you work with these types of assets!

How it works...

As you can see, adding a bit of randomness to identical models is not a complicated technique. Starting with the **PerInstanceRandom** node, this little helper assigns and exposes a random value to our **Instanced Static Mesh** components, which we can use to drive different types of interesting functionality – such as modifying the value of the **Base Color** property as we've done in this recipe. Just like we've done that, we could have also modified the **Roughness** parameter, or any other that we wanted for that matter. The other node we studied, the **Camera Depth Fade** node, allows us to hide certain meshes depending on how far they are from us.

Apart from those two, there are other nodes that you might find useful that work in a similar way. For instance, another cool node we could use is the **Per Instance Fade Amount** node, which is very similar to the **Per Instance Random**; instead of assigning a random float value to each instance of the Instanced Static Mesh component, this time the number we are given relies on the position of the instance itself with regards to our camera. We can use this to hide the models that are farther away from us, instead of the ones that are closer like we did in this recipe. Be sure to check that out!

See also

Even though we've used nodes that only work when applied to instances of a mesh, it's good to note that the **Instanced Static Mesh** component isn't the only one that can take advantage of them. There's another component that can also benefit from the techniques we've shown in the previous pages: the **Hierarchical Instanced Static Mesh** component (**HISM** for short). It's good to learn about both of them as they can be used to increase performance in our games. They are especially helpful when we need to scatter multiple instances of the same mesh, such as railings or trees, as the computer only stores information about one and repeats it across the world multiple times.

They are both very similar, but the HISM component allows us to pair the gains in performance given by the instancing feature with the creation of hierarchical levels of detail, something we studied in the *Combining multiple meshes with the HLOD tool* recipe back in *Chapter 6*. Be sure to check out that recipe out to learn more about that feature!

Adding dirt to occluded areas

Creating complex materials for 3D models requires artists to learn about the intricacies of the objects they work with. For instance, replicating the look of a dirty surface often makes the artist think about which areas are more likely to accumulate dirt. To aid with that type of work, render engines often include functionalities tailored to artists with the hopes that they will make their material-creating workflow easier.

One example of that comes in the form of the ambient occlusion render pass: the engine can easily compute which areas are closer to each other, and it allows us to use that information to drive the dirtiness of our shaders. This is what we'll explore in the next few pages, so buckle up for the ride!

Getting ready

The scene that you can open if you want to follow along working with the same assets I'll be using is called `07_02_Start`, and you can find it within the following directory: **Content | Levels | Chapter 07**.

As always, you can use your own 3D models instead, but there are some things you should take into consideration when doing so. First, the level that you want to operate on needs to have computed baked lighting. This is crucial to this recipe, as the material node we'll use needs to tap into that information and won't work otherwise.

Secondly, and as a consequence of my first point, your lighting needs to be set to stationary or static, as those are the types of lights that can produce baked lighting data.

Finally, the models that you choose to work with need to have properly laid out UVs so that the lighting information can be properly stored. We've seen all of that earlier in this book, so be sure to revisit the recipe named *Using static lighting in our projects* in *Chapter 1* if you need a reminder on how to work with static lighting. Keep that in mind before continuing with the following pages!

How to do it...

The very first step we'll need to take in this chapter is building the lighting into our level. As we stated previously, this is a crucial step in our journey before we attempt to create the material we'll be applying, as the new node we'll use this time relies on the existence of shadow maps. As a result, the first few steps we are about to take are very similar to those we took in the *Using static lighting in our projects* recipe back in *Chapter 1*:

1. To begin with, start by heading over to the **Rendering** section of **Project Settings** and navigate to the **Global Illumination** area.

2. In there, change **Dynamic Global Illumination Method** from **Lumen** to **None**.

3. After that, head back to the main viewport and open the **World Settings** panel. If it isn't already open due to us working with it before, make sure to enable it by clicking on the **World Settings** option within the **Window** panel.

4. Once inside the **World Settings** menu, scroll down to the **Lightmass** section and expand the **Lightmass Settings** category. There are a couple of options there named **Use Ambient Occlusion** and **Generate Ambient Occlusion Material Mask** that we need to ensure are ticked if we are to use that information inside our materials.

5. With the preceding steps done, select the **Build Lighting Only** option from within the **Build** panel. This is a process that can take a bit of time to complete, so make sure to wait until that's done.

   The previous steps have ensured that shadow maps have been calculated for our level, with the appropriate **Ambient Occlusion** information ready to be used within our materials. The next step will be the creation of the material that can take advantage of that fact.

6. Let's start by creating a new material and giving it an appropriate name, something indicative of what we'll be creating — I've gone with **M_DirtyWalls**, as that's going to be the use of our material. Double-click on the new asset to open up the material graph.

7. Right-click anywhere within the material graph and search for a node called **Precomputed AO Mask**. Select it to add it to the graph.

8. After that, include a **Cheap Contrast** node and wire its **In (S)** input pin to the output of the previous **Precomputed AO Mask** node.

9. Next, create a **Scalar Parameter** node and connect it to the **Contrast(S)** input pin of the previous **Cheap Contrast** node. This will allow us to control the contrast between the occluded and the unoccluded areas in our material. Give it a value of **5**.

10. Having reached this point, feel free to connect the output of the **Cheap Contrast** node to the **Base Color** input pin in our main material node. Doing that and applying the material to the walls in our scene will allow us to see the effects that these nodes have in the level, as seen in the following screenshot:

Figure 7.7 – The effects of our new material in the scene

> **Important note**
>
> Similar to what happened with the **Per Instance Random** node we studied in the previous recipe, the **Precomputed AO Mask** node won't show its effects in the material preview window. This is because the data that it taps into isn't available there – only in the specific models with which we work. Keep that in mind when working with this recipe, as you'll often need to refer to the main viewport whenever you want to see any results happening on screen.

The **Precomputed AO Mask** node will be the star of this recipe, as it is the node that will allow us to tap into the baked ambient occlusion information that we can use for our purposes. A common practice when working with this node is to use it to blend the base color of our material with a darker or dirtier version of it in order to show the dust that often settles in those crevices captured by the new node we are using. Let's take care of that next.

11. Create a couple of **Constant3Vector** nodes and assign them two different values. Seeing as we will apply this material to the walls in our scene, we can go for a shade of white and another shade of brown – the latter to drive the appearance of the dirty areas of the material.

12. Interpolate between the previous two nodes by creating a **Lerp** node and placing it after the previous two **Constant** nodes. Connect the white value to the **A** input pin and the brown color to the **B** input pin.

13. Connect the **Alpha** input pin of the previous **Lerp** node to the output of the **Cheap Contrast** node we created in *step 8*.

Tip

It can be difficult to know which input pin we need to connect our constants to in the Lerp node. As a rule of thumb, think of **A** and **B** as "black and white"—**A** will affect the black areas of the texture provided as the **Alpha** input pin, while **B** will affect the white ones.

Seeing as the **Precomputed AO Mask** node assigns lighter values to the areas that are occluded — perhaps a bit counterintuitively — the **A** input pin needs to be fed with the values that we want to have on the non-occluded areas, leaving the **B** input pin for the occluded ones. Setting the material to work in this way will leave us with a white and brown surface, which isn't too subtle a result. We can fine-tune our shader by introducing a second mask at this stage.

14. Add a **Texture Sample** node and select the texture called **T_ Water_ M** as its value (this is part of the Starter Content, in case you also want to use it on your own projects).

15. After that, include a **Texture Coordinate** node and hook it to the previous **Texture Sample** node. Set a value of **10** in both the **U Tiling** and **V Tiling** settings.

16. Next, create a new **Lerp** node and add it after the previous one we had in our graph. Connect the output of the previous **Texture Sample** node to the **Alpha** input pin of the new **Lerp** node.

17. As for the input pins, connect the **B** one to the output of the **Lerp** we created in *step 12*, and wire the **A** input pin to the output of the white **Constant3Vector** node.

18. Finally, connect the result of the latest **Lerp** node to the **Base Color** input pin on the main material node. The graph should now look something like this:

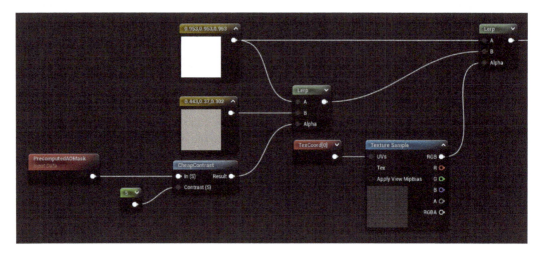

Figure 7.8 – The current state of the material graph

With that done, we can finally apply the changes we've made and save our material, only to go back to the main viewport and see the effects that the latest adjustments have made to the material:

Figure 7.9 – The look of the material we've just created when applied to the 3D model of the wall

As you can see, our wall now looks like it's seen some action in the past, with multiple marks of dirt on its surface. These are concentrated around the areas that are occluded, saving us from manually placing specific textures in those areas.

How it works...

The way the **Precomputed AO Mask** node works is very simple — it takes the lightmaps that have been calculated for our level and uses them as a value that we can tap into inside the Material Editor. Having said that, we usually want to apply some kind of modification to the information that we work with. This is because the data that we are presented with usually turns out to be a very smooth gradient, something that might not be ideal when we want to use it as a mask. More often than not, we'll want to apply some kind of transformation, such as the **Cheap Contrast** node we used in this recipe. Both have the ability to alter the base information, either by augmenting the contrast or by increasing the values of each pixel. This process leaves us with a more defined image, where the areas that are occluded and those that aren't can be easily differentiated — something that's usually good for our purposes.

See also

Even though we've managed to apply the material to the walls in the scene successfully, keep in mind that you'll need to build the lighting every time that you want to use this effect. As the name of the node itself implies, we are dealing with a precomputed effect here, so we won't be able to take advantage of it when we work with dynamic lighting. For those scenarios, make sure to check out the technique described in the *Taking advantage of mesh distance fields in our materials* recipe back in *Chapter 5*, which will allow you to create similar-looking effects in those circumstances.

Matching texture coordinates across different models

How we apply materials and textures to our models varies a lot, and it is something that depends on multiple factors: for instance, are we working on small props or are we texturing large surfaces? Do we need to accurately depict real-life objects, or can we include procedural-creation techniques in our workflow?

Those are some of the questions we need to ask ourselves before we begin working on an asset, and we've answered some of them in multiple different recipes in this book. However, all of those lessons had one thing in common: we were always looking at individual objects. This time around, we will explore how to tackle multiple objects all at once, ensuring that the materials we create look good on all of them regardless of their individual UV layouts.

Make sure to continue reading to find out how!

Getting ready

The key to tackling this recipe is to have two or more different models that look different when applying the same material on them. You can see this behavior in the following screenshot, which shows two toy tank track meshes that display this behavior:

Figure 7.10 – The initial state of the level

If you want to follow along working on the same level I'll be working on, open the level called 07_03_Start, which you can find inside the **Content | Levels | Chapter 07** folder.

How to do it...

Let's start this recipe by creating and working on the new material we'll be dealing with in the next few pages:

1. Create a new material and assign it to the two toy circuit tracks present in our scene. I've gone with the name **M_ Wood_ Track** this time.

2. Jump straight into the Material Editor and create the star of this recipe: the **World Position** node.

> **Important note**
>
> It's important to note that the name you should type when searching for this node is the one mentioned before, World Position. Having said that, you'll see another name once you add it to the material graph — **Absolute World Position**. That prefix depends on the actual properties defined on its **Details** panel, which can be changed.

3. Next up, create three **Component Mask** nodes and set them to use the following channels: the red and green channel on the first, green and blue for the second, and red and blue for the third.

4. Connect all the input pins of the previous masks to the output of the **World Position** node.

5. After that, add a couple of **Lerp** nodes to the graph to interpolate between the previous three masks.

6. Connect the output of the red and green mask to the **A** input pin of the first **Lerp** node and the output of the green and blue mask to the **B** input pin of the same **Lerp** node.

7. Once sorted, connect the output of the previous **Lerp** node to the **A** input pin of the second one, and connect the output of the red and blue masks to the **B** input pin of the second **Lerp** node.

You can take a look at the state of the material graph in the next screenshot:

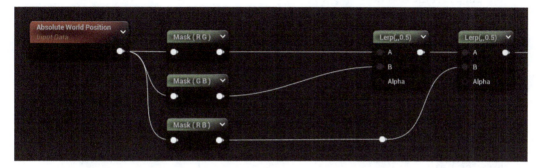

Figure 7.11 – The current state of the material graph

The previous set of nodes will allow us to use the **World Position** node on different surfaces of the models on which we apply this material. At its core, this node acts as a planar projection method, and we need to specify different projection planes for the effect to work on all possible surfaces, not just one. You can read more about this in the *How it works…* section of this recipe.

The next bit of logic we need to implement involves identifying the direction in which the faces that make up a given 3D model are oriented. Having that information will allow us to implement the appropriate projection method on those faces.

8. To start, create a **Pixel Normal WS** node. This node will give us the direction that each pixel is facing in world space coordinates.

9. After that, include a couple of **Component Masks** nodes right after the previous node. Select the red channel on one of them and the green channel on the other.

10. Next, connect the output of the **Pixel Normal WS** node to the input pins of both of the previous masks. Given how we've chosen the red and green channels as per *step 9* and seeing how the output of the **Pixel Normal WS** node is a vector, that means that we are getting the values in the X (red) and Y (green) directions.

11. Create a couple of **Abs** nodes and place each immediately after each of the previous **Component Mask** nodes, to which they should also be connected.

The **Abs** node gives us the absolute value of whatever figure reaches it, so it discards the sign of the incoming value. This is important to us at this point as we are interested in the axis the pixel is facing and not whether it is looking at the positive or negative end.

12. Next, look for an **If** node and add it to the graph.

13. Add three **Constants** and assign them the following **0.5**, **0**, and **1** values. The first one (**0.5**) should be connected to the **A** input pin of the **If** node; the second one (**0**) to the **A > B** input pin, while the third one (**1**) to both the **A == B** and **A < B** input pins.

14. With the preceding steps in place, position the **If** node and the three **Constants** next to the **Abs** node that's being driven by the red channel mask, and connect the output of **Abs** to the **B** input pin of the **If** node.

The previous set of steps performs an operation that determines whether a given pixel is facing in the direction that we want to analyze. In this case, we chose to study the X axis, as masking the red channel from the **Pixel Normal WS** node gives us those values. The **If** node then allows us to output a true or a false statement in the form of the **0** and **1 Constants** we used, based on whether the pixel is predominantly facing in the direction that we've chosen. We know that by comparing the incoming value against the **0.5** preset floating-point number – if the input is greater, it means that the pixel is mostly oriented in that axis.

Having said that, let's now duplicate the previous setup in order to analyze one of the other axes, the Y axis:

15. Create a copy of the previous **If** node and the three **Constants** and place those after the **Abs** node being driven by the green mask, just like we did in the previous step when working with the red mask.

16. Remember to connect the **Abs** to the **B** input pin of the **If** node. This part of the material graph should look something like the following screenshot, if only for reference:

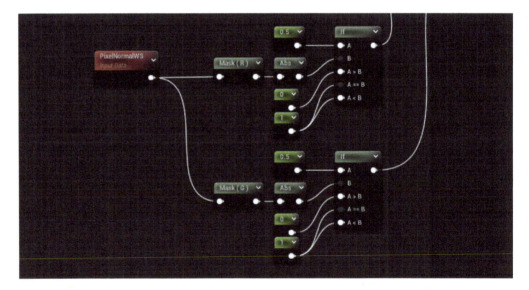

Figure 7.12 – The previous set of nodes we created, from steps 8 through to 16

The previous part of the graph creates a conditional statement that will assign a specific value depending on the direction the pixels on our models are facing, which is useful to drive the **Alpha** input pins of the previous **Lerp** nodes we created back in *step 5*.

17. Connect the output of the **If** node being driven by the red mask to the first of the two **Lerp** nodes; that is, the one being driven by the red/green and green/blue masks.

18. Connect the output of the other **If** node to the **Alpha** input pin of the remaining **Lerp** node.

Now that we've set up that logic, we finally have a material that can detect the direction that each pixel is facing and assign a value according to it. Next, we need to do a couple of extra things within our graph, that is, assign a texture and adjust its tiling. Let's do that now.

19. Include a **Divide** node and place it after the last **Lerp** node, connecting its **A** input pin to the output of that **Lerp** node.

20. Create a **Scalar Parameter** node and assign it a name and a value. I've chosen **UVing** as the name, as it will control the tiling of the textures, and **5** as a default value.

21. Add a **Texture Sample** node and select any texture that you'd like to see on the tank tracks. Connect the output of that to the **Base Color** input pin on our main material node. I've chosen one sample from the Starter Content for this node, called **T_ Concrete_ Grime_ D**.

22. Next, let's adjust the previous **Texture Sample** node so that it looks a bit different. We can do that by creating a **Multiply** node. Assign a **0.5** value to the **B** input pin and connect the output of the previous **Texture Sample** node to its **A** input pin.

23. Finally, feel free to add any extra nodes to the material to spice it up a little more, such as some parameters to control **Roughness** or any other properties that you'd like to change. I've created a **Constant** to drive the **Roughness** setting and set its value to **0.7**.

The material graph should now look something like the following screenshot:

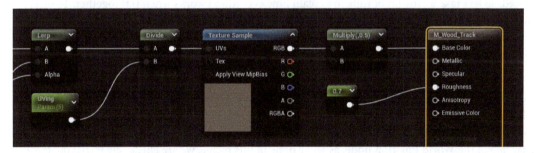

Figure 7.13 – A recollection of the last few nodes we've created

24. Finally, remember to save and apply the material to the toy track model in the scene.

As you can see, this method is a very useful whenever we want to match the texture coordinates of several models that don't share the same UVs. We can apply this to many different models, from roads to pavements to these toy tracks we've just created and many others. It's also a good technique to employ when dealing with models that don't have nicely laid out UVs, saving you the time of adjusting those in external software or even creating them altogether! Having said so, let's take a final look at the material in action:

Figure 7.14 – A final look at the material being rendered

How it works...

The way the material we've just created works is a bit hidden behind the network of nodes we've just placed, so let's take a little bit of time to explain how everything works when put together. To tackle that, we need to know that there are two basic parts to this material — the section where we specify how the projection works and the part where we determine the direction in which the pixels of our models are facing.

The first of those two is handled by the **World Position** node, which basically gives us the location of each pixel in a global coordinate system. Using that information instead of the UVs of the object is what allows us to planar project any texture that we later add, but we need to mask the node in order to get the specific plane in which we want that projection to happen. By masking the red and green channels of the **World Position** node, as we've done in this recipe, we are effectively retrieving the X and Y coordinates of a given pixel, something that allows us to project textures along that XY plane – something that we can call a top-down planar projection. It's good to note that we can translate each channel into one of the coordinate's axes — red being X, green being Y, and blue being Z. This is also indicated by the color-coded gizmo that Unreal uses when displaying this information.

The second part of our material deals with the identification of the direction in which our pixels are facing. We checked whether they were facing in the X or the Y direction, as we needed different projection planes for those cases. The pixels that are aligned mostly along the Z axis are also using a separate projection, but we don't check those as that's the base assumption. For the other two, X and Y (or red and green, using the material's terminology), we need to perform a detection that can be seen on the **If** nodes we place. What we perform in those nodes is a basic comparison — if the pixels are within a certain threshold, we assume that they are facing in a specific direction and proceed to mask them.

As for the final steps, they are all quite straightforward. We use the information we've gathered so far to drive the appearance of a texture, which in turn defines the look of the material. A little bit of spicing in the shape of a **Roughness** property, et voilà — we have a material!

See also

Even though we've used our material in a 3D object, planar projections are more often than not used on planar surfaces. We've mentioned roads and pavements already in this recipe, which are two examples that see this technique being used a lot. The calculations and the complexity of the resulting material in those cases are usually much simpler than what we've seen, as we only need to take one plane into account. Take a look at the following screenshot to see the resulting graph:

Figure 7.15 – The material graph needed when we want to project a texture from a top-down perspective

As you can see, things are much simpler and lighter, so make sure to use this option when you don't need to take all the axes into consideration!

Using interior cubemaps to texture the interior of a building

Working with a 3D model of a building comes with its own set of challenges. Many of them stem from the fact that these meshes are comprised of both large and small-scale pieces. This dual nature wreaks havoc among the artists that have to deal with them, and a great example of that problem can be seen when we look at their windows. Do we, as artists, allow users to see what's inside, with the repercussions of needing to model the interiors? Or do we simply make the glass reflective and opaque, limiting the views of those rooms?

In the following pages, we'll take a look at a more contemporary technique that brings windows to life thanks to a special type of texture called interior cubemaps, which allow us to add realism to our buildings without needing to model all of the rooms inside them.

Getting ready

Unlike most of our previous recipes, where we can usually jump straight into them, you'll actually need to have a specific type of asset to tackle this one. We will be working with a texture commonly known as a cubemap, and even though Unreal includes some as part of the engine, they aren't really suited to be used in interior shots. For that reason, I've downloaded one provided by Epic Games, which you can find through the following link:

```
https://docs.unrealengine.com/4.26/Images/RenderingAndGraphics/
Textures/Cubemaps/CreatingCubemaps/CubeMapNvidiaLayout.png
```

On top of that, you can also open the level provided alongside the Unreal Engine project for this book, named 07_04_Start, and available within the **Content** | **Levels** | **Chapter 07** folder.

How to do it...

The first thing we will do in this recipe is look at the map we have in front of us, which will help us understand what we are trying to achieve — plus which elements you'll need to have if you want to use your own textures and assets. Let's take a look at the scene:

Figure 7.16 – The initial state of the scene

As you can see, we have a basic block, which we'll use to represent a house of sorts — our goal will be to have some windows in its walls and show what's inside without actually creating any more geometry. If you want to bring your own models, something like a building without an interior will work great. Let's start working on the material that will allow us to achieve this effect:

1. Begin by creating a new material and giving it a name—something like **M_InteriorCubemap** would work well!

2. Next, double-click on the new asset and add your first node to the material graph: a **Texture Sample** node. Choose **T_ House_ Color** as its parameter, one of the textures I've created for this building, available in the Unreal Engine project accompanying this book.

> **Important note**
>
> The textures that we are using in this recipe have been created specifically for the model that we are applying them on, taking its UVs into account. Make sure to bring your own images if you plan on using your own models.

The previous texture has given us the look we will be applying to the exterior of our model, which you can check straight away if you compile, save, and apply the material on our model. The next step we'll need to take will be the creation of the appearance of the interior, and the blending between those and the exterior we've already created.

3. Continue by creating a **Lerp** node and place it after the previous **Texture Sample** node. Connect its **A** input pin to the output of the **Texture Sample** node.

4. After that, add another **Texture Sample** node and set it to use the **T_Default InteriorCubemap** image.

5. Then, connect the output of the previous node to the **B** input pin of the **Lerp** node we created in *step 3*.

6. Next, right-click and look for a specific node named **Interior Cubemap**. Place it before the **Texture Sample** node storing the cubemap, the one we created in *step 4*, and connect its **UVW** output pin to the **UVs** input pin of the **Texture Sample** node.

7. Next, add a **Constant2Vector** node and connect it to the **Tiling (V2)** input pin of the **Interior Cubemap** node. Assign it a value of **2** and **4** on the **R** and **G** channels, respectively.

8. Add another **Texture Sample** node to use as a mask and assign the **T_ House_ Mask** texture to it. Connect its output to the **Alpha** input pin of the **Lerp** node.

9.   Connect the output of the **Lerp** node to the **Base Color** input pin on our material. The graph should now look something like the following screenshot:

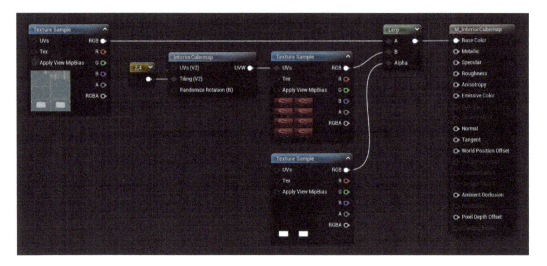

Figure 7.17 – The current state of the material graph

The previous set of nodes are the implementation of the cubemap logic inside the material. Everything is condensed in the **InteriorCubemap** node, which is a handy function that packs a much larger set of nodes inside. You can check out how that logic is constructed if you double-click on the node itself, or learn more about it in the *How it works…* section. In any case, we've already established the base logic that we need to implement when we use these types of assets within a material by taking a cubemap, applying it to a specific area that we have masked, and blending between that and the other parts of the material. The next bit we need to tackle is masking and creating the rest of the nodes.

10. Add two more **Texture Sample** nodes into the graph, as we'll need those to drive the normal and metallic properties of our material.

11. Next, assign the **T_House_ AORM** texture to the first of them and the **T_ House_ Normal** texture to the second.

12. Connect the output of the **Normal** map straight into the **Normal** input pin of our material.

13. After that, create a **Lerp** node and hook the output of the blue channel in the **T_House_ AORM** texture to its **A** input pin.

14. On completion, create a **Constant** and assign a value of **0** to it, and plug it into the **B** input pin of the previous **Lerp** node.

15. Connect the same texture we used as a mask to the Alpha channel of the new **Lerp** node. That would be the **Texture Sample** node we created back in *step 8*.

16. Finally, wire the output of the **Lerp** node into the **Metallic** input pin on our material.

All we've done in the previous steps was assign some extra information to our shader — in particular, to the **Normal** and **Metallic** channels. The Constant we've used could have been replaced with the actual **Roughness** values for the interior cubemap via a texture, just like we did in the **Base Color** section of the graph. This is the same setup that we would follow should if we decide to add any other information, such as emissive lighting or translucency. The **Emissive Color** channel is one that gets used quite often in interior cubemaps as there are often lights within the rooms that we see — especially at nighttime. The last thing we need to do is assign the material and see how it looks!

Figure 7.18 – A final look at the material we've created in this recipe

How it works...

As we've seen in this recipe, this interior cubemap technique can prove very useful when dealing with building interiors. That being the case, the cubemaps we'll probably want to use will be our own instead of the default one that we've used in this recipe. So, how would you go about creating one?

The key is to create a texture that fits within Unreal's expected formatting for cubemaps, which you can learn more about through the link found at the end of this recipe. The basics that we need to know is that said texture needs to adjust to fit a specific pattern that you can see in the following screenshot:

Figure 7.19 – The position where each texture needs to be placed for the interior cubemap to work

As you can see, that texture is made up of six different pieces, where each matches a specific camera direction. This is the key to capturing cubemaps—making sure that we align the camera appropriately. If we translate that into a real-world scenario, this would mean that you need to point the camera in each of the possible axis directions – the positive X axis, negative X axis, positive Y axis, negative Y axis, positive Z axis, and negative Z axis – and render the scene like that. You would then bring those six images into an image editing program, such as Photoshop, and arrange the pictures from left to right according to the order that Unreal expects, as shown in the previous screenshot.

Saving that file into the correct format is another matter, one that you need specific tools for, as provided by different vendors, such as NVIDIA or AMD. With that in mind, let me leave you with an interesting link regarding how to create, export, import, and use your own cubemaps:

```
https://docs.unrealengine.com/4.26/en-US/RenderingAndGraphics/
Textures/Cubemaps/
```

Using fully procedural noise patterns

In this recipe, we will talk about a technique that is both powerful and flexible but also quite demanding in terms of computing power. It is best used as a means of creating other assets, and it's one that isn't really suited for real-time apps and games. We are talking about the **Noise** node — a fully procedural mathematical system that allows you to create many different non-repetitive textures and assets.

Similar to the technique we studied in *Chapter 2*, in the recipe named *Creating semi-procedural materials*, this node takes things a bit further and enables you to use an effect that's very widespread in offline renderers, giving you the ability to create materials where repetition is not a concern. Let's see how it's done!

Getting ready

Unlike in the previous recipe, we won't need anything else apart from what the engine offers us to tackle this recipe. However, I've prepared a small scene that you can use should you choose to do so — its name is 07_05_Start and you can find it within the **Content** | **Levels** | **Chapter 07** folder. It contains a single plane onto which we'll apply the material we are about to create, so keep that in mind if you want to create your own level.

How to do it…

As always, let's start by creating the material that we will apply to the plane in our scene. In this case, we'll create one that will work as a cartoon ocean material, with animated waves driven by the **Noise** node we'll be introducing. Something good to note before we start is the fact that this material is going to have several different parts—one for the sea foam, one for the normal sea color, and another for a slightly darker variation on this last parameter just to spice things up a bit. We'll be referring to these different parts throughout this recipe, which will be good to remember. So, follow these steps to create the material:

1. Let's start by creating a new material and giving it an appropriate name, something like **M_CartoonOcean**.

2. Next, assign it to the plane in our scene and double-click on the material to bring up the Material Editor.

3. The first section we'll tackle will the different sea color variation. Start by creating two **Constant3Vector** nodes and adding them to the graph.

4. After that, assign two different shades of blue to the previous **Constant3Vector** nodes, keeping in mind that these colors will drive the **Base Color** property of our material.

5. With the previous two nodes in place, create a **Lerp** node and connect both of the previous **Constant3Vector** nodes to its **A** and **B** input pins.

    The next part that we need to create is the mask that will drive the **Alpha** input pin of our previous **Lerp** node. Instead of relying on a static texture, we'll use this opportunity to introduce our new procedural friend — the **Noise** node!

6. First, add a **Texture Coordinate** node and an **Append** node to our node graph, and connect the output of the former to the input of the latter.

7. Next, create a **Constant** and connect it to the **B** input pin of the previous **Append** node.

8. When the previous steps are done, right-click and look for the **Noise** node. Add it to our graph, and hook the output of the previous **Append** node into its **Position** input pin.

> **Important note**
> The reason why we are adding a Constant with a **0** value to the **Texture Coordinate** node is that the **Position** input pin of the **Noise** node expects a three-dimensional vector.

9. With the **Noise** node selected, look at the **Details** panel and set the following values: choose the **Fast Gradient - 3D Texture** method as the function and set the **Output Min** value to **-0.25**. All of the other properties should be left with their default values.

10. Once that's sorted, connect the output of the **Noise** node to the **Alpha** input of our **Lerp** node.

Seeing how we have created a few nodes thus far, now is a good time to look at the current state of the material graph:

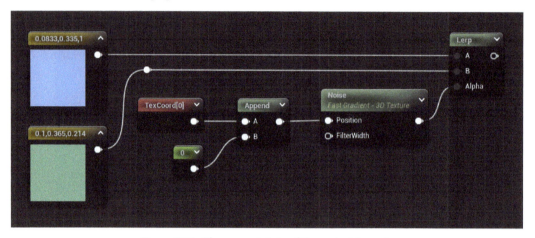

Figure 7.20 – The current state of the material graph

The previous steps have left us with a non-repeating pattern that allows us to mix two different colors without any kind of repetition. Of course, this first part is just introducing a very small color variation: a subtle effect that's used to randomize the look of the final shader. Let's continue and create the other parts of the material.

11. Continue by creating a third **Constant3Vector** node and giving it a value close to white—we'll use this to color the sea foam.

12. Add a new **Lerp** node and connect the previous **Constant3Vector** node to its **B** input pin.

13. After that, wire the output of the first **Lerp** node we created back in *step 5* to the **A** input pin of the new one.

With that done, we can now start creating the noise pattern that we'll use to mix between the normal color of the sea and the sea foam. We'll use the **Noise** node again, but we'll also want to animate it so that it looks even better in motion.

14. Start by creating a **World Position** node.

15. Next, add a **Panner** node and change the value of **Speed X** to **5** and **Speed Y** to **15**.

16. After that, create both an **Append** node and a **Constant** node with a **0** value, and place them after the previous **Panner** node.

17. Connect both the previous **Constant** and **Panner** nodes to the **Append** node—the output of the **Constant** node should go into the **B** input pin, while the output of the **Panner** node should be connected to the **A** input pin.

18. With these steps sorted, create an **Add** node and plug the output of the **World Position** node from *step 14* into its **A** input pin. As for the **B** input pin, connect the output of the previous **Append** node to that one.

19. Add a new **Noise** node and adjust the following settings in the **Details** panel: **Scale** set to **0.015**, **Quality** to **5**, **Function** to **Voronoi**, **Levels** set to **1**, and **Level Scale** set to **4**.

20. Finally, connect the output of the previous **Add** node from *step 18* to the **Position** input pin of the new **Noise** node.

Before we continue, it might be a good idea to go over some of the most recent nodes we have created. As mentioned before, the **Noise** node creates a non-repetitive pattern that we can adjust through the parameters exposed in the **Details** panel. Out of all of the settings that we can find there, perhaps the **Function** option is one of the most important ones, as it defines the shape of the noise that we get. We can further affect that shape through other different settings – for instance, we can animate it with the **Panner** node sequence we created before. Pay attention to the nodes we'll be creating in the following steps, as those also will adjust it, and make sure to also visit the *How it works…* section, where we'll go into greater detail about the different options available for the **Noise** node.

21. Continue by creating a **Constant** node and setting its value to **0.3**. Connect its output to the **Filter Width** input pin of the previous **Noise** node.

22. Next, include a **Power** node after the **Noise** node, which we'll use to push the black and white values apart. Connect the output of the **Noise** node to the **Base** input pin of the new **Power** node.

23. Add a **Constant** to drive the **Exp** input pin of the previous **Power** node, and give it a value of **4**.

24. After that, create a **Multiply** node and connect it's a input pin to the output of the previous **Power** node. Set the **B** parameter to **3**, which will increase the lighter values in our mask.

25. Connect the output of the previous **Multiply** node to the **Alpha** input pin of the **Lerp** node we created back in *step 12*.

These last sets of nodes have helped us define the **Noise** node we are using to drive the position and appearance of the sea foam color. With this done, we are almost finished tweaking the material. However, we can still adjust it a little bit more – after all, the sea foam pattern we've created is only being animated in one specific direction, which isn't great in terms of the final look it gives us. Let's add a bit more variation to that, not without checking the state of the material graph first:

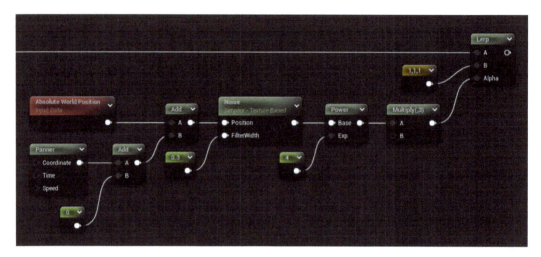

Figure 7.21 – The new nodes we have created since step 11

26. To start adding that variation we mentioned, copy the previous set of nodes we created from *steps 14* to *25* and paste them elsewhere in the material graph – we'll use them as the base to create a little extra variation on the sea foam front.

27. After that, change the values of the duplicated **Panner** node – **Speed X** should be set to **10**, and **Speed Y** should be set to **5**.

28. Next, modify the scale of the new **Noise** node and set it to **0.0125**.

29. With regards to the **Constant** driving the **Exp** input pin of the **Power** node, set it to **3**.

30. Connect the **A** input pin of the duplicated **Lerp** node to the output of the previous **Lerp** node we created in *step 12*.

31. Finally, connect the output of this last **Lerp** node to the **Base Color** input pin of our material, and apply the shader to the model in the scene. This is the final look of the material that we should be looking at:

Figure 7.22 – The cartoon ocean material being applied to the plane in our level

And there you go! That's what our material looks like in the end, without any repetition going on, no matter where you look. This is a very powerful technique that you can use to get rid of that problem, given that you are aware of the cost that it introduces to the rendering pipeline. That cost is something that you can check, either by looking at the shader instruction count or by relying on the tools that Unreal provides to test how well your app or game performs. As always, stay tuned for the next two sections, where we'll talk a little bit more about what's going on beyond this node's surface.

How it works...

Even though we've already used the **Noise** node a couple of times, it won't hurt us to take another look at it and see what we can accomplish by tweaking the different parameters that are exposed on the **Details** panel:

- The first of them, if we look at the different options in order, is the **Scale** setting. This is quite a straightforward parameter – the bigger the number, the smaller the noise pattern will be. This can be a bit confusing at first, as we will usually need to type in really small values if we want to use the noise more as a masking technique and not as a small grain effect. Values such as **0.01** and lower usually work better than the default **1** value, so be sure to remember this when you are on your own and can't figure out what's going on with your **Noise** function!

- The second parameter we can tweak is the **Quality** parameter, which will affect the end result in a subtle way – mostly by offering a smoother transition between the areas that are colored differently.

- The third parameter, the **Function** paramerer, is probably the most important one, as it controls the logic that generates the final pattern. We'll talk about these more in the *See also* section, as this is where the meat of the node is.

- **Turbulence** is the next option we can enable, which determines how jittered the pattern we create is. A good way to think of turbulence is to see it as variation within the already random pattern created by the **Noise** node.

- The next setting, **Levels**, also increases the variation of the end result that we get, making it richer in terms of finer detail.

- Further down, you can find the **Level Scale** setting, which lets you adjust the size of the levels. You can increase this value when you have a low number of levels in order to fake the detail you would get if you were using more, effectively boosting the effect to make it more efficient.

- Beyond those parameters we've already covered, we can also control the **Output Min** and **Output Max** settings: these control the minimum and maximum output values, with **0** being black and **1** being white. The default values are set to negative and positive, and it's useful to tweak them as if we had a scale: if we set a minimum of **-6** and a maximum of **1**, we'll have a greater range of values located in the negative, darker tones than lighter ones.

- Finally, the last settings are for **Tiling** and **Repeat Size**. Checking the first option will make the pattern repeat over the size that you specify, allowing you to bake a texture that you can then use as a noise generator at a much lower rendering cost. It's a handy feature when you want to create your own assets!

See also

The most important settings that we need to select within the **Noise** node are probably for the functions that drive this feature. They are different on many levels, but mainly in terms of their rendering cost and their final looks. Even though you might have seen them implemented in other 3D packages, one of the important things to note when choosing between the different types within Unreal is their limitations and the cost that they incur.

As an example, Epic's documentation tells us that certain types are better suited for specific situations. For instance, the **Simplex** function can't tile, and the **Fast Gradient** one is bad for bumps. That already signals to the user which one they should choose when creating a specific effect. Apart from that, the number of instructions is a really important element to keep in mind: for example, the **Gradient** function packs 61 instructions per level in its nontiled version, whereas the **Fast Gradient** option peaks at around 16.

Beyond that, an expensive but very useful function is the **Voronoi** one, which is one of the latest functions to be added to the engine. It is a really useful one when you're trying to recreate many different elements that we see in nature, as it comes close to reproducing stone or water, to name two examples. This is what we've used to recreate the ocean shader, but given different inputs, we could have just as well created a cracked terrain. Examples of this can be found in the following documentation:

```
https://www.unrealengine.com/en-US/tech-blog/getting-the-most-out-
of-noise-in-ue4
```

Adding detail with Detail Texturing

The way we create materials for a 3D model often changes based on the size of the object itself: the techniques we use to render small meshes aren't usually the same ones as the methods employed on big surfaces. In reality, size is not what really matters: more often than not, the difference lies in the distance at which the camera is positioned with regard to a given model. Big surfaces tend to be viewed at drastically different distances: for instance, we can stand right in front of a building or be 10 meters away from it. This circumstance creates the need to set up materials that look good both when viewed up close and when we are far away from them.

In this recipe, we will explore a handy material function that allows us to seamlessly add detail to our materials so that they look great upon close inspection: the **Detail Texturing** function. Let's take a look at how to work with it.

Getting ready

The material we are going to create in this recipe relies entirely on assets that you can find within the Starter Content, so you don't need to worry about incorporating any extra resources into your level. The only thing you'll need to have in your scene is a fairly big model that you can get really close to. If you don't want to spend time setting it up, feel free to open the level called 07_06_Start located within the **Content | Levels | Chapter 07** folder.

How to do it...

Let's get straight to the task at hand and create the new material we will be working on:

1. Start by creating a new material and giving it a name – **M_DetailTexturing** is a good option, considering the new node we are going to be using. Double-click on it to open the material graph.

2. Once inside the material graph, right-click inside it and search for the **Detail Texturing** function, and add it to the graph.

    The **Detail Texturing** material function contains several inputs that we'll need to adjust, as well as two outputs that we'll eventually wire to the main material node. In essence, we need to provide the function with the textures that we want to use both when viewing the material at a normal distance and when observing it closely. In return, the node will give us two output pins that are ready to be plugged directly into our material. Let's take care of setting it up now, starting with the information that we want to see at a standard distance.

3. Create a couple of **Texture Sample** nodes and assign them to whichever color and normal textures you want to use in this material. I've selected two that are part of the Starter Content, named **T_Brick_Clay_New_D** and **T_Brick_Clay_New_N**.

4. After that, add a **Texture Coordinate** node and connect its output pin to the **UVs** input pin of the two previous **Texture Sample** nodes. Since the new node will be driving the tiling of our material, set both its **UTiling** and **VTiling** properties to **10** – a value that works well once we apply the selected textures to the wall model that I have in my level. Make sure to fine-tune that value if you are working with your own assets.

5. With the preceding steps in place, connect the **RGB** output pins of the two **Texture Sample** nodes to the **Detail Texturing** function. The output of the **Base Color** texture should be applied to the **Diffuse (V3)** input, whilst the output of the **Normal** map should be connected to the **Normal (V3)** input pin.

    The previous set of nodes we have created (the two **Texture Sample** nodes and the **Texture Coordinate** node) is tasked with taking care of the textures that are applied as a base within the **Detail Texturing** node. Up next, we will be working with the textures that are going to enhance the look of the material, the ones that add that extra detail that gives its name to the material function we are using. Let's deal with those next.

6.   Create a couple of **Texture Object** nodes and assign the textures you want to use for the extra detail. In my case, I'm going to use the assets named **T_Rock_Sandstone_D** and **T_Wood_Pine_N**, which are a **Base Color** texture and a **Normal** map, respectively, that I think work well when enhancing the look of the previous brick textures we selected.

7.   Connect the output of the **Texture Object** node that contains the detail **Base Color** info to the **Detail Diffuse (T2d)** input pin of the **Detail Texturing** node, and connect the output of the other one to the **Detail Normal (T2d)** input.

Unlike the textures we used as the base information for the **Detail Texturing** material function, the ones used to add detail need to be provided as **Texture Object** nodes. This prevents us from using nodes such as the **Texture Coordinate** one to adjust their scale and placement. Instead, the **Detail Texturing** material function contains three input pins that we can tweak to adjust the scale and the intensity of the detailing effect, which we'll see now.

8.   Create three **Scalar Parameter** node, which we'll use to control the **Scale (S)**, **Diffuse Intensity (S)**, and **Normal Intensity (S)** input pins of the **Detail Texturing** material function. Feel free to name those parameters with the same name as the input pins.

9.   Assign **50** to **Scale**, set **Diffuse Intensity** to **1.75**, and adjust **Normal Strength** to read **10**. Then connect the output of each parameter to the appropriate input pin in the **Detail Texturing** node.

10. Finally, connect the **Diffuse** output pin of the **Detail Texturing** material function to the **Base Color** input pin in our material, and hook the **Normal** output to the **Normal** input pin on the main material node.

With that done, this is the look of the material graph we should have in front of us:

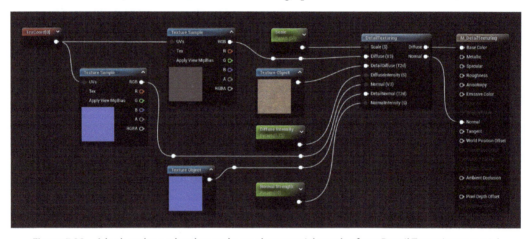

Figure 7.23 – A look at the nodes that make up the material graph of our Detail Texturing material

With that done, we can say that we have finished working on our new material! The last thing that I want to show is the results that we see on the screen when the node is working versus when it isn't. Let's take a look at the following screenshot to see those results in action:

Figure 7.24 – Comparison between using the Detail Texturing node (right) versus not using it (left)

To visualize the impact of the **Detail Texturing** node has on your material, simply set both the **Diffuse** and **Normal Intensity** parameters to **0**, which will effectively disable the effect that the detail textures we've chosen have on the material. While subtle, the results can make our materials much more flexible when they need to be applied on objects that need to look good both when viewed from far away as well as on close inspection. Be sure to give it a try whenever you find yourself needing to tackle one such asset!

How it works...

Even though we've used the **Detail Texturing** material function throughout the recipe, let's pause for a moment to see how it works. In essence, the function combines four different textures in a simple way: it adds the normal values of the texture we want to use to add detail to those ones already selected for the base layer. The same happens with the diffuse color information, and we are able to control the intensity of the blend thanks to the diffuse and normal intensity settings exposed in the material function. Finally, the scale can also be adjusted through its own dedicated parameter, which allows us to control the tiling of the textures used for detailing purposes.

See also

The **Detail Texturing** node is a very handy one, as we've seen in the previous pages, and the fact that it is a material function allows us to look inside its internal graph and learn how it works behind the scenes. I'll leave that for you to explore, but one thing that I would like to do before moving on is to replicate the internal logic as simple plain nodes within the material graph. I'll do that next:

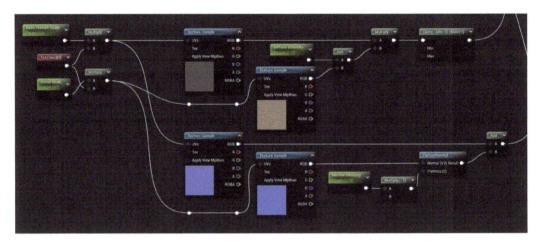

Figure 7.25 – Another way to implement the Detail Texturing function

As you see, we need a few more nodes in place if we want to replicate the same functionality we get out of the **Detail Texturing** function. Furthermore, as this is a technique that we might want to repeat across several materials, having things condensed into a function also works well in terms of keeping our code clean. If you want to explore those nodes in further detail, I've left them within the material graph I created for this recipe, a little bit further down from where the **Detail Texturing** node is located. Be sure to look at it if you want to pick it apart!

8
Going Beyond Traditional Materials

Most of the materials and techniques we've seen so far had one thing in common: they were designed to affect the look of our 3D models. Having said that, we can also use materials to drive other interesting effects; we can highlight interactive elements within the game world, or capture the scene from a particular point of view to then display it back to the user as if we were dealing with a CCTV feed.

We are going to take a look at a few particular cases in this chapter, including the following ones:

- Playing a video on an in-game TV
- Capturing the scene through a CCTV camera
- Highlighting interactive elements
- Creating snow on top of objects using layered materials
- Changing a sunny scene to a snowy one with a parameter collection asset
- Moving between seasons using curve atlases
- Blending landscape materials

As you can see, there are other interesting uses for materials that can help us with our game or app development. Let's explore them in the recipes!

Figure 8.1 – A quick look at some of the materials we'll create

Technical requirements

As usual, most of the materials that we'll explore in this recipe can be tackled using assets that you can find both as part of the Starter Content or using the Unreal Engine project provided alongside this book. With that in mind, let me leave you with a link to the location of the UE project:

`https://packt.link/20u7B`

Playing a video on an in-game TV

Something that amazed me back when I started to play video games was the fact that it was possible to display videos within those same games; it felt like the designers had woven another layer of interactivity to trick our brains into thinking that the worlds they had created were truly real. Videos within videos – that's something that left me starstruck back in the day!

Honoring that memory, we'll now proceed to tackle how to create a working television display and how to play back any videos that you might want to show using Unreal Engine's media player capabilities, which allow us to use videos as textures. Buckle up for the ride!

Getting ready

Seeing as the main goal of this recipe is going to be streaming a video on an in-game TV, the two things that we are going to need are a 3D model of a screen and a video file. As always, you have the option to use the assets provided alongside the Unreal Engine project bundled with this book, where you'll find everything that you'll need to complete this recipe. If going down that route, simply open the level called `08_01_Start` to start working.

If you prefer to work with your own assets, there are a few considerations that you'll need to take into account with regard to the 3D TV model and the format of the video file. To begin with, the UVs of the 3D model need to be set up in a very specific way if we want to make the most of the video file we will use; you can find more info regarding their setup in the *See also* section.

As for the video file, I recommend using an MP4 format, even though AVI files are also accepted. The file format is not the only thing that we need to worry about, as there is currently a quirk with Unreal 5, DirectX 12, and video files that sometimes prevents them from working. To solve that, make sure to implement the following steps:

1. Open the **Project Settings** panel and navigate to the **Platforms | Windows** section.
2. In the **Targeted RHIs** category, make sure that the **Default RHI** property is set to **Default**.

Video files also need to be stored in a specific folder called **Movies** if you want to package them alongside your game or app. Make sure that you create one with that name and place it within the main **Content** folder present in all Unreal 5 projects for things to work properly beyond the editor!

How to do it...

The first thing that we are going to do in this recipe is to set up a Blueprint that we can use to host both the 3D model of the TV and the logic that will allow us to display the video on it at a later stage. Let's start that process:

1. To begin with, right-click in the Content Browser and create a **Blueprint** asset of the **Actor** type. Remember to also assign a name to it – something such as **BP_TVScreen** could work!

2. Next, double-click on this new asset to open the Blueprint Editor. This is where we'll be operating throughout most of the recipe.

3. Once inside the Blueprint Editor, drag the static mesh for the TV display model into the Blueprint actor. You can do this by dragging and dropping the static mesh from the Content Browser into the Blueprint, or by selecting the **Add Component** option in the Blueprint itself and selecting a **Static Mesh Component** from the drop-down list. If you follow this second approach, make sure to set the new component to use the TV model that you want to use in the **Details** panel.

4. After that, make the new imported TV screen the default root of your Blueprint by dragging and dropping the name of the **Static Mesh Component** in the **Components** panel over the **Default Scene Root** one, as seen in the following figure:

Figure 8.2 – Making the Static Mesh Component the root of our Blueprint

> **Important note**
> Now that we have a Blueprint containing the model of the TV screen in it, make sure to replace the existing static mesh of the TV present on the main level if you are working on the map mentioned in the *Getting ready* section.

Before we go back to the Blueprint Editor to continue implementing different layers of functionality, we'll need to create a number of assets that will enable us to stream a video from disk and play it in our app. Let's do that now:

5. Start by creating a **Media Player** asset. You can do this by right-clicking in the Content Browser and looking under the **Media** tab.

6. As soon as you create the new **Media Player**, Unreal will also prompt you to create a **Media Texture** actor as well. Know that you can also create it by right-clicking within the Content Browser and looking inside the **Media** tab, just like we did in the previous step, in case you want to create another one at a later stage.

7. Open the new **Media Player** and look at its **Details** panel. Find the **Loop** option and check the checkbox beside it so that our video loops once we play it.

8. Next, create a **File Media Source** asset, which can be created from the same panel as **Media Player**. This new type of asset is meant to hold the video that we want to play, and that's what we'll need to reference next.

9. Now that we have the **File Media Source** asset, double-click on it so that we can adjust it. We need to set its **File Path** property to point to the location on your hard drive where the video that you want to use in this recipe is located.

   As mentioned in the *Getting ready* section, I've included an MP4 file for us to use in this recipe, called `SampleVideo.mp4`. You can find it within the **Content | Movies** folder of the Unreal Engine project provided alongside this book, so make sure to select either that video or one of your own that you want to use.

   Once all of the previous assets have been created and adjusted, we can get back to our TV Blueprint and continue implementing the different bits of functionality.

10. Jump to the **Event Graph** window of the TV Blueprint and create two new variables inside the **Variables** panel:

    - The first of those variables should be of the **Media Player Object Reference** type. You can name it **Media Player**, for example.

    - The second variable has to be the **File Media Source Object Reference** type. You can name it **Media Source**, for example. The point is to have recognizable names whenever we use these variables in the **Event Graph**!

    Now that those two variables have been created, we need to assign them some default values that match the previous assets we created in *steps 5* through *9*.

11. Select the **Media Source** variable and assign the **File Media Source** asset we created in *step 8* to it. You can do so by looking at the **Details** panel and choosing said element from the available drop-down menu.

12. Next, select the **Video Player** variable and assign the **Media Player** asset we created in *step 5* to it.

    We now have all of the basic ingredients that we need to implement the functionality within the Blueprint's **Event Graph**. Let's jump over there and start coding!

13. Start by getting a reference to the **Video Player** variable in the graph, something you can do if you drag it from the **Variables** panel and drop it into the main graph.

14. After that, drag a wire out of the output pin of the previous reference and start typing `Open Source` to create an **Open Source** node.

15. Next, get a reference to the **Media Source** variable and connect it to the **Media Source** input pin of the previous **Open Source** node.

16. With that done, duplicate the **Video Player** node and paste it after the previous ones we created, to the right of the **Open Source** function.

17. Drag a wire out of the new copy of the **Video Player** variable and type `Play` to create a **Play** node, which will automatically be connected to that variable.

18. With all of the previous sequence of nodes in place, connect the execution pin of the **Event Begin Play** node to the execution input pin of the **Open Source** node we created in *step 14*.

19. Finally, connect the output of the **Open Source** node to the input execution pin of the **Play** node.

Let's take a quick look at what the **Event Graph** should look like:

Figure 8.3 – The logic we've created within the Event Graph

At this point, it's probably a good idea to review what we've done. We basically start with two variables: **Video Player** and **Media Source**. You can think of these as the software that is going to play your video and the video that you want to play. At this point, we've only told the video player which video it should play thanks to the **Open Source** node, and then we went ahead and told it to play it through the **Play** node. All of those steps will see us playing the video that we've chosen, but if we want to see it in our screens, we still need to create a material and apply it to the static mesh of the TV present in the Blueprint. Let's do that now:

20. Let's continue by creating a new material and naming it something like **M_TV_ImageDisplay**.

21. Next, create a **Texture Sample** node and connect it to the **Emissive Color** input pin of our main material node.

22. After that, select **Texture Sample** and set the **Texture** parameter to use the **Media Texture** actor that was automatically created for us back in *step 6*.

23. With that done, save the material and apply it to the **Element 0** material slot present in the **Static Mesh Component** used to hold the TV model in the Blueprint with which we've worked.

If everything went according to plan, you should now be able to look at the video playing on the TV screen once you start playing the level! Let's take a final look at what that should look like in the next picture:

Figure 8.4 – A final look at the material being displayed on the model of the TV screen

How it works...

This recipe saw us using a video as a texture, and we were able to do so thanks to a few assets that allowed us to open, format, and play said file. Let's now take a brief detour and see how they work before moving on to the next recipe.

The first of the new assets we used was **Media Player**. You can think of it as a program within Unreal that allows us to play back videos, in a similar way as we would when using a video player outside of the engine.

Having a video player is only one part of the equation, and the next bit we need is the clip that we actually want to watch: that would be **File Media Source**. We can specify which file we want to see through that asset, so you can think of it as the movie that you want to watch.

Having said that, we need to keep in mind that watching a video is not the only thing that we want to do in Unreal. Our real objective is to use that file as a texture, and the way to do so is through the **Media Texture** asset – a special resource that allows us to parse videos as textures, allowing us to use them within the **Texture Sample** node in our materials as if they were any other standard 2D image.

See also

If you want to create your own 3D model of a screen to use in this recipe, there is one important thing you need to consider when adjusting its UVs. Instead of trying to lay them out in a way that matches the shape of the object, we want them to occupy the entirety of the 0-to-1 UV space. This will ensure that whatever image comes our way gets shown across the entirety of the display. You can see an example of how the UVs should be set up versus how they are usually created in the following figure:

Figure 8.5 – The way we should set up the UVs of our model (left) vs. how they are usually done (right)

This might sound counter-intuitive, as laying out the UVs in this way would traditionally lead to distortion across the textures applied to the model. This is not the case when using video files as textures, where we have to map the incoming file to the entirety of the UV space. Keep that in mind when creating your own models!

Capturing the scene through a CCTV camera

As we've seen by now, materials usually work hand in hand with textures to drive certain material parameters. Those usually come in different shapes and sizes: sometimes, textures are static 2D images that we use to drive the appearance of a model, while on other occasions we might rely on video files to create the effect of an animated sequence (as we saw in the last recipe).

Beyond those two examples, there is yet another type that doesn't rely on existing data, but one that allows us to capture what is happening inside the level. This is called a **Scene Capture** component, and it allows us to create interesting functionalities, such as displaying what a camera placed inside our scene would see – mimicking the behavior of a CCTV camera. In this recipe, we'll study how to work with this type of asset, known as **Scene Capture**, and how to put it to good use inside our materials.

Getting ready

Most of the assets we'll use are part of the engine, especially the ones that will enable us to capture the scene. We've also included some extra ones that will act as props, which, as always, are included as part of the project we'll provide. If you want to work with those, simply open the level named 08_02_Start, something you can find within the **Content** | **Levels** | *Chapter 8* folder.

How to do it...

Capturing a scene can be done for a multitude of reasons, but we've decided to center our efforts on a CCTV feed. That being the idea, what elements do you think are needed in such a system? Having an idea about that will help us visualize the whole process, as there are going to be multiple phases. If we think about it in real-life terms, we'll probably need at least a camera that can capture the scene, a display where we can play the video back, and some sort of storage unit or transfer mechanism that can move the data between the camera and the display.

As you'll see, those same actors will also be needed in Unreal, so let's start creating them:

1. Start by creating a new **Actor** Blueprint somewhere within the Content Browser and give it a name that denotes its intended nature as a capturing device – following the theme of the scene, you can probably go with something such as **BP_Webcam**. Incidentally, that's kind of how the static mesh we'll use looks!

2. Open the Blueprint Editor and create a **Static Mesh Component** by selecting one from the **Add Component** drop-down menu.

3. Look for another type of component in the same panel as before – this time, we'll need to look for **Scene Capture Component 2D**.

    With those assets in place, it's time to assign some default values to them.

4. Continue by making the **Static Mesh Component** the root of this Blueprint. You can do so by dragging and dropping its name into **Default Scene Root** on the **Components** panel.

5. With the **Static Mesh Component** still selected, look at the **Details** panel. The section named **Static Mesh** will let you assign the model that you want to use for this asset. If you want to use the same one I'll use, go ahead and select the **SM_WebCam** mesh included in the project.

6. Go back to the Content Browser and right-click on your desired directory to create a new asset of the **Render Target** type. You can create it by looking inside the **Textures** section of the drop-down menu that appears when you right-click within the Content Browser.

7. Give a name to the previously created asset, something such as **RT_Webcam**.

8. With the previous asset ready, go back to the Blueprint of the webcam and select **Scene Capture 2D**. Look inside its **Details** panel and scroll all the way down to the **Scene Capture** section, where we need to assign **RT_Webcam Render Target** as the value for the **Texture Target** property.

9. There's another setting we need to adjust a little bit further down the location of the **Texture Target** property. Its name is **Capture Every Frame** and we need to make sure that it is checked.

10. With **SceneCaptureComponent2D** still selected, adjust its location and rotation so that it sits in front of the static mesh we are using as the webcam. This will allow the component to record the scene properly, as it would otherwise look black if we didn't move it (as it would be inside the camera!). Take a look at the following figure for reference:

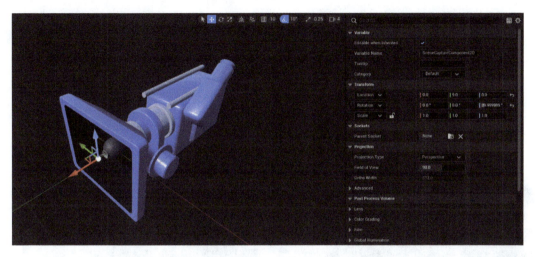

Figure 8.6 – A look at the position of SceneCaptureComponent2D with regard to the webcam

We now find ourselves at a point where we can finish working with this Blueprint, as we already have a setup that allows us to use it to record what's happening on the scene. Having said that, I think we can still spice things up a little more by making the camera rotate as if it were a surveillance device. Let's take care of that by adding a looping motion to the camera that will make it rotate between two defined angles.

11. In order to create the logic that will allow the camera to rotate, start by heading over to the **Event Graph** section of the Blueprint.

12. Once in there, create a **Timeline** asset right after the **Event Begin Play** node. You can do so by right-clicking anywhere on the graph and selecting the **Add Timeline** option.

13. Connect both of those nodes together – the output of the execution pin of **Event Begin Play** to the **Play** input pin on **Timeline**.

14. After that, double-click on the **Timeline** node to open Curve Editor.

15. Once inside Curve Editor, add a **Float Track** by clicking on the **+ Track** button located at the upper-left corner of the new window. While at it, go ahead and give it a name: something such as **Camera Rotation** could work!

16. After that, assign a duration to it by typing a number in the **Length** entry box, located a few spaces to the right of the previous + **Track** button. Set it to **10** and click the **Loop** button located a little bit to the right.

17. With the setup of the **Float Track** in place, let's create the different points that make up the curve: to do so, start by right-clicking in the main section of the graph and selecting the **Add key to CurveFloat0** option. A new point will appear.

18. Set the **Time** property to **0.0** and **Value** to **0.25**. You can modify those two settings once you have the point selected, as two small editable boxes will appear in the upper-left corner of the graph.

19. Create two more points, and give them the following values: **Time** of **5** and **Value** of **-0.25** for the first one, and **Time** set to **10** and **Value** to **0.25** for the last one.

20. With the points in place, select the three of them, click with your right mouse button on top of any, and select the **Auto** option from the **Key Interpolation** section of the new contextual menu.

Because this is the first time we work with a timeline in this book, make sure to head over to the *See also* section of this recipe if you want to learn more about this. Seeing as we've done quite a few things in the previous steps, let's take a look at what the float curve should look like:

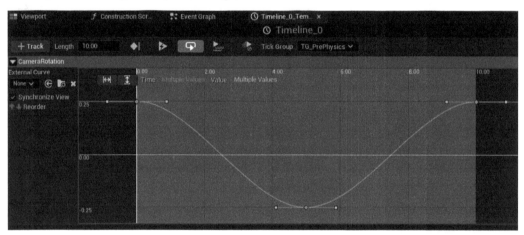

Figure 8.7 – A look at the resulting float track

With the timeline created, it's now time to go back to the **Event Graph** and create the logic that will allow us to animate the webcam. Let's do that now:

21. Start by dragging a reference to the **Static Mesh Component** used by our Blueprint into the **Event Graph**.

22. After that, create an **Add Relative Rotation** node by dragging a wire out of the **Static Mesh Component** reference and typing the name of the new node we want to create.

23. Next, right-click over the **Delta Rotation** input pin and select the **Split Struct Pin** option. Doing that will allow us to control the rotation on each individual axis independently.

24. Connect the output float pin of the timeline (which should have the name of the **Float Curve** asset you created in *step 15*) and connect it to the **Delta Rotation Z (Yaw)** input pin of the **Add Relative Rotation** node.

25. Finally, connect the **Update** output pin of the timeline to the execution input pin of the **Add Relative Rotation** node.

Our graph should now look something like the following figure:

Figure 8.8 – The look of the Event Graph for our Blueprint

As a quick summary of what we've achieved so far, know that we've managed to create an animated value that drives the rotation of our model. The **Timeline** asset is a tool we can use to animate a given value, such as a float in this case, by specifying different values at different times. Hooked to the **Yaw** rotation value of the **Add Relative Rotation** node, it will make the camera swing both ways at a fixed period. It will first turn one way, stop, and then rotate in the other direction, thanks to the way we've defined the curve. On top of that, it will also constantly loop, giving the appearance of non-stop motion.

Now that we have the Blueprint in place, we can proceed to store the camera's view on **Render Target** and use it within a material to drive the appearance of the display of the TV screen present on the level. The last part that we need to put in place to realize this vision is the material itself – which, as we are about to see, is fairly simple!

26. Create a new material anywhere you like within the Content Browser, and give it an appropriate name. I've gone with **M_TV_WebCamDisplay**.

27. Next, open the Material Editor and drag and drop the previously created **Render Target** into the graph.

28. After that, connect the output of **Render Target** to the **Emissive Color** input pin of our material.

29. Drag and drop the webcam Blueprint into the level and place it wherever you think makes the most sense.

30. Finally, apply the material we created back in *step 26* to the **Element 0** material slot of the TV screen model present in the level.

After doing that, let's take a moment to review what your scene should look like (more or less, depending on where you've placed your webcam!) at this point:

Figure 8.9 – A look at the level, with the image captured from the webcam displayed on the screen

You should now be able to hit **Play** and see the image on the TV screen update as the webcam rotates. With all of that done, we can finally say: mission accomplished!

How it works...

Let's take a moment to review the logic we've implemented in this recipe. Simply put, all we did was set up a webcam that captured the scene and passed that information to a texture, which could then be reused inside a material. Applying that material to the 3D model of the screen gave us the effect we were after, closing the circle of the whole setup. With this bird's eye view of the problem, let's quickly take a look at the responsibilities held by each of the assets we've used to tackle it.

The first of them is the webcam – a Blueprint that incorporates both **Static Mesh Component** and **Scene Capture 2D**. We needed the first of those to give the user a visual representation of the actual webcam, whilst the second one is there to actually do the capturing work. This **Scene Capture 2D** component works alongside **Render Target** to store the information that it captures, so you can think of the component as the video recorder and **Render Target** as the storage media. Finally, the material is simply the vessel that takes the information stored in the texture and displays it back to the user. That's the whole idea!

See also

As promised, here's some more info about timelines, which aims to get you up to speed with this topic. To start, let's take a quick look at the following screenshot, where we can see the timeline graph:

Figure 8.10 – A look at the timeline graph

The screenshot is broken down into the following parts:

- **A:** The **+ Track** button contains a series of options that allow you to create different types of timelines. We've used one based on the float type, which allows us to specify several values and times. Other types can be selected, and you'll choose them whenever it makes sense for your project. For example, we have access to **Vector** tracks, useful for defining positions for objects; **Event** tracks, which provide us with a different execution pin that triggers at certain frames that we can define; and **Color** tracks, useful for animating colors. We can also select an external curve we might have in our Content Browser to use that instead.

- **B:** This part of the editor grants us access to several important properties that control the behavior of the curve, such as its length, whether it should loop, or whether it should be replicated over the network. One important setting is the **Use Last Keyframe** button, which automatically adjusts the length of the timeline to match the position of the last key we create. This often matches what we want, so make sure to check this box if needed!

- **C**: Located in this corner are important buttons for daily work with timelines. Things such as framing the graph so that we can visualize its contents or adjusting the values of the different keys can be located here. Something to note is the context-sensitive nature of certain panels, such as the **Time** and **Value** ones, which will only show if a certain key is selected.

- **D**: This is the main window of the Curve Editor, where we can create the different keys and give them values. If we right-click on the key frames themselves, we'll be able to modify the interpolation between them.

Having covered all of these different parts, make sure to play around a little bit with all of them in order to get comfortable and know what everything does! It will boost your confidence when working with timelines, which can be a bit confusing initially. One important setting I'd like to point out again is **Use Last Keyframe**, which if left unchecked, will force you to define the length of the animation. This is something easy to forget, so keep that in mind!

Highlighting interactive elements

So far, we've had the opportunity to work with various materials; some of them more traditional, as seen in previous chapters, and some others more disruptive, as we are starting to see. Continuing with that trend, we are now going to take a look at a different type of shader that we haven't yet tackled in this book.

I'm talking about post process materials – a type of asset that taps into the post-processing pipeline that Unreal has in place, enabling us to create some very cool and interesting effects. We'll explore one such example in the following pages, using it to highlight interactive elements within our games and apps. Let's get started!

Getting ready

You'll soon find out that the material we are about to create relies largely on math nodes and scripting, which means that there are not a lot of custom assets, such as textures or models, needed to tackle this recipe. As a consequence, it'll be easier than ever to apply the knowledge we acquire to any custom level that you might have already with you.

Having said that, the reliance on those custom math nodes stems from the fact that we are going to apply a mathematical technique that might not be easy to understand if we only focus on how it is applied. Seeing as the main body of the recipe is more practical in nature, I'd like to refer you to the *How it works…* section, where we'll explore the theory behind the effect we are going to apply in greater detail.

In any event, and as always, we'll be providing you with a map that you can use to test the functionality we are about to introduce. Its name is `08_03_Start`, and you can find it in the **Content** | **Levels** | *Chapter 8* folder.

How to do it...

As we said in the last few lines, we'll be creating a new material of the post-process type which will let us highlight certain key gameplay elements. Post-process materials are inherently different from the ones we've used so far, and the techniques we are about to implement are better understood when looking at the theory behind them. Don't be afraid though, as I know this can sound a bit daunting! But do remember to head over to the aforementioned section, as it contains some nifty information. With that said, let's dive right in:

1. Let's start by creating a new material, as this is what we'll be using to highlight the different gameplay elements in our scene. As for its name, I'm going to go with **M_EdgeOutline**, as that's what the shader is going to be doing – highlighting the edges of models. Double-click on it to enter the Material Editor.

2. With the main material node selected, let's focus on the **Details** panel, as there are a couple of things we need to change there. The first of them is **Material Domain** – select the **Post Process** option instead of the default **Surface** one.

3. Next, adjust the **Blendable Location** setting of the material. You can find this option within the **Post Process Material** section of the **Details** panel, a little bit further down from where we were in the previous step. Make sure to change that setting from the default **After Tonemapping** to the **Before Tonemapping** option. Refer to the following screenshot in case of doubt:

Figure 8.11 – The first two settings we need to tweak in our new material

The previous option is one that is available on this type of material, which gives us the ability to decide at which point the shader gets inserted into Unreal's post-processing pipeline. Make sure to check out the *See also* section for more information about this process!

With that done, we can now start to create the logic within the node graph. Since we'll want to outline the shape of certain objects, the first bit we'll have to calculate is the part that deals with the detection of the edges. This is done in a similar way to how many other image-editing programs do it, and it is based on a convolution operation – something we'll explore in the *How it works…* section.

4. Continue by creating five different **Constant 2 Vector** nodes; we'll use these to sample between different pixels in order to be able to detect the edges of our models.

    Given a known pixel, we'll want to compare it against its neighbors located above, to the right, to the left, and below it. That fact is going to determine which values we need to use for the new **Constant 2 Vector** nodes we have just created, which we'll assign next.

5. Next, assign the previous **Constant 2 Vector** nodes the following values: **(-1,0)**, **(1,0)**, **(0,-1)**, **(0,1)**, and **(0,0)**. This will let us sample the left, right, bottom, top, and central pixels respectively, as mentioned in the previous step.

    The next bit of code that we need to have in place to actually sample those pixels is going to be repeated multiple times, and we can use this as an excuse to use **Material Functions**. These assets are basically chunks of the material graph that can be reused inside any other material where we want to have that same functionality, without the need to copy-paste the same nodes over and over multiple times.

6. Head back to the Content Browser and create a new **Material Function**. This can be done by right-clicking inside the browser and looking under the **Materials** section. Name it **MF_PixelDepthCalculation**.

7. After that, double-click on the new function to access its material graph. Once in there, right-click and type Function Input, the name of the first node that we need to place. This will let us have an input in our function once we place it inside the main material we previously created.

8. With that new node selected, head over to the **Details** panel and change **Input Type** to **Function Input Vector 2**. You can also give a name under the **Input Name** section, such as **Offset**, as the values that get connected to that pin are going to be used for that. Whilst in there, check the **Use Preview Value as Default** checkbox as well.

    Using the **Function Input Vector 2** option for **Input Type** allows us to provide the **Material Function** with **Constant 2 Vector** type values, which is what we added to our material back in *step 4*.

9. Next, include a **Multiply** node and connect one of its input pins to the output of the previous **Function Input**.

10. Add a **Scene Texel Size** node and connect it to the other input pin of the previous **Multiply** node.

    Scene Texel Size is a handy node that enables us to consider the different resolutions under which our app might be running. This is especially important in the material that we are trying to create, as it is intrinsically tied to the resolution at which we want to run the app or game. Think about the following situation: if the size of the outline we want to render was to be fixed at a value of 4 pixels, the relative dimension of said effect would be much more obvious in a low-resolution display as opposed to on a high-resolution one, where 4 pixels occupy less relative space. **Scene Texel Size** will allow us to account for that.

11. With that done, go ahead and create a **Texture Coordinate** node above the previous ones we've created.

12. Include an **Add** node after it and connect its **A** input pin to the output of the previous **Texture Coordinate** node. As for pin **B**, connect it to the output of the **Multiply** node we created in *step 9*.

13. Add a **Scene Depth** node after **Add**, connect both, and wire its output to the existing **Output Result** node. The graph of the **Material Function** should look something like the following image:

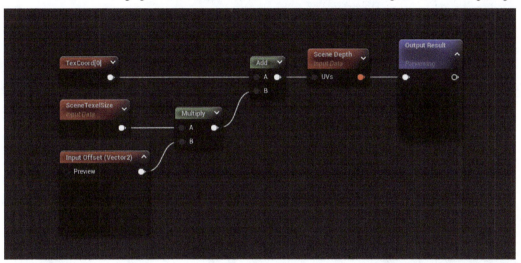

Figure 8.12 – A look at the material graph for the Material Function we've just created

> **Important note**
>
> Don't worry if you see a red warning that says that only transparent or post process materials can read from **Scene Depth**, as that's where we'll be using this function.

Feel free to close the graph for the **Material Function** once you are done placing the last of those nodes. We are now in a position where we can reuse all of that logic back within the main material we previously created. That means that we only need to include a small function call to perform the preceding bit of code, saving us from copy-pasting it multiple times across the graph. Even better, we can reuse that code in other materials should we need to. Pretty nifty!

Having said that, I'd like to take a moment to make sure that we understand what the previous function is doing. In essence, the function will give us the depth value of the pixel that we want to analyze. We start by giving it the relative location of the pixel that we want to study thanks to the **Function Input** node we added to the graph. Said value is then adjusted thanks to the **Screen Texel Size** node, which modulates the incoming value taking into account the screen resolution. This is then added to the **Texture Coordinate** node so that we can locate the pixel on the screen, and then read its depth value thanks to the **Scene Depth** node, completing the analysis. Seeing as we now understand what the function is doing, let's go ahead and implement it:

14. With the previous **Material Function** sorted, go back to the material graph of the **M_EdgeOutline** material and create a **Material Function Call** node inside it.

15. With the new node selected, head to the **Details** panel and assign the previously created **Material Function (MF_PixelDepthCalculations)** in the selection slot.

16. After that, copy/paste **Material Function Call** four times, positioning each instance at the side of the **Constant 2 Vector** nodes we created in *step 4*.

17. Next, copy **Material Function Call** a fifth time and place it below the other four. We'll need to connect this one to the output of the **Constant 2 Vector** node to which we assigned the value of **(0,0)** back in *step 5*.

18. Create a **Scalar Parameter** node and place it before the **Constant 2 Vector** nodes. This will allow us to control the width of the outline effect, so let's give it an appropriate name. I've gone with **Line Width** and set its initial value to **5**.

19. Create four **Multiply** nodes and place them between each of the first four **Constant 2 Vector** nodes and the **Material Function Calls**. Connect pin **A** to the output of the **Constant 2 Vector** nodes, and the output to the input of the **Material Function Calls**.

20. Connect every input pin **B** of the **Multiply** nodes to the output of the **Scalar Parameter** node we just created that controls the line width.

Seeing as we have created quite a few nodes, let's take a quick look at the state of the graph:

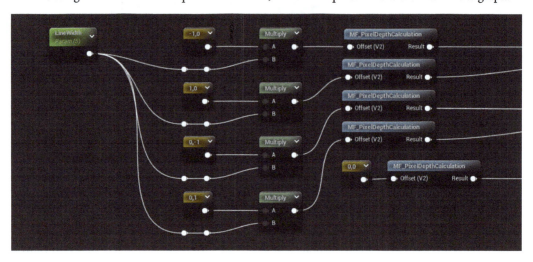

Figure 8.13 – A look at the previous set of nodes we have created

So far, we've managed to provide our depth calculation function with the values that it has to analyze: we told it to look to the pixel to the left, to the right, below, and above thanks to the **Constant 2 Vector** nodes we used, but we adjusted the actual distance that it has to look in those directions thanks to the **Scalar Parameter** node called **Line Width**.

We finally fed that information to our **Material Functions**, which are now ready to give us some results. The next steps will involve working with those values, so let's tackle that next:

21. Continue by creating an **Add** node and placing it after the first two **Material Functions**, whilst using those as the input.

22. Repeat the previous operation by creating another **Add** node, but this time, use the remaining **Material Functions** as the inputs.

23. After that, add a **Multiply** node right after the fifth **Material Function**, and connect the output of that one to the **A** input pint.

24. As for input pin **B**, we'll need a **Constant** node connected to it, so create one and assign it a value of -**4**.

    The reason behind using that number under **Constant** is directly related to the convolution operation we are performing, so be sure to head over to the *How it works...* section to learn more about that.

25. Next, create another **Add** node and connect its two input pins to the output of the two **Add** nodes we created in *steps 21* and *22*.

26. After that, create another **Add** node and connect its **A** input pin to the output of the previous **Add** node we created in *step 25*. As for input pin **B**, connect it to the output of the **Multiply** node we created in *step 23*.

27. With that done, create an **Abs** node, something that will give us the absolute value of anything we connect to it. We need this to ensure that every value from now on is positive.

    Seeing as we've created quite a few of the same nodes, let's quickly review the graph to ensure that we don't get lost:

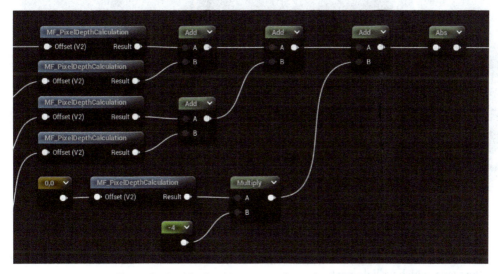

Figure 8.14 – A look at the new nodes we've just created

The reason why we've used so many **Add** nodes is that we want to sum all of the depth values for the pixels around the current one we are analyzing. Doing that will allow us to know whether the pixels surrounding the one we are studying present big changes in their depth values – if all the pixels share similar depth results, the sum of those calculations will tend to cancel out or be close to zero; if the opposite is true and certain pixels present widely different depth figures, then the result of the sum will be something other than zero. We'll use this information to tackle the next few steps, where we'll perform that comparison.

28. Continue by creating an **If** node, and place it after the previous **Abs** node, connecting its **A** input pin to the output of that. This will create a conditional branch that we will populate next.

29. Next, create a **Scalar Parameter** node, assign a value of 4 to it, and connect its output to input pin **B** of the previous **If** node. We'll use this value to compare against the data in input pin **A**, which is the value of the depth calculation we performed earlier. The number that we choose here will determine the jump in depth value that the **If** node will use in order to distinguish between the areas that it needs to highlight and the ones that it doesn't. As such, let's name the **Scalar Parameter** node something such as **Threshold**.

30. After that, create two **Constant** nodes and connect the first to the **A > B** input pin of the **If** node. As for the second, connect that one to the **A < B** pin. The first one should be given a value of **1**, and the second one a value of **0**.

Seeing as we've done quite a bit, let's pause for a moment and take a look at the last few nodes we've created, as well as a quick look at how the scene would look if we were to apply this material as a post process effect at this stage:

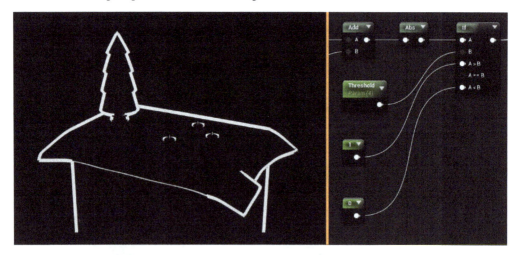

Figure 8.15 – A look at the last few nodes and the current state of the material

Now that we've completed the previous steps, we can finally say that we have an edge-detection system in place. You can check this out if you apply the material to the post process volume in the scene, which we'll do next:

31. In order for the material to work at all, connect the output of the **If** node to the **Emissive Color** input on the main material node.

32. After that, exit the Material Editor for a moment, go back to the main level, and select the **Post Process Volume** actor present there. Looking at its **Details** panel, scroll down to the **Rendering Features** section and expand the category named **Post Process Materials**.

33. Once in there, click on the + icon next to the **Array** word.

34. In the drop-down menu that will appear after clicking the + button, select **Asset**.

35. With the previous step implemented, select the material we've created to see how it is impacting the scene.

    At this point you should see a black-and-white image in front of you, displaying most of the edges that we have in our scene, as we saw in *Figure 8.15*. We'll use this binary output as a mask to create a highlight effect in some of the objects in our scene. Let's do that now!

36. Go back to the post-process material and create a **Lerp** node. Connect its **Alpha** input pin to the output of the **If** node created in *step 28*.

37. After that, create a **Scene Texture** node and connect it to pin **A** of **Lerp**.

38. With this new asset selected, look at its **Details** panel and select **Post Process Input 0** under **Scene Texture Id**. This is a pass that we can use to get hold of the scene color, which means how the base scene looks without this new post-process material being applied.

39. Next, add a **Vector Parameter** node and name it something such as **Outline Color**, as this is what it's going to be used for. Assign the color that you want to it.

40. Include a **Make Float 4** node and connect each of its input pins to the **R**, **G**, **B**, and **A** channels of the previous **Vector Parameter**.

41. Get a hold of a **Multiply** node, place it right after the previous **Make Float 4** node, and connect its **A** input pin to the output of the **Make Float 4** node.

42. Next, introduce a **Scalar Parameter** node and name it **Outline Intensity** since it's going to work as a multiplier for the chosen outline's color intensity. Start giving it a value of something such as **5**, and connect it to the free available input pin of the previous **Multiply** node.

43. Connect the **Multiply** node from *step 40* to the **B** input pin of the **Lerp** node created in *step 36*.

The previous set of nodes can be seen in the next image. The **Outline Color Vector Parameter** controls the color of the effect, whilst **Outline Intensity** controls the brightness:

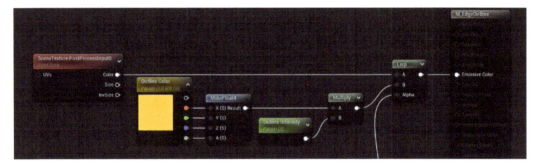

Figure 8.16 – A look at the last few nodes we have created

Completing these steps and applying the material like that to the scene will give us a level where every object has a noticeable outline effect. While we are definitely on the right track, this is not exactly what we want – we need a way to control which objects get affected and which ones don't. This is going to be the last thing we tackle in this material, so let's get to it!

44. Create a new **Lerp** node and place it after everything we've done before. Connect its **B** input pin to the output of the previous **Lerp** node created in *step 36*.

45. Copy the **Scene Texture** node we created in *step 37* and paste it again close to the **A** input pin of the new **Lerp** node, as that's where it's going to be connected. Wire it up now!

46. After that, create another **Scene Texture**, but select the **Custom Depth** option as **Scene Texture Id**.

47. Next, connect the previous **Scene Texture** to a new **Multiply** node that we can now create, and choose a low value for the **B** input pin – **0.01** works well in our level.

48. Throw a **Frac** node after the previous **Multiply** node. This asset outputs the fractional portion of the original values we feed it, useful in this case, as we really need a very low value and we wouldn't be able to type it without Unreal rounding up to 0.

49. Finally, connect the output of the previous **Frac** node to the **Alpha** input pin of the **Lerp** node created in *step 44* and connect that to the **Emissive Color** input pin of our material.

In the previous steps, we blended the original scene (which we get as a render pass through the **Post Process Input 0 Scene Texture** node) and the outline effect according to the **Custom Depth** of the scene. **Custom Depth** is a property that has to be activated for each model we have in our level, allowing us to determine which meshes we want to see. This can be done in many different ways, and it lends itself to being tweaked interactively via Blueprint commands. The final step we need to take is to actually set up the objects that we want to highlight!

50. Select the object that you want to adjust and scroll to the **Rendering** section of its **Details** panel.

51. In there, we should be able to see a setting called **Render Custom Depth Pass**. Check the checkbox next to it and see the magic happen!

With all of that done, our scene should now look something like the following figure:

Figure 8.17 – A final look at the scene

And that's it! You can control the **Custom Depth** property of a **Static Mesh Component** through Blueprint functionality, by calling a reference to the specific objects through the **Level Blueprint**, for example. This is something that can enhance any of your projects, so make sure to give it a go!

How it works...

The edge-detection logic applied in this recipe is based on a mathematical operation called convolution. That being the case, it's better if we tackle what that is at this point instead of bogging you down with the details while you are trying to complete the recipe.

First of all, convolution is the name given to an operation performed on two groups of numbers aiming to produce a single third one. In image-based operations, those two groups of numbers can be broken down like this:

- We have a grid of defined values, known as the kernel, acting as the first group.

- The second one is going to be the actual values of the corresponding pixel to the previous grid. The size of that grid is often a 3x3 pixel matrix, just a little thing where the center pixel is the protagonist.

The way we use it to calculate the third value – that is, the result – is by multiplying the values of the kernel by the values of the pixels underneath and adding all of them to produce a final number that gets assigned to the center pixel. By moving and applying this multiplication across the image, we get a final result that we can use for different purposes, depending on the values of the kernel grid: to sharpen edges, to blur them, or to actually perform edge detection as we did in this recipe. The kernel values we can use in that case are the ones we saw in this recipe, which follow the Laplacian edge-detection system, and which are the following:

Figure 8.18 – A look at the kernel values used in this recipe, as seen in the purple matrix

The whole concept was performed on this recipe in the initial stages, roughly from *steps 4* to *30*. We did it in different places—first, we got the left, right, upper, and lower pixels of the grid by creating the **Constant2Vector** nodes and using those as input for the **Material Functions**, where we multiplied the values of the kernel by the pixel underneath. After that, we got the value of the center pixel by multiplying that by the value of the kernel, the **-4 Constant** node we created in *step 24*. Finally, we added everything together to get the right result. There was no need for us to calculate the pixels in the corner as the kernel uses values of **0** in those regions.

See also

Before we finish, I'd like to leave you with some additional thoughts that can be helpful to understand how post process materials and the post-process pipeline work in Unreal Engine. So far, we've had the opportunity to become familiar with many different types of shaders – opaque ones such as wood or concrete, translucent ones such as glass or water, and things in between, such as wax. One of the ways we can define them all is through their interaction with the lights in our scene, as we can sort them depending on how that happens. This is because those types of materials are meant to be applied to the objects that live within our levels, just so that the renderer can calculate the final look of our scenes.

This is not true for post-process materials, because they are not meant to be applied to 3D models. They are there to be inserted into the engine's post-process pipeline, contributing to the scene in a different way, mainly by operating on any of the different render passes. Unreal, as a render, stores the information of the scene in multiple layers, such as the depth of field, the temporal anti-aliasing, the eye adaptation, and the tone mapper. Having access to those multiple render passes can be very helpful in order to achieve certain effects, just like we've done in the present recipe. This is very powerful, as we are affecting how the engine works at a fundamental level – giving us the ability to create multiple visual styles and rendering effects. The possibilities are almost endless!

Creating snow on top of objects using layered materials

Introduced in a later version of Unreal Engine 4, material layers are a technique that takes different types of materials and puts them together using an easy-to-use interface. The layering mechanism comes with no additional pixel shader instructions, so relying on this technique doesn't hinder our apps or games' performance.

To see just how easy it is to put this together, we are going to see how we can use this concept to add snow to the top of a material based on the world. Let's see how next!

Getting ready

Everything we'll use in this recipe is part of the engine or provided as part of the Starter Content, so make sure that you have that in your project. And as always, feel free to open the level called `08_04_Start` if you want to continue using the same assets I'll employ. See you in a sec!

How to do it...

Creating a material layer is going to be very easy, so let's take care of that next:

1. Start by right-clicking anywhere inside the Content Browser, and select the **Material Layer** option located under the **Materials** section.

2. As for its name, choose whatever you think works best – I'll go with **Layer_Base** for this particular instance. Double-click on the newly created material to open up the Material Editor.

   This editor should look fairly familiar to you if you've used **Material Functions** before, the only difference being that on the right-hand side, there's a new node called **Output Material Attributes**. This result is what will be used within our material later on.

3. To the left of the **Input Material Attributes** node, right-click and create a **Make Material Attributes** node by searching for its name and selecting it.

   This node should look similar to the other materials we created previously. We can create this material just like any of the ones we've created previously, but for the sake of simplicity, we will only use the **Base Color** and **Normal** channels.

4. Next, navigate to the **Starter Content | Textures** folder within the Content Browser. From there, select the **Diffuse** and **Normal** textures to bring into the **Material Layer** editor, and drag and drop them into there. For this example, I used the assets named **T_Brick_Clay_New_D** and **T_Brick_Clay_New_N**.

5. Connect the textures to the **Base Color** and **Normal** channels of the **Make Material Attributes** node.

6. After that, connect the output of the **Make Material Attributes** node to the **Preview** input of the **Input Material Attributes (Material Attributes)** node.

7. We want to be able to support having other objects covered in snow easily, so right-click on both **Texture Sample** nodes and select the **Convert to Parameter** option. Name the **Diffuse** texture **Base Diffuse** and the **Normal** texture **Base Normal**. If all went well, you should have something that looks like the following figure:

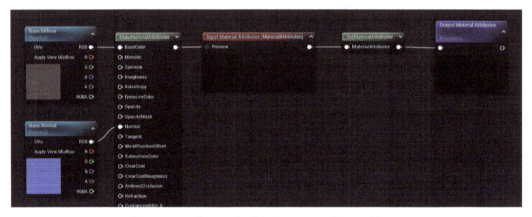

Figure 8.19 – The graph of the Layer_Base Material Layer

8. Click on the **Apply** and **Save** buttons from the editor and exit this layer.

   Now that we have our base layer, we need another material layer for the snow.

9. In the same folder as your **Layer_Base** material layer, right-click on the Content Browser and create another such asset, naming it **Layer_Snow** this time. Double-click on it to enter the **Material Layer** editor.

10. This material will be built in the same way as the previous material layer, so repeat *steps 3* to *7* within the new asset. Instead of using the brick texture, we will use something that looks like snow. In our case, the Starter Content does not include a specific texture that covers that material, but it does have another one called **T_Concrete_Poured_D**, which looks pretty close. Let's use that one and its **Normal** map companion in this material layer.

    Now that we have two material layers, it is time to put them to good use. Material layers cannot be applied to an object directly, so we will need to create a material that uses the material layers we have created instead.

11. Create a material and name it something along the lines of **M_SnowLayeredMaterial**. Double-click on it to enter the Material Editor.

12. To the left of the default channels, right-click and add a **Material Attribute Layers** node. This is the element that we'll use to combine the two layers we have created.

    Since this is a standard material, you'll notice that the main material node contains all of the traditional channels that we've used throughout this book – the **Base Color** property, **Roughness**, the **Normal** channel, **Emissive**, and so on. Working with material layers means that we define those properties in the material layer assets themselves, so we have no use for them in the main material used to combine them. For that reason, we'll need to adjust the look of the main material node, something we'll do next:

13. From the **Details** tab of the material, scroll down to the **Use Material Attributes** property and check it. You'll see all of the standard inputs combined into one property named **Material Attributes**.

14. Next, connect the output of **Layer Stack** to the **Material Attributes** property of the **M_SnowLayeredMaterial** node.

15. Now that we have the connections created, we can start adding to our **Material Attribute Layers** node. To do so, select that node from the **Details** tab and click the arrow next to the **Default Layers** property. Expand the **Background** subsection and, from the **Layer Asset** dropdown, select our **Layer_Base** material layer.

16. Afterward, click on the + icon next to **Default Layers** to add a new layer to the stack. Select and expand the newly created **Layer 1** section and set its **Layer Asset** property to the **Layer_Snow Material Function**.

    We should now have something like this before us:

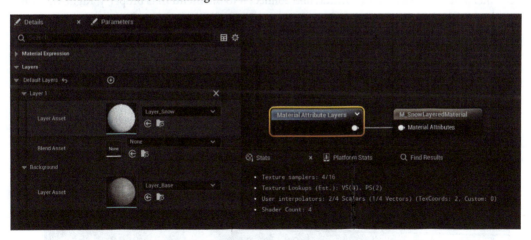

Figure 8.20 – A look at the Material Attribute Layers node and its setup

You'll notice that the snow is now covering the original background. Layers work in a similar way to how image-editing software packages such as Photoshop handle them, with new layers being drawn on top of each other.

To control that blending process, we have access to the **Blend Asset** property, which defines what parts of **Layer 1** should be drawn on top of **Background**, so we will implement that next.

17. Hit **Apply** and **Save**. Then, close the editor.

18. Right-click inside the Content Browser to bring up the contextual menu and look inside the **Materials** section to find a **Material Layer Blend** asset, the next item we want to create.

19. Name it whatever you want—I'll go with **Snow_Blend** for this particular instance. Double-click on the new asset to open the Material Editor for it.

As you can see, we get two default nodes by default, which can be seen on the left side of the previous screenshot: **Input Top Layer** (which we'll use for the snow) and **Input Bottom Layer** (the bricks). The output of those nodes are being passed to a third one called **Blend Material Attributes**. The **Alpha** property there dictates how the two layers should be mixed, much like in a **Lerp** node. For this particular recipe, we want to blend the two materials we've created based on the alignment of the vertices based on the world: we want the snow to appear on the top. Let's take care of that next.

20. Somewhere below the two input nodes, add a **World Aligned Blend** function. Then, connect the **w/Vertex Normals** output to the **Alpha** property of the **Blend Material Attributes** node.

21. To the left of the **World Aligned Blend** node, create a **Constant** node and connect it to the **Blend Sharpness (S)** input. As for the value, set it to **10**.

22. Afterward, create another **Constant** and connect it to the **Blend Bias** property. Right-click on this node, select the **Convert to Parameter** option, and name it **Snow Bias**.

Seeing as we have created a few nodes, let's look at the Material Editor before moving forward:

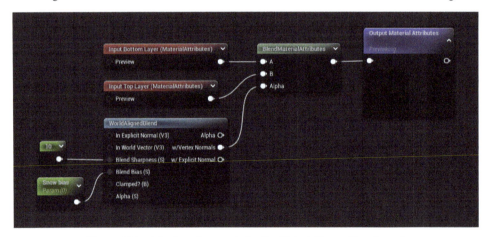

Figure 8.21 – A look at the graph for the Material Layer Blend asset

23. After that, save your **Material Layer Blend** asset and exit its editor.

24. Back in **M_SnowLayeredMaterial**, select the **Material Attribute Layers** node and set the **Blend Asset** property to be the **Snow_Blend** asset we have just created. If all goes well, you should see the top half of the preview image covered with our snow layer!

25. To make this easier to see with traditional assets, go ahead and create a scene with a more complex mesh, such as **SM_MatPreviewMesh_02**, and drag and drop our material onto it.

    With all of those changes in place, this is the material that we should see:

Figure 8.22 – The material is applied to the model in the scene

Now that we have the basic material, we can instantiate it very easily, as we'll see next.

26. Right-click on **M_SnowLayeredMaterial** and select the **Create Material Instance** option. Name this new asset **MI_SnowLayeredGold** and double-click on it to open the editor.

27. Unlike before, you'll notice that the menu has been streamlined. Navigate to the **Layer Parameters** panel and open the **Background** property.

28. After that, click on the arrow next to **Layer Asset** and then **Texture Parameters**. There, you should see our previously created parameters.

29. Check the checkboxes next to the **Base Diffuse** and **Base Normal** parameters. You should be able to assign new texture values to those, so go ahead and choose a gold-like texture. The Start Content includes one asset that can work well, named T_Metal_Gold_D and T_Metal_Gold_N.

30. After completing the previous step, create a duplicate of the **Material Preview Mesh** actor present in the level and assign the new Material Instance we've created to that model.

31. With that done, go back to the Material Editor and, under **Layer 1**, expand the **Blend Asset** and **Scalar Parameters** arrows, and then play with the **Snow Bias** property. Note that **Material Preview Mesh** will be modified in real time.

We should now see a material akin to the one depicted in the following figure:

Figure 8.23 – A look at the new Material Instance we've created

This idea can also be easily expanded so that you can work with any material that you would want to tweak often, such as color schemes for character skins.

How it works...

A material layer acts in a similar way to a **Material Function** but has the ability to create children, just like when you create Material Instances.

The **Material Layer Blend** node provides instructions on how to draw the different layers in the material on top of each other. In our case, we used the result of the **World Aligned Blend** node to determine where the snow layer should be drawn, which is on the top of the surface it is placed on.

When we create an instance of the parent material, you'll notice that, when opened, the material displays the **Layer Parameters** menu instead of the default editor. This menu only shows aspects of the material that are parameters, with the idea of making it very simple to add and/or modify aspects of the shader. It's important to note that you still can access the **Details** tab as well if that is what you'd prefer.

See also

For more information on layered materials, check out the following link: `https://docs.unrealengine.com/5.1/en-US/layering-materials-in-unreal-engine/`

Changing a sunny scene to a snowy one with a parameter collection asset

One common issue faced by artists is making multiple materials change at the same time. We've already learned about parameters and how we can change them at runtime using Blueprints, in the recipe called *Orienting ourselves with a compass* back in *Chapter 5*. Having said so, we had to change each material parameter individually, which isn't ideal when dealing with a large number of materials.

Material Parameter Collection actors allow us to create special variables that can be referenced across multiple materials and then modified either inside the editor or at runtime, through both Blueprints and/or C++. To see just how easily those can be used, we'll see how we can make multiple materials change at the same time in this recipe, giving the impression that it has snowed in the level with which we'll be working.

Getting ready

Material Parameter Collections are a type of asset that, as their name implies, let us adjust several different parameters or variables that we create within it. As a consequence, this will be a data-oriented recipe, where we won't need much in terms of external assets in order to execute it: no custom textures or models are needed in this one! The few external references will be locked to assets present in the Starter Content package, so no need to look elsewhere for those.

One thing you'll want to have is a level containing a few different models where we can apply different versions of the material we'll be creating. We've created one such map for you, named 08_05_Start, which is part of the project provided alongside this book.

How to do it...

Keeping things simple, let's start the recipe by creating and adjusting the asset that we want to demonstrate, **Material Parameter Collection**:

1. Start by right-clicking inside of the Content Browser and, under the **Materials** section, select the **Material Parameter Collection** option. Name this new asset whatever you wish; I went with a simple name, **RustCollection**.

2. Double-click on the new collection asset to open a window that will allow you to adjust its properties. You'll see two main categories there: **Scalar** and **Vector Parameters**.

3. After that, click on the + button next to the **Scalar Parameters** option. Once a new entry gets created, expand the **Index [0]** option to discover two properties there – **Default Value** and **Parameter Name**. Set the latter to **Rust Amount**.

4. Hit the **Save** button and return to Unreal Editor.

As this is the first time that we are dealing with **Material Parameter Collection**, here is what the previous options should have looked like:

Figure 8.24 – A look at the setup we've just created

With that out of the way, we could now start to use our new asset inside a material, or even inside the **Material Layer Blend** asset we created in the previous recipe. For simplicity purposes, let's focus on a material this time around.

5. Continue by creating a new material where we can demonstrate this effect, and double-click on it to open the Material Editor. I'll name the new asset simply **M_BasicMetallicMaterial**.

6. Once inside the Material Editor, start by creating a **Vector Parameter** node and setting its name to **Base Color**. This is one of the properties that will drive that attribute of the material.

7. Next, create a **Texture Sample** node and assign the **T_Metal_Rust_D** texture to it. This asset is part of the Starter Content, so make sure that you've added it to your project in case you haven't done so yet.

    With those two nodes in place, the idea now will be to blend between the two according to the **Rust Amount** value we defined in the **Material Parameter Collection** asset. Let's do that next.

8. Continue by creating a **Lerp** node and connecting the two previous ones to its **A** and **B** input pins. Input pin **A** should be connected to the **Vector Parameter** node we named **Base Color**, whilst input pin **B** should be wired to the **Texture Sample** node.

9. After that, create a **Collection Parameter** node by right-clicking on the material graph and typing its name. Once created, select the node and, from the **Details** tab, set the **Collection** property to be the **RustCollection** asset we created in *step 1*.

10. With the new node in place, connect its output to the **Alpha** input pin of the previous **Lerp**.

11. Finally, connect the output of the **Lerp** node to the **Base Color** property of the material.

    Seeing as we are trying to create a basic metallic material, let's also adjust its **Metallic** property next.

12. Create a couple of **Constant** nodes and assign a value of **1** and **0** to them. One will make the material fully metallic, which we want to see when there's no rust, whilst the other will make it non-metal, similar to what happens when rust accumulates on a material.

13. Add a new **Lerp** node and connect the **Constant** node that has a value of **1** to its **A** input pin. Connect the other one to its **B** input pin.

14. As for **Alpha**, connect it to the output of the **Collection Parameter** node, just like we did back in *step 10*.

15. After that, connect the output of this **Lerp** to the **Metallic** input pin of our material.

Completing the previous steps will have left us with a material that we can instantiate and use to demonstrate the parameter collection asset. Before doing that, let's take a moment to review the material graph we should have in front of us:

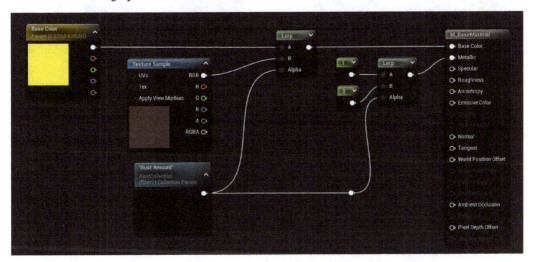

Figure 8.25 – A look at the material graph

As mentioned in the previous paragraph, let's now create several instances of this material so that we can demonstrate the effect:

16. Go back to the Content Browser and right-click on the material we've just created. Select the **Create Material Instance** option and create four instances.

17. After that, adjust each new Material Instance by exposing the **Base Color** variable and modifying its default value. Seeing as we have four instances, assign a different color to each of them, and apply them to the different models that should be on your level.

18. With that done, double-click on the **Material Parameter Collection** asset and adjust the **Default Value** property of the **Rust Amount Scalar Parameter** node.

Assigning different values to the previous parameter should see all of the materials that you've just applied to the scene change at the same time, demonstrating the effect we were trying to achieve in this recipe, as seen in the following figure. Make sure to get a hold of this effect when you need to enact this type of change across your scenes!

Figure 8.26 – A look at the same materials after changing the Rust
Amount setting (0.5 on the left and 1 on the right)

How it works...

As we've seen in this recipe, material collections allow us to modify the properties of multiple materials at once. This is great both in terms of usability and performance: it's much easier and more efficient to tweak a single **Material Parameter Collection** asset than it is to adjust multiple material parameters. Dealing with a single centralized item instead of multiple individual ones also reduces the rendering overhead, allowing for smoother gameplay experiences.

With that in mind, there are also some limitations that we need to be aware of when deploying this technique. For instance, materials are limited to referencing two **Material Parameter Collection** assets, so we need to make sure that we don't cross that threshold. Furthermore, each of those collections can store up to 1,024 **Scalar Parameter** and 1,024 **Vector Parameter** entries. These are very big numbers, which we are unlikely to surpass, but it is a limitation nonetheless.

One issue that we are more likely to face is the need to recompile shaders that are affected by modified parameter collections. This is something that happens every time we change the number of parameters within a given collection: doing so will trigger the recompile process, which can take a bit of time depending on your hardware and the number of materials that are affected. In light of this, one solution is to create more parameters than strictly needed, giving us a bit of headroom in case we ever want to go back and add more settings.

See also

If you want a bit more info regarding the **Material Parameter Collection** asset, please check the following link:

```
https://docs.unrealengine.com/5.1/en-US/using-material-parameter-
collections-in-unreal-engine/
```

Moving between seasons using curve atlases

In our journey through Unreal's material repertoire, we sometimes find ourselves using different types of parameters. Just in the last recipe, we looked at how we could garner the strength of both the **Scalar** and **Vector** types through the use of a **Material Parameter Collection** asset. Having said so, the two types of variables available there are best suited to drive certain fixed properties, such as the **Roughness** value used by a shader or the color needed to be displayed on a specific material.

More nuanced types of data can also be parameterized in Unreal thanks to the use of **curves**. These are special types of assets that can hold different values, meant to give a range of options as opposed to a single one. In this recipe, we'll take advantage of that characteristic by creating a material that can sample different values from a single curve. We'll use that feature to create a material that can display the most common colors shown on tree leaves throughout the year, creating the illusion of a shader that can adapt to seasonal changes.

Getting ready

Just like in the previous recipe, this one will also rely on assets that can be created directly within Unreal. We will use some other resources available through the Starter Content, so make sure to include it in your project if you haven't done so already.

If working with the Unreal Engine project provided alongside this book, make sure to open the level called `08_06_Start` to follow along using the same examples you'll see me employing. With that out of the way, let's jump right to it!

How to do it...

Seeing as the star of this recipe is going to be curves, let's start by creating one such asset:

1. Start by navigating to the Content Browser and creating a **Curve** asset. You can do this by right-clicking and looking inside the **Miscellaneous** category.

2. You'll be prompted to choose the type of curve that you want to create immediately after completing the previous step: from the **Pick Curve Class** window, select the **Curve Linear Color** type.

3. As with every new asset that we create, we'll need to give it a name. Let's go with **Summer Curve** this time around, as we'll use it to drive the colors that our models should have during summertime later.

4. With that out of the way, let's continue by double-clicking on the new asset to open its editor. This is what that window should look like at this point:

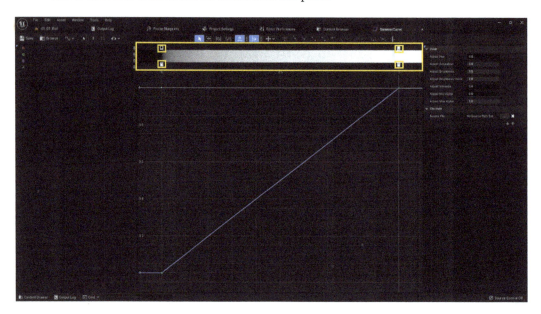

Figure 8.27 – A look at the Curve Editor

> **Important note**
>
> If this is the first time you work with the Curve Editor, make sure to head over to the *See also* section to learn how to operate it.

The area where we need to focus is going to be **Curve Gradient Result**, which I've highlighted in *Figure 8.27*. This is the black-and-white gradient that appears toward the top of the editor, currently displaying four keys in it: two at the top and two at the bottom. The bottom keys adjust the **Alpha** values, whereas the top ones affect the color. The next bit we are going to do is create some values for this curve, making it look greener, like the summer tones shown by tree leaves in the northern hemisphere.

5. Double-click on the black key (the one currently located at the top left corner of the gradient) to open up the **Color Picker** menu. Once in there, click on an intense green value, and see how the gradient has been updated.

6. Next, add more points to the curve by right-clicking on the graph and selecting the **Add key** option. Assign other similar green values, making the curve look something like the following figure:

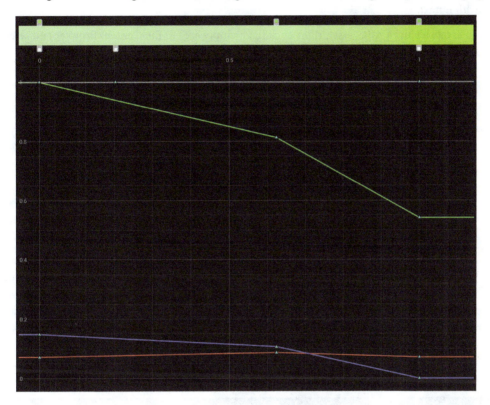

Figure 8.28 – A look at the curve we've just created

7. After that, click on the **Save** button and return to Unreal Editor.

Doing this will have granted us a curve that contains several tones and can be used to represent the summer colors usually seen in northern hemisphere vegetation. Just as we've set up this one, we'll proceed to create three more that depict the tonalities seen in the other three seasons.

8. Continue by creating three new curves of the **Linear Color** type and naming them something such as **SpringCurve**, **FallCurve**, and **WinterCurve**. You can replicate *steps 1* to *3* to achieve that goal.

9. After that, adjust the internal values to reflect the tones seen in those seasons. You can use warmer colors for the curve controlling the fall season, white tones for the one controlling winter, and more vibrant values for the one representing spring.

As an alternative to creating those three curves, you can use the ones contained in the following folder of the Unreal Engine project provided alongside this book: **Content | Assets | Chapter08 | 08_06**.

Now that we have the curves in place, the next few steps are going to deal with the creation of an asset that we can use to sample between them. Its name is **Curve Atlas**, and we'll deal with it next.

10. To create a **Curve Atlas** asset, right-click in an empty area of the Content Browser and select the **Curve Atlas** option within the **Miscellaneous** section.

11. Next, give it a name (I've gone with **Season Atlas** this time around) and double-click on it to open the editor.

12. Once inside the editor, look at the **Details** tab and focus on the **Gradient Curves** property. Whilst there, click on the + button to add an element to that array, and set that item to the first curve (**SpringCurve**).

13. Continuing from this, continue clicking the + button to add the next curve (**SummerCurve**), and then a couple more times to add the **FallCurve** and **WinterCurve** assets.

14. Uncheck the **Square Resolution** checkbox and set the **Texture Height** property to **4**.

15. Hit the **Save** button and return to Unreal Editor.

As you've seen, we set the **Texture Height** property to a value of **4**, something that matches the number of curves contained in the atlas, whilst leaving **Texture Width** set to its default value of **256**. This is important in that we remove the empty space not occupied by the curves in the vertical axis while leaving enough horizontal resolution for the gradients to be correctly represented. The horizontal resolution is a parameter that you might need to play around with: simple curves can be represented with fewer pixels, while complex ones might need more. In any event, make sure that both the height and the width are multiples of two. With all of those changes in place, we should be looking at a **Curve Atlas** asset similar to the following one:

Figure 8.29 – A look at the setup of the new Curve Atlas asset

Now that we have a **Curve Atlas** asset, we can create a material that uses it:

16. Continue by creating a new material and naming it something akin to **M_SeasonGround**, and double-click on it to open the Material Editor.

17. Once inside the editor, create a new **Texture Sample** node and assign the **T_Ground_Moss_D** texture to it. This is an asset available in the Starter Content.

18. Next, create a **Desaturation** node right after the previous **Texture Sample** node.

19. With the last node selected, look for the **Luminance Factors** property inside the **Details** panel and change the **R, G,** and **B** values to **1**.

20. After that, connect the **RGB** output pin from **Texture Sample** node to the input of the **Desaturation** node, and connect the output of **Desaturation** to the **Base Color** property in our main material node.

    Implementing the previous set of steps has effectively removed the color from the texture we chose to use, giving us a black-and-white version of it. We've done that in order to use the colors stored in the **Curve Atlas** asset we created earlier, something we'll implement next.

21. Right-click in the graph and add a **Curve Atlas Row Parameter** node.

22. You'll be prompted to assign a name to the new node as soon as you try to create it. I've used the word **Season** this time around.

23. With the new node selected, look at the **Details** tab and set the **Atlas** parameter to be the **Curve Atlas** asset we created back in *step 10*, the one we named **Season Atlas**. As for the **Curve** property, set it to be the **SpringCurve** asset.

24. With the previous changes in place. Create a **Blend_Overlay** node to the right of the previous ones we've placed.

25. Next, connect the output of the **Desaturation** node to the **Base (V3)** property of the **Blend_Overlay** node, and connect the output pin of the **Season** node to the **Blend (V3)** property of the **Blend_Overlay** node.

26. Lastly, connect the output of the **Blend_Overlay** node to the **Base Color** input of our material. The material graph should now look something like this:

Figure 8.30 – A look at the graph of the material we have just created

We are now in a position where we have a material that has access to the **Curve Atlas** asset we've created, but one that can't change the curve that it is currently using. That's something that we want to change, and we'll rely on blueprints to do that, as we'll see next:

27. Let's continue by opening the **Level Blueprint** for your level.

28. Once inside the graph, grab a reference to the model to which you want to apply the material. You can do this by dragging the name of the object from the **Outliner** panel and dropping it inside the graph.

29. With the reference to the object in place, drag a wire out of its only pin and type `Create Dynamic Material Instance` to add one such node to the graph. You'll notice that this won't be the only node created when performing that action, as another one called **Static Mesh Component** will also be automatically created. That one is an intermediary node needed to complete this operation.

30. As you'll soon see, the new node contains a few parameters that we can tweak. The first one we'll focus on will be **Source Material**: set that property to be the material we created in this recipe, **M_SeasonGround**.

31. To the right of the **Create Dynamic Material Instance** node, create a **Set Scalar Parameter Value** node. Ensure that the **Context Sensitive** option is disabled and use the version that has **Target** set to **Material Instance Dynamic**.

> Tip
>
> You can check the target of a given function before you create it by hovering over its name. You'll see several functions appear on the contextual menu once you start typing their name – simply hover over them to see what their target is.

32. Under **Parameter Name**, type the name of the parameter we created in *steps 21* and *22* – this should be called **Season** if you've used the same name I used. Under **Value**, type **0.5**.

33. Finally, connect the **Return Value** output pin of the **Create Dynamic Material Instance** node to the **Target** property of the **Set Scalar Parameter Value** node. Seeing as we now have the whole logic set up, let's review the contents of the Event Graph:

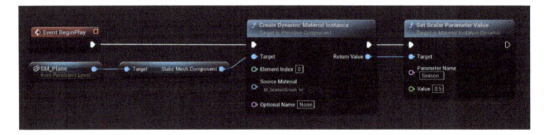

Figure 8.31 – A look at the contents of the Event Graph

As you can see if you go back to the main editor viewport and hit **Play**, the ground plane now has a new material that is modified by the curves we created earlier! Try changing the value to anywhere between **0** and **1** to notice how it changes.

Figure 8.32 – The material in action!

How it works...

Before we move on, let's take a moment to review how to operate inside the Curve Editor. To do so, let's focus for a moment on the editor itself, as seen in the next screenshot:

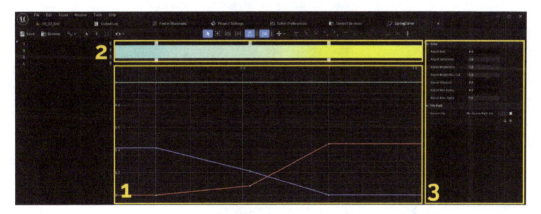

Figure 8.33 – A look at the Curve Editor

As you see in the preceding figure, I've highlighted three different areas, which correspond to those places where you'll be doing most of the work when adjusting curves. Let's see what each of them does:

- The first section, and the biggest area of them all, is **Curve Graph**: a section of the editor where you'll create new keys, delete any existing ones you may want, modify their values, or in general tweak any of the data related to those points.

- The second section is known as **Curve Gradient Result**. This is the area where we can see the result of all of the keys we've added to the curve.

- The third area is **Color Panel**. This is the place where you can adjust curve-wide parameters – settings that affect all keys of your curve.

As mentioned, most of the work will be done in **Curve Graph**, which takes up most of the space in the Curve Editor. In it, you'll be able to add keys to all of the curves by right-clicking in an empty space. Alternatively, you can also hold the *Shift* key and click on an individual curve to add a single key there.

See also

As you've seen in this recipe, the final material allowed us to sample one of the values stored in the **Curve Atlas** asset we created. With that in mind, something else we could do is create a bit of extra code within our **Level Blueprint** to allow us to see some of the other values stored in our curves.

Taking our base recipe, we can easily expand on this by modifying the value we are sampling through the **Set Scalar Parameter Value** node. Instead of using a fixed number, we can start at 0 and gradually increase the value to see the changes that makes over time. This can be done by creating a variable and then incrementing its value over time with a **Delay** node before calling the **Set Scalar Parameter Value** node again. You can see this in action in the 08_ 06_ End map provided alongside the Unreal Engine project accompanying this book, of which you can see a snippet next:

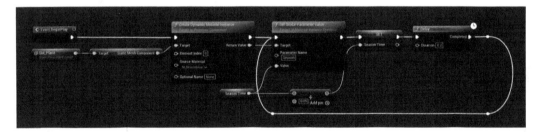

Figure 8.34 – An adjustment to the logic that will make the material iterate over the different curves

With that said, adjusting the value of the **Season** parameter will only allow us to move between the different curves stored in our atlas. As you might recall, the **Season** parameter is the **Curve Atlas Row Parameter** node we created back in *step 21*. Adjusting its value moves us along the *Y* axis of the atlas, and seeing as that texture is as tall as the amount of curve stored in it, adjusting the value of the parameter basically means changing from one curve to another.

Something else we could do is move along the X axis of the atlas instead, which would allow us to traverse the contents of the individual curves stored in it. To do so, you can do the following:

1. Go back to the **M_SeasonGround** material and create a **Scalar Parameter** node named something like **X Position**.

2. Connect the output of the previous parameter to the **Curve Time** input pin of the **Season Curve Atlas Row Parameter** node.

3. Head back to the **Level Blueprint** and adjust that setting instead. This will allow you to control the position of the curve in which you want to be.

You can review the **M_SeasonGround** material I've left for you in **Content | Assets | Chapter08 | 08_06** if you want to explore things further – be sure to take a look!

Blending landscape materials

As we've seen throughout this book, materials can be used in plenty of places: in 3D models, in UI elements, as rendering passes, and so on. Another place where they are extremely useful is in landscape objects, a special category of 3D models that differ from standard 3D meshes in the way they are created. Unlike traditional static meshes, landscapes are usually huge in size, so the way we texture them needs to be inherently different just because of that reason.

To tackle that problem, Unreal comes bundled with a very powerful landscape system that allows you to have a single material that can blend between various textures through the use of the **Landscape Layer Blend** node. In this recipe, we will explore how to create and apply such a material.

Getting ready

Seeing how we'll be working with landscapes, make sure that the level you decide to work on contains one such asset. If you don't want to create one, feel free to open the level named 08_07_Start, where you'll be able to find a scene already set up for you to apply the technique we'll discuss in this recipe.

How to do it...

Seeing as we want to apply a material to a landscape actor, let's start the recipe by creating said material!

1. Create a material and call it something akin to **M_Landscape**, and double-click on it to enter the Material Editor.

2. After that, continue by right-clicking in an empty area of the graph and adding a **Landscape Layer Blend** node.

3. With the previous node selected, focus on the **Details** panel and, under the **Layers** property, click on the + button to add the first layer to your blend.

4. Click on the arrow next to the newly added **Index [0]** property and expand the options for the layer. Change the value of the **Layer Name** setting to **Grass**.

5. After that, click on the same + button we clicked on *step 3* to add an additional layer. Under the new **Index [1]** entry, change the value of **Layer Name** to **Rock**.

6. With that done, connect the output pin of the **Layer Blend** node to the **Base Color** property of the main material node.

7. Afterward, create a couple of **Texture Sample** nodes and set a texture for each of them. We'll connect these to the two layers we created in the **Landscape Layer Blend** node, so choose an appropriate image to represent a patch of grass and rocks. You can use the ones contained in the Starter Content, named **T_Ground_Grass_D** and **T_Rock_Slate_D** respectively.

8. With the new **Texture Sample** node in place, connect the output of the one containing the **Grass** texture to the **Layer Grass** input pin of the **Landscape Layer Blend** node, and connect the one containing **Rock** to the **Layer Rock** input.

The material graph should look something like the following figure:

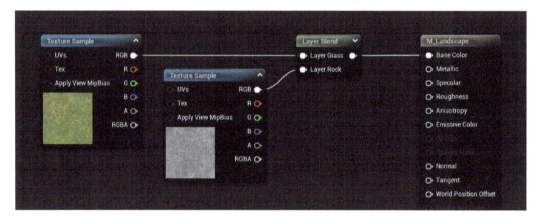

Figure 8.35 – A look at the current state of the material graph

Now that we've taken care of the **Base Color** property of the material, let's focus on the **Normal** attribute.

9. Select the three nodes we created thus far, and proceed to copy-paste them by hitting *Ctrl + C* and then *Ctrl + V* on your keyboard.

10. With the new duplicate nodes in place, connect the output of the new **Layer Blend** node to the **Normal** property of our material.

11. After that, select the two new **Texture Sample** nodes that we have just duplicated and change the textures used by them. Seeing as we were using the **Diffuse** version of the **Grass** and **Rock** textures before, choose their **Normal** counterparts now: **T_Ground_Grass_N** and **T_Rock_Slate_N**.

12. As the final step in the material creation process, click on the **Apply** and **Save** buttons and exit the Material Editor.

    Now that we have the material ready, the next bit we are going to focus on is going to be on the creation of a simple landscape object that we can use to display our new material. Let's take care of that next.

13. Create a level and add a landscape object to it. You can do this by going to the **Modes** drop-down menu in the main editor viewport and choosing the **Landscape Mode** option. Alternatively, if you want to work with an already existing landscape actor, feel free to open the level mentioned in the *Getting ready* section of this recipe, 08_07_Start.

14. After that, assign the **M_Landscape** material to our new landscape.

15. Select the landscape in your scene and focus on the **Details** panel. In there, find the option called **Landscape Material** and assign the **M_Landscape** shader we created earlier.

16. Next, click on the **Paint** button and scroll down to the **Target Layers** section. From there, open the **Layers** property and click on the + button next to each layer to add a **Landscape Layer Info** Object Instance for the layer. From the menu that pops up, select **Weight-Blended Layer (normal)**, as seen here:

Figure 8.36 – Adding a Weight-Blended Layer (normal) info object to our landscape

17. Select a folder to hold the information in and, once selected, hit the **OK** button.

    After waiting a few seconds, you should notice the landscape change so that it uses the first layer to fill the entire area.

18. Seeing as we also configured the landscape material to use a rock asset, repeat *steps 16* and *17* for the **Rock** layer this time around.

19. Now, with the **Rock** layer selected, click and drag it within the scene window. You should notice the **Rock** layer is painted on the scene wherever your mouse cursor passes over it, as seen in the following figure:

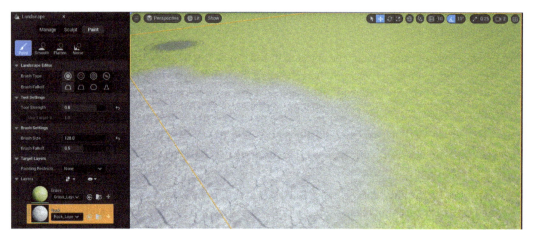

Figure 8.37 – The new material in action

And with that, we have seen how we can blend between layers within landscapes!

How it works...

The **Layer Blend** node allows us to blend multiple textures or materials together so that they can be used as layers within the **Landscape Mode**. Each layer has a number of properties that dictate how it will be drawn. We are using the default **Blend Mode**, **LB Weight Blend**, which allows us to paint layers on top of others.

From the **Paint** section of the **Landscape** menu, you can make use of the **Brush Size and Brush Falloff** properties to change how much is being painted at once. We can also add as many layers as we'd like to get whatever look we are looking for.

See also

Even though we are blending between the two layers contained in our landscape material using the **LB Weight Blend Mode**, you can learn about the other blend types at the following link:

https://docs.unrealengine.com/5.1/en-US/landscape-material-layer-blending-in-unreal-engine/#landscapelayerblendtypes

On top of that, you can learn more about landscapes and how to create, sculpt, and edit them at the following link:

https://docs.unrealengine.com/5.1/en-US/creating-landscapes-in-unreal-engine/

9

Adding Post-Processing Effects

Welcome to the last chapter of this book! We are going to continue talking about shaders in Unreal Engine 5 but, this time, we are going to focus our attention on a very specific variety: post process materials. What makes them unique is the realm in which they operate: instead of applying this type of material to 3D objects as we've done throughout most of this book, these shaders are meant to be used as part of the rendering pipeline. So, what is that?

Unreal, like any other rendering engine, contains a component that takes care of depicting what we've placed on our levels, called the **renderer**. The work done by this system involves calculating how lighting interacts with 3D models, understanding the depth of the scene, and sorting out which elements are visible and which ones are occluded. Together, these operations are known as the **rendering pipeline**. Post process materials are shaders that get injected into it, which can be used to create image-wide effects, such as color grading or setting up custom tones for your levels.

In this chapter, we'll learn how to work with them and how to apply them to our scenes, by studying the following recipes:

- Using a Post Process Volume
- Changing the mood of a scene through color grading
- Creating a horror movie feeling using post process materials
- Working with a cinematic camera
- Rendering realistic shots with Sequencer
- Creating a cartoon shader effect

And just as a teaser, here is a quick look at some of the things we'll be implementing:

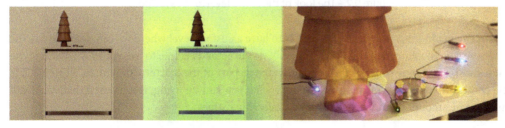

Figure 9.1 – A look at some of the effects we'll create in this chapter

Technical requirements

All of the materials that we'll tackle rely on assets contained in the Unreal Engine, so nothing that you'll need beyond access to the software. In any event, let me leave you with a download link to the Unreal Engine project I'll be using: `https://packt.link/20u7B`

Using a Post Process Volume

In order to access the different post process effects that Unreal has in store for us, we will need to place a specific actor in our level. This actor receives the name of **Post Process Volume**, which is a container in the shape of a box that specifies its area of influence. In this recipe, we'll learn how to enable it in our scenes and how to work with it – something that will allow us to access many of the settings that we are going to be exploring later in this chapter.

Getting ready

I've prepared a scene for you to use as you traverse the different recipes of this chapter — it is a very simple one, but it should help demonstrate the different post process effects that we are going to be studying in the next few pages. You can locate the file by navigating to the following directory inside the Unreal project we are providing alongside this book: **Content | Levels | Chapter09**. The name of the scene is `09_01_Start`, and as soon as you open it, you'll be greeted by the next environment:

Figure 9.2 – The look of the demo level for this recipe

The purpose behind this little still-life scene is to have a custom level where we can demonstrate the different settings contained in the **Post Process Volume** actor. Feel free to use your own assets if you want instead, or employ some of the ones provided as part of the Starter Content, as all we need is some content to look at at this point. No matter what you end up using, rest assured that you'll still gain the same amount of knowledge from the following recipes.

How to do it...

Let's start the recipe by locating the actor we are going to be playing with:

1. Navigate to the **Quickly add to the project** menu, then in the **Volumes** category, select **Post Process Volume**:

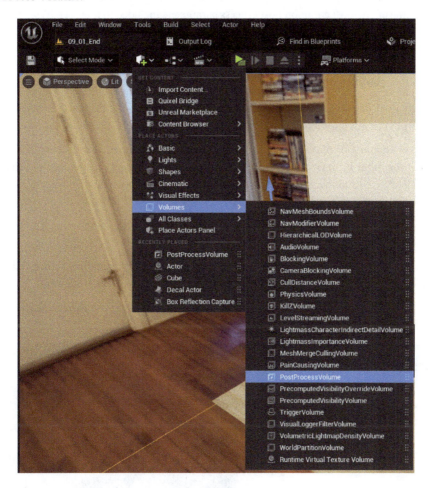

Figure 9.3 – A fast way to add a Post Process Volume actor to our levels

---

Tip

You can also start typing the name of the actor you want to place immediately after clicking on the **Quickly add to the project** menu, which will allow you to narrow down your search.

2. Drag and drop the **Post Process Volume** actor into the level. To ensure that we can always see the effects of this new actor, head over to the **Details** panel and check both the **Enabled** and the **Infinite Extent (Unbound)** options available within the **Post Process Volume** category. This will propagate the effects that we enable within this actor throughout the whole scene.

    With those initial steps out of the way, we've managed to place an actor that will allow us to tweak some of the most common post process parameters included in most rendering programs. Having said that, we haven't actually changed any settings in our new actor, so we won't see any difference at this stage, no matter whether stepping inside or outside of the volume's bounds. For demonstration purposes, we'll now proceed to enable a film grain effect, just so that we can see how the volume works.

3. With the **Post Process Volume** actor selected, scroll down to the **Film Grain** section in the **Details** panel. Check the **Film Grain Intensity** parameter there, and set its value to **2**.

    On top of enabling or disabling the film grain effect, we can also tweak its behavior through many of the other **Film Grain** settings. For instance, we can adjust certain options that affect how visible the effect is over certain areas of the image, such as over the shadows, the mid-tones, and the highlights, or how big the grain used in the effect is. Let's take care of those things next.

4. Continue by checking the box next to the **Film Grain Intensity Shadows** setting and increasing the value to **4**. This will make the film grain effect pop a bit more in the darker areas of the scene.

5. After that, adjust the size of the grain through the **Film Grain Texel Size** setting. Check the box next to it and set the value to something greater than the default **1**, such as **1.5** instead.

    With all of that done, this is how the **Film Grain** section of the **Post Process Volume** actor should look:

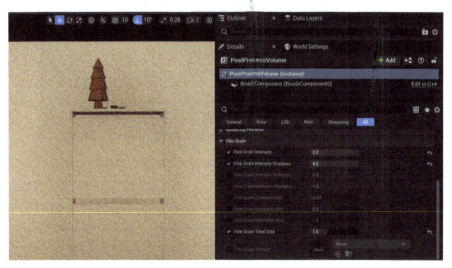

Figure 9.4 – The settings used in the Film Grain section of the Post Process Volume

> **Tip**
>
> On top of changing the default look of the **Film Grain** effect as we've done in the previous steps, you can also change its shape by selecting your own texture in the **Film Grain Texture** setting.

Implementing all of the previous steps has given our scene a look that definitely deviates from the one we had at the beginning. We can easily compare both by enabling or disabling the **Post Process Volume** actor, as we did in *step 2*, which will allow us to compare the before and after states of the scene, as seen next:

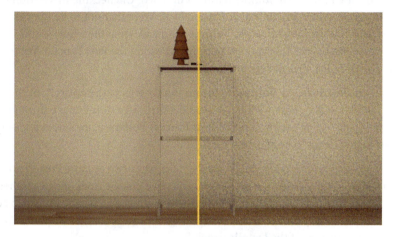

Figure 9.5 – Comparison shot showing the differences between
enabling and disabling the Film Grain effect

Something we can also do through the **Post Process Volume** actor is to control the harshness of the transition between being inside the volume or being outside of it. That is done through the **Post Process Volume Settings** section in the actor's **Details** panel, which contains certain parameters that allow us to control the smoothness of the transition. We'll look at those in more detail in the *How it works…* section of this recipe, as they are a bit more theoretical in nature.

And that is it! We should now be ready to work with **Post Process Volume** actors and implement them across the scenes with which we work – something that we'll revisit in some of the next recipes. Be sure to stick around for a while longer in the next sections if you want to learn a bit more about the settings we've just tweaked but know that you can move on to the next recipe if you'd like to learn about other effects you can implement thanks to **Post Process Volume** actors.

How it works...

Something that we've seen in this recipe is how to make sure that our **Post Process Volume** actor works across the entirety of the level. This is something very useful, as we typically want the settings that we tweak within that actor to extend beyond its boundary box. We often do this so that we can forget about the size of the selected volume when we are only using one.

Convenient as that might be when we only have to deal with an instance of that actor, we can't forget that the situation I've just described is not always the one we'll face. From time to time, especially in complex environments, we'll want to place several **Post Process Volume** actors throughout the scene. Think, for example, about a level that contains outdoor areas as well as interior ones — we may want to emphasize the differences between both ambiances by applying different post process effects. Let's see which settings we can use to adjust their behavior:

- The first thing that we need to check when working with multiple **Post Process Volume** actors is that the **Infinite Extent (Unbound)** setting is disabled. Unchecking this option will ensure that only the area within the volume is affected by it. Were that not to be the case, we would have potentially multiple **Post Process Volume** actors affecting the whole world – something that is sure to wreak havoc! This option can be found within the **Post Process Volume Settings** section of the **Details** panel, as seen in *step 2*.

- Seeing as each **Post Process Volume** actor no longer affects the entirety of the level after tweaking the previous option, we'll need to adjust their area of influence through the **Scale** value – tweaking it so that each individual **Post Process Volume** actor surrounds only the part of the level that we want to be affecting through it.

- Despite the previous options, there might still be times when two or more **Post Process Volume** actors partially overlap – especially in the areas where we change from one to another. In those situations, make sure to pay attention to the **Priority** setting, which controls which actor takes precedence in precisely those scenarios. This setting can also be found within the **Post Process Volume Settings** section of the **Details** panel.

- The **Blend Radius** property, located immediately after the previous one, controls the area around the bounding box in which the settings we choose for the **Post Process Volume** actor get blended with any others that might be applied to the scene – be they the default ones or those of a different volume.

There's more...

As you've probably seen for yourself when working with the **Post Process Volume** actor in this recipe, this type of actor contains many sections that we can tweak inside of its **Details** panel.

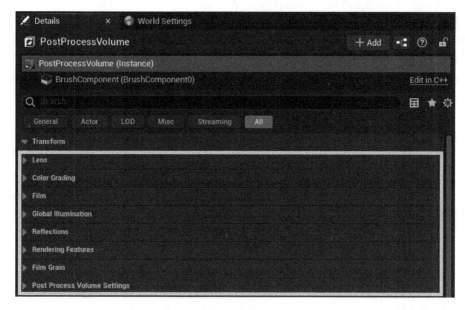

Figure 9.6 – A look at the Details panel of the Post Process Volume actor

Even though we will look at some of them in upcoming recipes, let's take a moment now to go for a high-level view of what those areas affect:

- **Lens**: This contains controls that affect certain camera effects, such as the exposure, chromatic aberration, bloom, or lens flares.

- **Color Grading**: This is the place to go when we want to modify the color and balance of our scene. Operations such as color correction, creating an intentional mood for the level, or affecting the overall contrast of the scene are performed primarily here.

- **Film**: This enables the modification of the tonemapper in UE5. You can think of the tonemapper as the part of the engine tasked with mapping the brightness values of the on-screen pixels to what can be displayed by our screens, if we simplify the explanation a little bit. The parameters found in this panel should not need constant adjustment, as they define how the brightness conversion should be happening.

> **Important note**
>
> The tonemapper can be a complex topic, so be sure to check the official documentation if you need more info: https://docs.unrealengine.com/5.1/en-US/color-grading-and-the-filmic-tonemapper-in-unreal-engine/.

- **Global Illumination**: This allows us to specify which rendering approach should be used to tackle the global illumination of the scene. On top of that, we can configure certain specific aspects of each individual method in some of its subsections.

- **Reflections**: This is very similar to **Global Illumination** in that it lets us specify the rendering method used by Unreal when tackling reflections – as well as the intricacies of each method that we choose.

- **Rendering Features**: This controls certain effects that apply to the 3D world. Examples of that are the use of post process materials, the ambient occlusion that is happening across the level, or the quality of the motion blur.

- **Film Grain**: As we saw in this recipe, this allows us to apply and control that effect should we choose to use it in our levels.

- **Post Process Volume Settings**: This controls how the **Post Process Volume** actor behaves in the world. Examples are defining which overlapping volume should be affecting the world when there are multiple ones in place or the distance in which two different volumes blend.

Changing the mood of a scene through color grading

After taking some time to familiarize ourselves with **Post Process Volume** actors, it is now time to start looking at how to use the different functionalities we can find within them. The first section that we will cover is the first one that we can find if we look at the **Details** panel with one of those actors selected: the **Color Grading** section.

Color grading tools provide artists with a series of options that they can use to alter the look of the final image they render. Similar techniques have been prominent in motion pictures, for example, where the captured footage is adjusted to satisfy a particular need; be that the establishment of a stylized look or the ensuring of color continuity. What we are going to be doing in the following pages is exactly that, tweaking the default camera values to modify the look of our scene.

Getting ready

Seeing as we are going to be focusing on changing the look of our scene in this recipe, something that we'll need to help us tackle that is to have a level populated with some 3d models. We'll use a copy of the one we worked on in the previous recipe, so feel free to open the level called 09_02_Start contained in the Unreal Engine project provided alongside this book if you want to follow along using the same examples you'll see me employing. The only difference between this map and the one we started with in the previous recipe is that this one contains a **Post Process Volume** actor already placed in it, so make sure that you also have one if working with your own assets.

How to do it...

Seeing as we want to modify the look of the level, let's start by taking note of the current state of the scene:

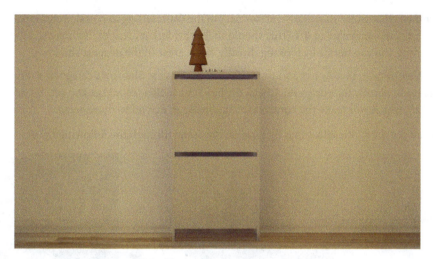

Figure 9.7 – The initial appearance of our level

Having noticed the warm tint present in the level (mostly given by the lighting used in it, as well as the color palette used in the materials), let's start to adjust it to make it a bit more neutral:

1. Select the **Post Process Volume** actor present in the level and focus on its **Details** panel. We are going to be working in the **Color Grading** section, so expand that category.

2. Within the **Color Grading** section, you'll notice that the first area that we can adjust is the **Temperature** category. Expand it and make sure that the **Temperature Type** property is set to **White Balance**, which should be the default value for that field and the one we want to adjust.

3. Next, check the box next to the **Temp** setting, which will allow us to adjust the field next to it. Set the value there to **4500**, which will immediately make the scene look a bit cooler.

    As you can probably already tell by looking at the new state of the scene in front of you, the level already feels much colder. Be sure to check the *How it works…* section to better understand how modifying the **White Balance** setting affects our scene.

    The next categories we can affect within the **Color Grading** section are **Global**, **Shadows**, **Midtones**, **Highlights**, and **Misc**. Ignoring that last one for a moment, you'll soon notice that the other four contain the same settings within them: basically, controls that affect the saturation, the contrast, the gamma, and so on.

    The twist here is that, even though they are the same controls, they each affect the area of the image under which they are grouped, with the **Global** category controlling the entire image while **Shadows**, **Midtones**, and **Highlights** focus on the dark, neutral, and bright areas of the image, respectively. With that in mind, let's tweak some of the settings there a bit next.

4. Something we can do with these controls is to make the darker areas of the image a bit more prominent, as the scene can look a bit too white as it stands. To change that, expand the **Shadows** section and check the box next to the **Gain** property.

5. After that, click on the drop-down arrow next to the **Gain** setting to reveal even more controls, and set the default value to **0.1**. This should make the darker areas of the image a bit darker – something you can notice by looking directly at the areas in the deepest shadows.

6. With that done, let's now expand the tones affected by the **Shadows** category, as the default value is too narrow for our scene. Check the box next to **Shadows Max** and set the value there to **0.125**, which will make the controls for the shadows extend to more areas.

You can see the subtle effect that this has on the scene through the following figure:

Figure 9.8 – The changes made to the Gain and Shadows Max
settings and how they made our tree a bit darker

> **Tip**
> The changes we'll be making throughout this recipe are subtle in nature, so it's always a good idea to make these adjustments yourself and see the results in your monitor to really appreciate the difference tweaking these parameters is making.

Now that we've adjusted the shadowy areas a little bit, we can continue to focus on the neutral tones, or midtones, next. These are predominant through the level with which we are working, so let's adjust them a bit in order to make the scene feel a bit more vibrant.

7. Expand the **Midtones** subsection and check the box next to the **Saturation** parameter in order to be able to adjust it – something we'll do in the next step.

8. After that, expand the **Saturation** setting to adjust the values contained within it. This time, instead of assigning a singular value as we did in *step 5*, adjust the different channels in an independent manner so as to make the yellow and red tones more prominent. You can do this by assigning a value of **1.21** to the **Red** channel and **1.4** to the **Yellow** channel, while leaving the **Green** and **Blue** fields untouched.

You can take a look at that in the following figure:

Figure 9.9 – A look at the changes made through the Saturation property of Midtones

With all of those changes in place, we can now compare the initial scene we were looking at against the current state it is in. If you want to do so, just select the **Post Process Volume** you have on your scene and uncheck the **Enabled** option under the **Post Process Volume Settings** section inside the **Details** panel. Ticking that box on and off will let you toggle the **Post Process Volume** that is affecting our scene. Let's take a look at the results:

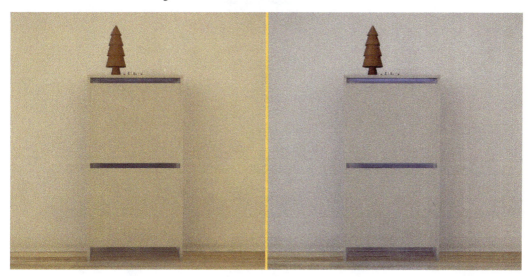

Figure 9.10 – A look at the changes we've made to the scene so far

They feel quite different, don't they? The changes might be subtle, but they are definitely there – we are now looking at a much whiter scene, which feels colder overall and is a bit more saturated. Whereas the first image can make you feel warm and cozy, the second one is a bit more neutral.

However, let's not stop with the adjustments just yet! Now that we know the basics, let's get a bit crazy. What if we gave the scene a bit of a horror touch?

9. To do so, let's tweak the value of **Saturation** under the **Highlights** section. Let's set that to something really high – I've chosen **1000** for this particular scene. Doing so should really make the brightest parts of the image stand out in an unnatural way, which could make the viewer feel a bit uneasy and on edge.

10. Finally, let's apply a color tint to the scene. Horror movies and psychological thrillers often make use of a greenish tint when they want to highlight the fact that something is wrong or just to create a bit of an atmosphere. If you head over to the **Misc** section of the **Color Grading** tab, you will find a setting called **Scene Color Tint**. Changing that to a greenish pastel color should give us the result we are after:

Figure 9.11 – The final look of the scene, after applying a horror-like effect

Look at that! With that final touch, we've covered most of the settings that we can tweak inside of the **Color Grading** post process category. Now that you have an idea of what they do and what they affect, you may want to play around a little bit more with them to get more familiar and confident. The job of a color grader requires patience, as tweaks and corrections need to constantly be applied until you get to the look you are after. In spite of that, the tools that we have at our disposal are quite powerful and should serve us well. Have fun with them!

How it works...

The color grading tools inside Unreal are quite similar to what can be found in other software packages such as Photoshop or Gimp. They allow users to modify certain image parameters, such as the saturation or the contrast of the scene – something that we've tackled in this recipe from a practical point of view. Now that we have a bit more time, let's take a further look at some of those settings to try to understand the theory behind them.

The first one I want to talk about is the initial **White Balance** property we saw and tweaked. That property is controlled via the **Temp** value (which is just short for **temperature**). The amount we set in that field controls one characteristic of the light we see – and that is its color. Light temperature is a physical property of the lights with which we work, and each type has a different typical value that is expressed in Kelvin. For reference, light emitted by candles usually stays around the 1,800 **Kelvin** (**K**) mark, while clear daylight is often expressed with a value of 6,500 K. Warmer light types (those with a reddish or orange tone to them) have a lower temperature value, with the opposite being true for colder emitters (sunlight or fluorescent lights).

Despite what I've just said, the **Temp** setting works in the opposite way. According to the previous explanation, lights that have a high color temperature are considered "cold"; that is, they have a blueish tint to them. Meanwhile, lower values get us in the range of reds, yellows, and oranges; for instance, a value of 3,000 K would be in the yellow range. That being the case, you can probably infer that the higher the temperature, the bluer the light color. However, why did we have to decrease the temp value of the white balance if we wanted to get colder values back in *step 3*?

The answer is that we are not tweaking the temperature values of the lights in the scene; instead, we are defining the base temperature against which all lights should be compared (which is, by default, 6,500 K). Imagine that you have a light of 3,500 K because the default white balance is calculated against that value of 6,500 K; that means that your light is quite warm. If we decrease the default of 6,500 to something like 4,500, the value of your light is still lower than the standard, but not by as much as before. That means that it will look closer to white than it previously did. This is how the **White Balance** temp setting works.

Something else that I'd also like to mention is the different options that we have in the **Global**, **Shadows**, **Midtones**, and **Highlights** categories. When we looked at them in this recipe, we only adjusted the overall multiplier that we can find in each subsection. Here's an example to refresh your mind:

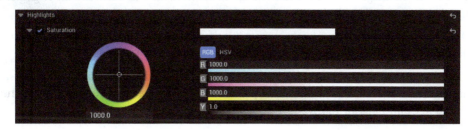

Figure 9.12 – A look at the Saturation selection panel

We usually tweak the value below the color wheel – that is effectively a multiplier that affects all of the RGB values equally. However, we can have even finer control if we know that we only want to be affecting a specific channel. In those cases, feel free to adjust each of the **R, G, B**, and **Y** values separately, which will in turn adjust the red, green, blue, and yellow tones independently from each other. Additionally, we can even change between the **RGB** mode and the **HSV** one, which will have us modify the hue, the saturation, or the value instead of a specific color channel.

> Tip
>
> Color temperature can be quite an extensive topic, so feel free to read more about it on the following site: `https://en.wikipedia.org/wiki/Color_temperature`.

See also

Before moving on to the next lesson, I wanted to spend some time talking about **Look-Up Tables** (or **LUTs** for short). They are a type of asset that allows us to match the post process tweaks we make in other software, such as Photoshop or Gimp, in Unreal, using a texture as opposed to dialing in different settings as we've done in this recipe.

Think about this – you have an environment in Unreal that you want to tweak, just like we did in the previous pages. You can do so through the methods we've seen before, by modifying and tweaking the **Post Process Volume** and playing with its settings until you get the effect you are after. However convenient that can be, some users might feel more comfortable taking that image, pasting it into an image editor, and adjusting it there. Fortunately for us, this process can be replicated thanks to LUTs.

The way we would do so is by following the next steps:

1. Export a sample screenshot out of Unreal that clearly shows your scene.

2. Load that sample shot in your image editor of choice, alongside a neutral LUT sample. One such asset is provided by Epic, which you can find at the following website: `https://docs.unrealengine.com/5.1/en-US/using-look-up-tables-for-color-grading-in-unreal-engine/`. The idea is that all of the changes that you will implement in the next step are applied to both your sample shot from Unreal and the neutral LUT.

3. Next, adjust the image however you like: change the contrast, the saturation, and the brightness... any changes you make will also affect the LUT texture.

4. Once you are happy with your image, export just the adjusted LUT file. The modifications that the image has undergone are going to enable us to replicate that same effect in Unreal.

5. After that, import the modified LUT back into Unreal and double-click on it once it is inside the engine. In the **Details** panel, set the **Mip Gen Settings** option to **No Mipmaps**, and **Texture Group** to **Color Lookup Table**.

6. Finally, head back to your **Post Process Volume** actor and, under the **Color Grading | Misc** section, tick the **Color Grading LUT** checkbox and select your recently imported texture. That should leave you with a scene looking exactly like the one you modified in your image-editing software.

As you can imagine, this technique can be quite useful – especially for those users who are more versed in adjusting these types of settings outside Unreal. However great that is, it also has some shortcomings that we should be aware of. Even though we won't cite them all, they all revolve around the notion that LUTs don't translate well across several screens. Even though you might be doing a correction to your particular scene and you might be happy with it, due to the nature of these textures operating in a low dynamic range, those changes might not translate well into other displays – especially those using a different color space. Keep that in mind when you work with them!

Creating a horror movie feeling using post process materials

Welcome back to another recipe on post-processing effects! I'm quite excited about this one, as we are about to start creating our own personal effects instead of using the ones available through the engine. To do so, we'll take advantage of a particular type of shader called **Post Process Materials**, which will allow us to adjust the scene as a whole. They are a peculiar kind of material, as they get applied inside of a **Post Process Volume** actor – so the knowledge we acquired in the previous recipes is going to come in handy. In this recipe, we'll put this post process material to use in order to create a red pulsating effect, with the intention to create more tension in our scene. Let's see how to do it!

Getting ready

Getting ready for this recipe means having a scene at hand where you can test the post process material we'll be creating. If you don't have one that you can use, feel free to open the map called `09_03_Start` contained in the **Content | Levels | Chapter09** folder of the Unreal Engine project provided alongside this book, which is a duplicate of the horror movie-style level we worked on in the last recipe.

We'll use that map as a base in order to enhance the feel of the map by introducing a pulsating effect through a post process material, just to make it look a bit more menacing and sinister.

How to do it...

Seeing as we want to apply a post process material to the scene, let's start creating that asset:

1. Create a new material, give it a name (I've named it **M_PostProcessSample**), and open the Material Editor.
2. With the new material in place, the first thing we need to do is specify that it is of the **Post Process** type. To do so, select the main material node and head over to the **Details** panel — the first property inside the **Material** section is called **Material Domain**, and we need to set that to **Post Process**.

Now that we have the preliminary setup of the shader out of the way, we can start programming the behavior of the material. Seeing as we want to create a pulsating effect, where the color of the scene blends between its standard one and a redder version in a pulsating manner, let's start retrieving the color of the scene first.

3. Right-click inside the main material graph and look for the **Scene Texture** node.

4. Once created, head over to the **Details** panel and change the default value from **Scene Color** to **PostProcessInput0**. Doing that will let us access the unadulterated color of the scene, which is what we want at this stage, as we'll be modifying it later.

Important note

The **Scene Texture** node is a key component of many post process materials. Because we are not applying these materials to a 3D model, but instead to the scene we are looking at, we need to specify which part of the rendering pipeline we want to affect. The **Scene Texture** node lets us do that.

---

5. Next, create a **Component Mask** node by right-clicking and typing that name.

6. With the previous node selected, make sure that the **R**, **G**, and **B** values are ticked in the **Details** panel for our new **Component Mask**.

7. After that, connect the **Color** output pin from the **Scene Texture** node to the input pin of our new mask.

    The reason why we are masking the color of the scene is that we only want to play with the RGB values provided by the **Scene Texture** node, and we want to discard any other information that might be provided by that node. With those two nodes in place, the **Scene Texture** and the **Component Mask**, we have secured ourselves access to the image that is being rendered, and what we are now going to do is overlay certain other information on top of that.

8. Create a **Texture Sample** node and assign the **T_PanningDistortion** asset to it. This image is part of the Unreal Engine project provided alongside this book, and it contains a reddish gradient that we are going to overlay with the base rendered image of the scene. Make sure you are using a similar asset if working with your own content.

9. Let's now create a **Lerp** node that we can use to overlay the previous **Texture Sample** node with the existing **Scene Texture** node.

10. With the new **Lerp** node in place, connect the output of **Component Mask** to its **A** input pin. As for input pin **B**, connect the **RGB** output of **Texture Sample** to that one.

11. With regards to the **Alpha** pin of the **Lerp** node, connect it to the **Alpha** output pin of **Texture Sample**.

12. Now that the previous sequence of nodes is in place, connect the output of the **Lerp** node to the **Emissive Color** property of our material, and apply and save it.

13. Add a new **Lerp** node and connect the **Constant** node that has a value of **1** to its **A** input pin. Connect the other one to its **B** input pin.

14. As for **Alpha**, connect it to the output of the **Collection Parameter** node, just like we did back in *step 10*.

15. After that, connect the output of this **Lerp** to the **Metallic** input pin of our material.

    Completing the previous steps will have left us with a material that we can instantiate and use to demonstrate the parameter collection asset. Before doing that, let's take a moment to review the material graph we should have in front of us:

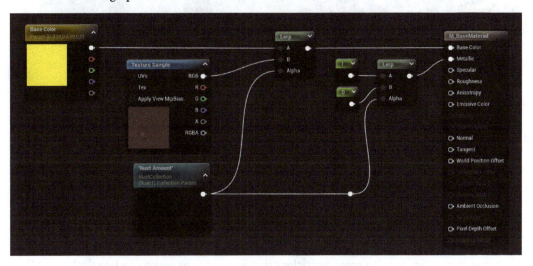

Figure 8.25 – A look at the material graph

As mentioned in the previous paragraph, let's now create several instances of this material so that we can demonstrate the effect:

16. Go back to the Content Browser and right-click on the material we've just created. Select the **Create Material Instance** option and create four instances.

17. After that, adjust each new Material Instance by exposing the **Base Color** variable and modifying its default value. Seeing as we have four instances, assign a different color to each of them, and apply them to the different models that should be on your level.

18. With that done, double-click on the **Material Parameter Collection** asset and adjust the **Default Value** property of the **Rust Amount Scalar Parameter** node.

Assigning different values to the previous parameter should see all of the materials that you've just applied to the scene change at the same time, demonstrating the effect we were trying to achieve in this recipe, as seen in the following figure. Make sure to get a hold of this effect when you need to enact this type of change across your scenes!

Figure 8.26 – A look at the same materials after changing the Rust
Amount setting (0.5 on the left and 1 on the right)

How it works...

As we've seen in this recipe, material collections allow us to modify the properties of multiple materials at once. This is great both in terms of usability and performance: it's much easier and more efficient to tweak a single **Material Parameter Collection** asset than it is to adjust multiple material parameters. Dealing with a single centralized item instead of multiple individual ones also reduces the rendering overhead, allowing for smoother gameplay experiences.

With that in mind, there are also some limitations that we need to be aware of when deploying this technique. For instance, materials are limited to referencing two **Material Parameter Collection** assets, so we need to make sure that we don't cross that threshold. Furthermore, each of those collections can store up to 1,024 **Scalar Parameter** and 1,024 **Vector Parameter** entries. These are very big numbers, which we are unlikely to surpass, but it is a limitation nonetheless.

One issue that we are more likely to face is the need to recompile shaders that are affected by modified parameter collections. This is something that happens every time we change the number of parameters within a given collection: doing so will trigger the recompile process, which can take a bit of time depending on your hardware and the number of materials that are affected. In light of this, one solution is to create more parameters than strictly needed, giving us a bit of headroom in case we ever want to go back and add more settings.

See also

If you want a bit more info regarding the **Material Parameter Collection** asset, please check the following link:

```
https://docs.unrealengine.com/5.1/en-US/using-material-parameter-
collections-in-unreal-engine/
```

Moving between seasons using curve atlases

In our journey through Unreal's material repertoire, we sometimes find ourselves using different types of parameters. Just in the last recipe, we looked at how we could garner the strength of both the **Scalar** and **Vector** types through the use of a **Material Parameter Collection** asset. Having said so, the two types of variables available there are best suited to drive certain fixed properties, such as the **Roughness** value used by a shader or the color needed to be displayed on a specific material.

More nuanced types of data can also be parameterized in Unreal thanks to the use of **curves**. These are special types of assets that can hold different values, meant to give a range of options as opposed to a single one. In this recipe, we'll take advantage of that characteristic by creating a material that can sample different values from a single curve. We'll use that feature to create a material that can display the most common colors shown on tree leaves throughout the year, creating the illusion of a shader that can adapt to seasonal changes.

Getting ready

Just like in the previous recipe, this one will also rely on assets that can be created directly within Unreal. We will use some other resources available through the Starter Content, so make sure to include it in your project if you haven't done so already.

If working with the Unreal Engine project provided alongside this book, make sure to open the level called 08_06_Start to follow along using the same examples you'll see me employing. With that out of the way, let's jump right to it!

How to do it...

Seeing as the star of this recipe is going to be curves, let's start by creating one such asset:

1. Start by navigating to the Content Browser and creating a **Curve** asset. You can do this by right-clicking and looking inside the **Miscellaneous** category.

2. You'll be prompted to choose the type of curve that you want to create immediately after completing the previous step: from the **Pick Curve Class** window, select the **Curve Linear Color** type.

How it works...

Post process materials are a bit different from the ones we apply to 3D models. Just as we can take advantage of the UVs in our meshes to indicate how textures should wrap around those models, we can use certain render passes to aid in the creation of post process materials. This is because of the domain in which they operate: instead of working with geometry, post process materials work on the scene we can see as a whole.

As a result, the logic we need to apply when working with this type of material is different. Working with 3D models and the PBR workflow usually means that we need to pay close attention to the way light interacts with the objects it touches. We don't need to focus on that as much when working on post process shaders: instead, we usually need to think about how we want to tweak the current look of the scene.

Instead of UVs, the information that we need from the scene usually comes in the shape of different scene textures, such as the one we dealt with before, named **Post Process Input 0**. There are many others, which we can use according to our needs. For instance, we can also access the **Subsurface** color of the level, the **Ambient Occlusion** pass, or **Scene Depth** to create interesting effects that take those inputs into account. The point is that we have access to many render passes, and we can use them to create a material that suits our needs.

See also

Before we go, something else that can be of interest is the position in which the post process material we create is inserted within the post-processing pipeline. If we take a look back at the **Details** panel for our material, we can see that there are several options available when we scroll down to the setting called **Blendable Location**: **After Tonemapping**, **Before Tonemapping**, **Before Translucency**, **Replacing the Tonemapper**, and **SSR Input**. Let's see what each of those does:

- Selecting the **Before Tonemapping** option means that the effect that we are trying to create will get inserted into the rendering pipeline before the color grading and tonemapping operations take place. That can be of interest if we need to access the raw pixel values in our map, in case any of the tonemapping adjustments cause issues with the effects that we want to create.

- In contrast to the previous option, choosing **After Tonemapping** means that the effects of our post process material get implemented after the color grading and tonemapping pass.

- In a similar way to the previous two options, the **Before Translucency** one means that the effect is calculated before the translucency pass is rendered.

- **Replacing the Tonemapper** is the option to choose if you want the shader you are creating to take care of the tonemapping operation instead of relying on the one used by Unreal.

- Finally, **SSR Input** is there if you need access to the pixels before screen space reflections are added to the image.

It's always good to know what the previous options do before implementing a post process effect, so do play with those settings if you are experiencing any issues with the materials you create.

Working with a cinematic camera

In this recipe, we are going to take a break from **Post Process Volume** actors to focus instead on the **Cine Camera** actor. That doesn't mean that we are leaving behind the world of post-processing effects – far from it, in fact! In this recipe, we are going to learn how to work with cameras in Unreal and how to take advantage of some of the post process effects available through them – such as depth of field, bloom, and lens flares. As you'll see, working with virtual cameras is very similar to working with real ones, so jump on board to learn all about them!

Getting ready

We are going to continue using the same scene we worked on in previous recipes, so feel free to open the level called `09_04_Start` contained in the **Content | Levels | Chapter09** folder if you want to follow along using the same assets you'll see me employing.

As usual, feel free to continue to use your own scenes if that's what you prefer to do! If you had something suitable for the previous recipes, chances are that will work here as well. Make sure you at least have multiple objects scattered at different positions; this will help us highlight the depth of field effects we are going to be exploring. That's it for now!

How to do it...

Seeing as we want to study different effects we can enable using a camera, let's start this recipe by creating one such asset:

1. Start by creating a **Cine Camera** actor by selecting that option from the **Quickly add to the project | Cinematic** menu.

2. Next, navigate to the **Outliner** panel and right-click on the newly created camera. From the options available, choose the **Pilot 'CineCameraActor'** one.

    Completing the previous step will allow us to control the camera in the same way that we usually traverse the scene, making the task of positioning it much easier than if using standard move and rotate controls.

3. Now that you have control over the actor, make sure to position it looking at some of the existing objects in the level – for instance, make it look at the Christmas tree and the small candles. We should be looking at a similar view as the one depicted in the next figure:

Figure 9.17 – The current view of the camera

At this stage, you might notice that everything that the camera sees is completely blurred – as shown in *Figure 9.17*. The next bit we are going to tackle is getting the objects into focus, while also applying a bit of depth of field to the scene. Let's do that next.

4. With the camera still selected, look at its **Details** panel and expand the **Current Camera Settings** section. In there, scroll down to the **Focus Settings** area and check the box next to the **Draw Debug Focus Plane** option.

   With that setting enabled, we should now be seeing a purple plane in front of us. That element is there as a visual cue for us to know where the focus plane is, which corresponds to the area that is going to be in focus. We need to modify its position so that the elements that we want to see clearly are at the same distance as that plane from the camera.

5. Play around with the value next to the **Manual Focus Distance** field and set it to something that allows you to clearly see the objects that you have in front of you. As a reference, try to focus on one of the candles. For me, the value I needed to use was **65**.

6. When you are happy with the results, go ahead and uncheck the **Draw Debug Focus Plane** option.

   Now that we have the candle in focus, we can try to blur the other objects a bit more by playing around with some of the other camera options. Let's see how next.

7. Scroll down to the **Current Focal Length** parameter and change the default value of **35.0** to something greater, such as **50**. This will make the candle take a more prominent role in the viewport.

8. Next, move to the next setting, called **Current Aperture**, and increase the value from the default **2.8** to something like **5**. This will smooth out the transition between the areas in focus and those out of it.

You can see what the view from the camera is at the moment and compare it to what it used to be in the next figure:

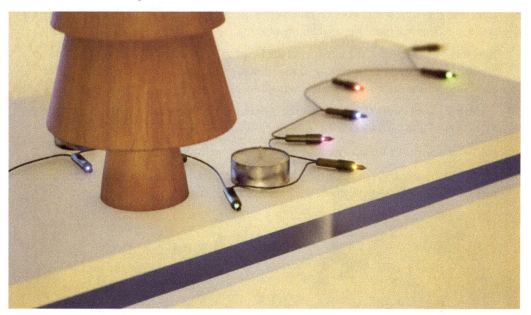

Figure 9.18 – The new view from the camera

> **Tip**
>
> All of the parameters that we've tweaked before in our **Cine Camera** actor are identical to what you can find when working with real cameras, so the knowledge that you gain here can also be applied there. For more information, be sure to check out the following link: `https://www.paragon-press.com/lens/lenchart.htm`.

With that, we've pretty much covered all of the settings that we need to be aware of in order to manually set up a cinematic camera. The main settings that we need to adjust are **Focal Length**, **Aperture**, and **Focus Distance**, much like we would in a real-world camera. Now that we've looked at those properties, let's continue applying other effects to the scene. The next ones we are going to look at are usually deployed when trying to tackle a specific artistic vision: I'm talking about bloom and lens flares. Let's see how to enable them in our camera.

9. With the camera still selected, scroll down to the **Post Process** section of the **Details** panel and expand the **Post Process | Lens | Bloom** category. This is the area that controls the halo effect that happens around bright objects.

10. In there, change the **Method** setting from **Standard** to the more advanced **Convolution**. This is a more expensive way of calculating bloom but is also more visually appealing.

11. After that, check the box next to the **Intensity** parameter and increase its value from the default **0.675** to **3**. This will make this effect more obvious.

12. Now that we've increased the intensity of the bloom, scroll down to the **Lens Flares** category. Similar to what we saw happening in the **Bloom** section, lens flares control the perceived scattering of light when working with cameras.

13. Once in the **Lens Flares** section, check the box next to the **Intensity** property and increase the value to something greater than the default, such as **3**.

14. After that, enable the **Bokeh Size** property and increase the value to something like **5**.

With all of those changes in place, the Christmas lights in our scene should have become much more prominent. Let's quickly review the view from the camera:

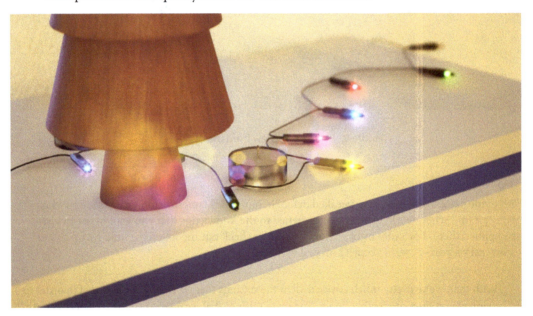

Figure 9.19 – The look from the camera with the new bloom and lens flares settings in action

Finally, we can still add one extra effect to the scene in order to make the Christmas lights even more prominent.

15. While still in the **Post Process** category, expand the **Image Effects** section and check the box next to **Vignette Intensity**. In there, increase the value from the default **0.4** to something like **0.6**.

The scene should have now become darker around the edges, making the objects in the center of the frame stand out a bit more. We can quickly review all of the changes we've made by looking at a comparison image between the original state of the scene versus the now final one:

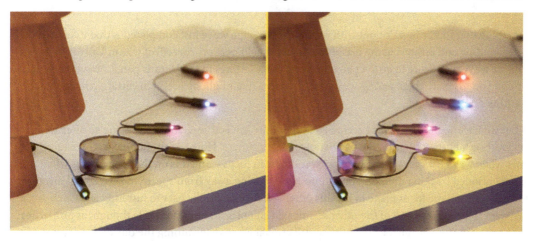

Figure 9.20 – Comparison between the initial versus the final state of the camera

The changes are subtle but noticeable and, when used moderately, these effects can enhance the look of a scene by making it a bit more believable. Make sure to play around with different values to add that little bit of extra realism to your levels!

How it works...

We've spent most of our time adjusting different post process effects inside the **Cine Camera** actor in order to achieve a particular look in the scene. Seeing as we have two different actors that can take care of those effects (the **Post Process Volume** and the **Cine Camera** actor itself), how do we decide when we should be using each one of them?

The answer lies in the specific goals that you want to reach with your project. Sometimes, one of the two actors is more useful or better suited to satisfy some of your needs. We can explore this through different examples.

Imagine, for instance, that you are playing a multiplayer shooter. If you were to receive any damage, you might be expecting to see some effects pop up on your screen – such as a red tint to the whole image or some other similar visual cue that indicates damage. This situation asks for the said effect to be local to the specific player that is suffering the effect, so it makes sense to adjust any post process effects through the player camera.

Another example where we could instead benefit from using a **Post Process Volume** is when we want to add post process effects to a whole open-world game. If we want to artistically tweak how the colors look or modify the feel of the scene when the climatology changes, we might find a good ally in this type of actor.

See also

Something that I'd like to point out with regard to the recipe we've just completed is that it focused on understanding and mastering the use of the depth of field post process effect within Unreal. In order to do so, we've made sure to go over all of the pertinent settings in the **Details** panel. However, I'd like to point you to a particular setting we haven't tweaked that might be useful in certain circumstances: **Tracking Focus Method**.

So far, we have spent the current recipe tweaking different settings, and we've done so after we had previously established **Manual Focus Distance** of the camera. However useful that is, there are other times when we know the specific actor we want to be in focus – and it is on those occasions when the aforementioned tracking method comes into play. Moving away from specifying the actual distance at which we want the focus plane to be, this other system enables the user to just indicate which actor they want to be in focus. All of the other settings do still play an important role though, as properties such as the aperture or the focal length are still crucial when determining the out-of-focus areas.

However, changing to this method when you do know which actor needs to be in focus can save you time from figuring out the distance value you should be typing. If you want to check it out, be sure to head over to the **Current Camera Settings** section of your camera's **Details** panel and look under **Focus Settings** | **Focus Method**.

Rendering realistic shots with Sequencer

Seeing as we've had the chance to work on great-looking scenes in the past few recipes, it would be a shame to leave those to dry out and not make the most out of them. Given how we've put quite a bit of effort to make them shine, why not use the opportunity to create realistic-looking renders?

In this recipe, we'll take a look at some of the options that Unreal puts at our disposal that allow us to create high-quality renditions of our levels. We'll look at both the **High Resolution Screenshot** feature as well as the automated batch-rendering process we can perform through **Sequencer** – Unreal's cinematic editor. On top of that, we'll discover how to tap into Unreal's built-in **Path Tracer**, an offline tool that will allow us to use the most realistic rendering system included in the engine. Let's see how to do all of those things next!

Getting ready

If you've completed any of the scenes we tackled in the previous recipes of this chapter, chances are you'll already know the drill: we are going to be working on a level that contains a few meshes in it, which is the same one we've seen in the other recipes of this chapter. The level itself is called 09_05_Start, and you can find it within the **Content | Levels | Chapter09** folder of the Unreal Engine project provided alongside this book. If working with your own assets, simply make sure to have a scene containing multiple 3D assets that we can use for visualization purposes.

Something to note this time around is the use of **Path Tracer**, which we'll employ at the beginning of the recipe to make realistic renders. This feature of the Unreal Engine requires access to hardware-accelerated ray tracing graphics cards and the latest version of the Windows DirectX 12 graphics API, so keep that in mind before tackling the next few pages. Apart from that, we'll also be dipping our toes into Sequencer and the Movie Render Queue plugin, so make sure to head over to the *See also* section if you've never worked with those editors before.

How to do it...

Seeing as how we want to create realistic renders in this recipe, the first few steps we are going to take will involve enabling the Path Tracer in Unreal. Let's take care of that first:

1. Enabling the **Path Tracer** requires Unreal to use the DirectX 12 graphics API. To enable it, head over to the **Project Settings** panel and navigate to the **Rendering** section.

2. Once in there, scroll down to the **Hardware Ray Tracing** area and make sure that both the **Support Hardware Ray Tracing** and **Path Tracing** options are enabled.

3. With the previous settings turned on, and while still inside **Project Settings**, change from the **Rendering** section to the **Platform | Windows** one.

4. In there, focus on the **Targeted RHIs** category (which, luckily for us, is at the very top) and set the **Default RHI** setting to **DirectX 12**.

> **Important note**
> You might be prompted to restart the editor at this stage – something you'll have to do if you want the changes to take effect. Wait for the engine to restart and open the scene with which you were working to continue with the recipe.

With all of the previous steps completed, we are now in a position where we can finally see the Path Tracer in action. Let's take a look at how to do that, as well as how to adjust certain parameters that control its behavior.

5. From the **View Modes** panel, scroll down to the **Path Tracing** option and select it. The engine will proceed to render the scene using this new approach to rendering, of which you can see a sample in the next figure:

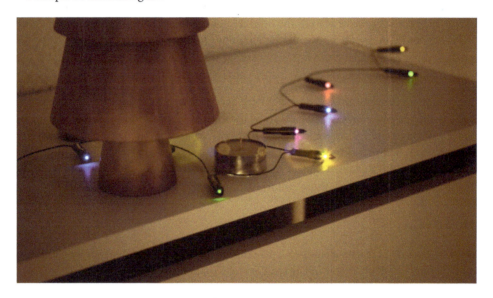

Figure 9.22 – The scene being rendered using the Path Tracer

The engine will now proceed to render the scene using a path tracing approach, which is a more expensive and accurate form of ray tracing. Make sure to visit the *How it works…* section if you want to learn more about the inner workings of this approach to rendering a scene.

Even though you'll find more information in that section of this recipe, you will have noticed, upon completing *step 5*, that the scene looked quite noisy immediately after clicking the **Path Tracing** button. As time goes on (and as long as you don't change the position of the camera!), the view will start to become more and more clear, and the quality of the rendered image will consequently increase. This is because path tracing is a more complex form of ray tracing, where more rays are used to produce the final look of the scene.

This is a very taxing operation for our GPUs, which need to shoot multiple rays over a period of time before being able to create a final render. This operation can be adjusted using a **Post Process Volume** actor, which will allow us to modify the number of rays that the Path Tracer shoots before producing a final image. Let's take care of that next.

6. Create a **Post Process Volume** actor by selecting one such actor from the **Quickly add to the project** drop-down menu.

7. With that new actor selected, focus on its **Details** panel and scroll down to the **Post Process Volume Settings** section. In there, check the **Infinite Extent (Unbound)** option, which will propagate any changes we make to that actor to the entirety of the level.

8. Once that's done, and while still looking at the **Details** panel for the **Post Process Volume** actor, navigate to the **Rendering Features** section and expand the **Path Tracing** area. This is the place that contains all of the settings related to the **Path Tracer** operation.

Two of the settings contained in the **Path Tracing** area that control the quality of the rendered image are the **Max. Bounces** and **Samples Per Pixel** options. The first one controls the number of times that a ray is allowed to bounce before being used to produce the final pixel that we see on screen. The logic here is simple: more bounces equals better results. The second option works in a similar way, but in this case, we can specify the number of samples being used to define the look that a given pixel should have. A third interesting option in this panel is the **Filter Width** setting, which effectively controls the anti-aliasing quality of the scene, with larger values making edges less aliased. Make sure to play around with the values here until you are happy with the results!

9. To create some sample renders, check the **Max. Bounces** setting and increase the default value from **32** to **64**.

10. Do the same for the **Samples Per Pixel** parameter, going as high as **32768** this time around.

11. As for the **Filter Width** setting, increase it to **5** from the default value of **3**. This will ensure that the edges of the models are properly anti-aliased.

With all of the preceding steps in place, let your computer process the scene for a while and look at the results once the render clears out. As a reference, here is a side-by-side comparison of the before and after results:

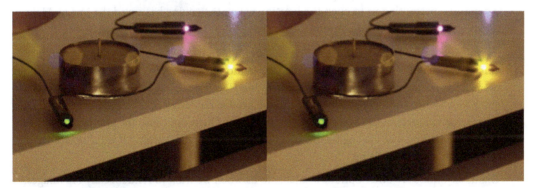

Figure 9.22 – A render taken with the Path Tracer default settings and the ones we've set in the last steps

Even though the differences can be subtle, you might notice how there are fewer artifacts underneath the reflections on some of the lights. Be sure to check things out for yourself on your screen, as things are bound to become more obvious there.

> **Important note**
>
> The Path Tracer defaults to using the fallback mesh on models that have **Nanite** enabled, which is probably something you don't want to happen. To mitigate that, make sure to enter the following console command when working with those models: `r.RayTracing.Nanite.Mode 1`. If you haven't used console commands in the past, know that you can access the console where you'll be able to type them by hitting the *tilde* (~) key.

Now that we know how to create realistic images using the Path Tracer, we need a way to export those creations outside of the engine. To do so, we'll study how to export images from Unreal, using both a quick and easy method and another, more professional one. Let's start with the easy option.

12. To create a quick screenshot of your scene, simply click on the **Viewport Options** button (the one located in the upper-left corner of the viewport) and select the **High Resolution Screenshot…** button.

13. A new panel should have appeared upon clicking on the previous button, allowing us to choose certain parameters: the one we want to tweak now is **Screenshot Size Multiplier**, which we'll increase from the default **1** to **2**. This will make the resolution of the resulting screenshot twice as big as the current viewport resolution, so take that into consideration when typing a value into that field.

14. After that, click on the **Capture** button, which will see Unreal taking a high-resolution screenshot of the scene. You'll see a prompt appear in the lower-right corner of the editor once the operation is complete, prompting us to open the location where the engine has saved the image. Make sure to go there and check the results for yourself.

As you've seen, the **High Resolution Screenshot** tool is a handy feature that allows us to take quick renders out of the current camera view. On top of that, it also offers us the option to tweak a few more settings, which we'll study in the *How it works…* section. Even though it's a quick tool, it's not entirely perfect, as it can have trouble rendering really large images. In fact, using large numbers in the **Screenshot Size Multiplier** field can cause the graphics driver to crash – something that the engine makes us aware of through a warning if we choose a value of **3** or higher for that setting.

Thankfully, there are other options available to us when we want to render really big images. Unreal has recently introduced the **Movie Render Queue** plugin as part of its suite of extensions, which will allow us to take high-resolution renders without needing to worry about the resolution at which we want to shoot them. Furthermore, this new approach to rendering images will allow us to specify which vistas we want to render without having to manually click on a button every time we want to take a screenshot, while also giving us the option to tweak certain parameters, which will ensure that our images are shot at the highest possible quality.

Let's see how to enable and work with this new feature next:

15. Because **Movie Render Queue** is a plugin, we'll first have to enable it. To do so, click on **Edit | Plugins** to access the appropriate panel.

16. In the search box at the top, start typing Movie Render Queue until you see the plugin appear. Check the box next to it to enable it.

17. You will be prompted to restart the editor at this stage. Do so to be able to use the plugin and reload the same level with which we are working once the editor restarts.

Now that we have the plugin in place, we are almost ready to start shooting our renders. The way we operate the Movie Render Queue is a bit different than what we saw with the **High Resolution Screenshot** option, as the new method gets triggered using **Sequencer**. With that in mind, we'll have to create a new sequence before we can use it, as that is the element that specifies what needs to be rendered. Let's see how to do that next.

18. Create a new **Level Sequence** asset by right-clicking and selecting the **Level Sequence** option available within the **Animation** category. As for the name of the new actor, let's go for something like **LS_Renders**.

19. Upon creating the new asset, drag and drop it anywhere within the level with which we are working and double-click on **Level Sequence** to open its editor – called **Sequencer**.

20. With the main level viewport and Sequencer open side by side, drag and drop any cameras that you want to render from the level into Sequencer. I'll just use one of the cameras present in the level, **Camera 02**.

21. A new **Camera Cut** will be automatically created upon adding a camera to the sequence. Feel free to modify its length so that it only occupies a frame, as we only want to take a screenshot and not a video. This is what we should have in front of us at this stage:

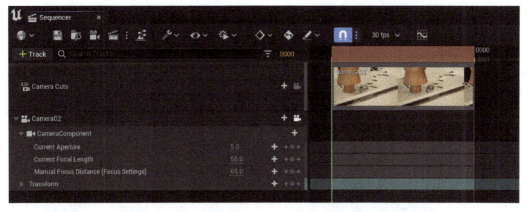

Figure 9.23 – A look at the new Level Sequence we've created

Now that we've specified which camera we want to render, the next steps we'll take will see us invoking the **Movie Render Queue** plugin and adjusting the parameters that we want to use to render the level. Let's tackle that now.

22. Click on the clapperboard icon to bring up the **Movie Render Queue** panel. This is the fifth button available in Sequencer if we start counting from the upper-left corner.

23. After that, click on the **Unsaved Config*** button to bring up the **Preset Selection** panel. This is the area where we can configure the settings that we want Unreal to use when shooting our renders.

24. Once inside that panel, start by unchecking the **Deferred Rendering** slider.

25. After that, click on the + **Setting** button and choose the **Path Tracer** option located toward the bottom of the list.

26. Next, select the **Output** option present in the left-hand side area of the panel and adjust the parameters contained therein to your liking. In my case, I've set **Output Resolution** to 3840 x 2160 (4K) and configured **Output Directory** to point to one of the folders on my computer.

27. In case we want to reuse this configuration later, click on the **Load / Save Preset** button in the top-right corner of the panel and select the **Save As Preset** option, and save it wherever you fancy.

28. Finally, click on **Accept** to go back to the **Movie Render Queue** panel.

29. With that done, click on the **Render (Local)** button to start the rendering process.

If all goes well, the computer will start to render the scene, a process that can take a bit of time seeing as we are using the Path Tracer and a high-quality preset. Despite that, choosing to render a single frame as we've done before should greatly speed things up, so it won't be long until you have something that you can take a look at. In our case, here is the result:

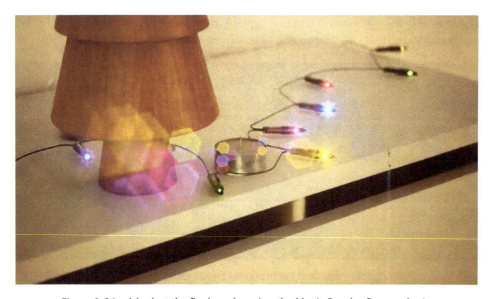

Figure 9. 24 – A look at the final render using the Movie Render Queue plugin

As you've seen, working with Sequencer and Movie Render Queue isn't as immediate as dealing with the **High Resolution Screenshot** tool, but the possibilities in terms of scheduling a render or obtaining higher resolution images more than make up for that difference in convenience. All in all, both tools have their uses, so make sure to keep both in mind when you need to produce a render!

How it works...

Seeing as we've spent a little bit of time going over how to work with Unreal's Path Tracer, let's now take a moment to learn more about this new rendering system.

As we've mentioned in the main body of the recipe, path tracing and ray tracing aren't that different – in fact, the two names mean the same thing. The difference between both is an artificial construct, something needed by the real-time rendering community to differentiate between the ray tracing implementation that can be achieved within real-time engines against that achieved by offline renderers. Even though the approach to rendering a level is similar in both cases, the fidelity that offline renderers can achieve thanks to not being constrained to producing multiple frames per second can't currently be matched by real-time implementations. In light of that, the industry has chosen to differentiate both systems by calling the real-time implementation ray tracing and the offline version path tracing.

With that in mind, we can think of the Path Tracer as a beefed-up alternative to the standard ray tracing techniques we've seen earlier in this book, specifically in the recipe called *Using software and hardware ray tracing* back in *Chapter 4*. You might remember how the quality of the ray tracing effects we studied back then could be adjusted using certain toggles – with path tracing, we don't need to worry about adjusting the quality of the effect, as that is always calculated in a physically correct way. The trade-off is the time it takes our computers to calculate the final render, which makes this technique impractical for real-time use under current hardware limitations.

Having said that, the benefits of the Path Tracer become obvious when we don't need to produce real-time graphics. As an example of that, certain sectors that require realistic graphics can benefit from using the Path Tracer, such as architectural visualization or the movie industries. Furthermore, realistic graphics can also be beneficial in gaming workflows, where we can use the Path Tracer to inform us about how a level should look so that we can then adjust the non-path-traced version to look as close as possible.

See also

As promised in the *Getting ready* section, let's now take a quick look at both the **Sequencer** and the **Movie Render Queue** editors. Let's start with the first one:

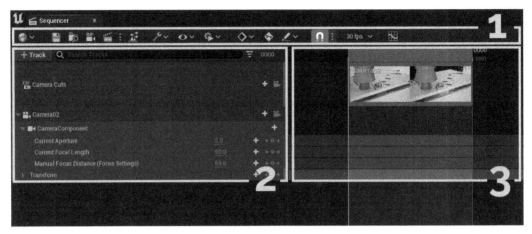

Figure 9.25 – A look at the Sequencer editor and its different components

As you can see in the previous figure, I've highlighted the three different areas where we'll spend most of our time when working with a sequence. Here is what you can do in each of them:

- The area of the editor marked with the number **1** contains buttons that allow us to manage and adjust the sequence. There are plenty of options at our disposal there: from saving any changes we make to rendering the sequence into a video or adjusting the framerate of the clip. As you can see, these are all high-level settings that control the sequence.

- The second section I've highlighted contains a list of all of the actors that the sequence is currently tracking, as well as the different camera cuts. There's an important button located in the top-left corner, named **+ Track**, which will allow us to add different actors to the sequence. On top of that, we can also control certain parameters that affect the objects that we add, and create different keys that allow us to animate those same parameters – something that works in combination with the next area that we'll see.

- The third section of the editor is the area where we can adjust the different keyframes that we've created before. This will allow us to animate any of the parameters that we are tracking. We should use this section in combination with the previous one: we can control the location of the keys in this one, and we can modify the values in the previous section.

Sequencer is a very flexible and robust system that allows us to create very intricate animations. Even though we've now covered the basics, the truth is that we could create a small book just dedicated to that topic. Make sure to check out the official docs if you want to learn more: https://docs.unrealengine. com/5.1/en-US/cinematics-and-movie-making-in-unreal-engine/.

Now that we've looked at the **Sequencer** panel, let's focus on **Movie Render Queue** next:

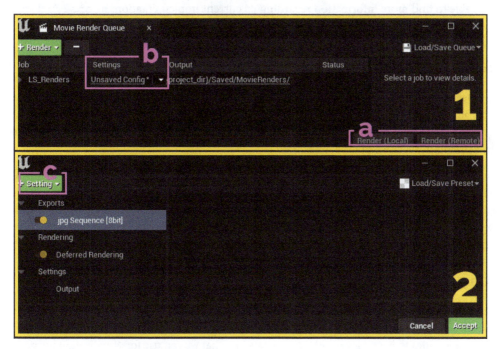

Figure 9.26 – A look at the Movie Render Queue and the Movie Render Queue settings panels

We access it through one of the buttons present in the **Sequencer** editor: specifically, the clapperboard button we clicked back in *step 22*. The panel is a very straightforward one, and it works in tandem with the **Sequencer** one, which allows us to configure the different presets we want to use when rendering a clip. We can call this second area the Preset selection panel, even though it doesn't have any specific name. Let's briefly go over the different options available in those two panels:

- The first panel I've highlighted is the **Movie Render Queue** one. There are not many options available to us there, with the most important ones being probably the **Render (Local)** and **Render (Remote)** options (**a**). These are the two buttons that will trigger the rendering of the scene, either in the machine that is running Unreal or in a remote one, depending on the option we choose. The settings used for the rendering process are defined in the **Settings** tab (**b**), which is the button that we can click to bring up the second panel, which we'll cover next, the Preset selection one.

- The Preset selection panel is the place that allows us to define the different parameters that should be used when rendering the scene. To do so, simply click on the + **Setting** button (**c**) to add a new parameter to the rendering process. Clicking on each of the selected parameters allows us to tweak the different settings that they control, which will appear in the right area of the panel.

Finally, let me leave you with a link to Epic's official documentation for the Path Tracer, where you'll be able to find more information regarding its current implementation: `https://docs.unrealengine.com/5.1/en-US/path-tracer-in-unreal-engine/`.

Creating a cartoon shader effect

As we've seen throughout this book, Unreal can be a very versatile real-time rendering engine. Proof of that is all of the different techniques we've been able to apply thus far, which allowed us to create a wide range of different materials. Many of them fell under the category of realistic shaders: assets that tried to mimic what we can see in real life.

Having said that, those are not the only types that the engine can tackle. In this recipe, we'll put Unreal's flexibility to the test by creating a post process material that allows us to render certain objects as if they were drawn in a cartoon style. This is commonly known as cel- or toon-shading, a non-photorealistic approach to displaying our models, which makes them look as if drawn by hand. Make sure to keep reading if you want to learn more!

Getting ready

There are not many requisites to tackle this recipe! Seeing as we are going to be working on a post process material, most of the heavy lifting will be done by the logic that we create. As always, it's probably a good idea to have a level with some assets in it that we can use for visualization purposes. If you are working with the Unreal Engine project provided with this book, make sure to open the level named `09_06_Start`, which you can find within the **Content** | **Levels** | **Chapter09** folder.

How to do it...

As we mentioned that we were going to apply a cartoonish effect to the scene by creating a post process material, let's go ahead and start creating the shader that will allow us to do so first:

1. Start by creating a new material and giving it a name. I've gone ahead and named the new asset **M_CelShadingPostProcess**.

2. With the new material in place, double-click on it to enter the Material Editor and focus on the **Details** panel. As with the other post process materials we've created, we'll have to adjust the **Material Domain** property: make sure to set that value to **Post Process**.

3. After that, and with our gaze still set on the **Details** panel, scroll down to the **Post Process Material** section and adjust the value of the **Blendable Location** property by setting it to **Before Tonemapping**.

The previous steps are necessary to have our shader work alongside Unreal's post-processing pipeline. Setting the **Material Domain** property to **Post Process** is something that we need to do if we want to use the shader within a **Post Process Volume** actor, and adjusting the **Blendable Location** lets us determine at which point our material gets inserted into Unreal's rendering pipeline. The **Before Tonemapping** option allows the shader to be inserted before any color correction steps happen so the values with which we work inside the material don't get affected by that process, which could throw a spanner into our calculations if we are not in control of those – for instance, if we later decide to apply some color correction options in a **Post Process Volume** actor as we've done in other recipes of this chapter.

Now that we have that under control, let's continue by creating the logic that will make our material unique. The first steps we are going to take will involve working with some of the scene buffers, which will allow us to gather information from the scene to tackle some of the calculations that we are about to make. We'll comment on those as we move forward, so buckle up and let's start coding!

4. Let's create a **Scene Texture** node and make it the first node that we add to this material.

5. With the previous node selected, focus on the **Details** panel and set the **Scene Texture Id** property to **Post Process Input 0**. This will grant us access to the look of the scene as if we weren't using a **Post Process Volume** actor.

6. After that, create a **Desaturation** node and connect it to the **Color** output of the previous **Scene Texture** function. This will give us access to a grayscale version of the previous information.

7. With those two nodes in place, copy and paste those functions, placing the duplicates a little bit below the originals.

8. Next, select the copy of the **Scene Texture** node and change its **Scene Texture Id** setting to **Diffuse Color**. This will give us very similar values to the ones offered by the **Post Process Input 0** option – the key difference being that the **Diffuse Color** option isn't affected by the lighting in the scene.

9. Now that we have those four nodes at our disposal, let's compare them by creating a **Divide** node and connecting the inputs of the two **Desaturation** nodes to their **A** and **B** input pins. The **Desaturation** node affecting **Post Process Input 0** should be connected to the **A** pin, whereas the **Desaturation** connected to **Diffuse Color** should be connected to input pin **B**. An illustration of this can be seen in the next figure, alongside the effect that applying the material as it currently stands has on the scene:

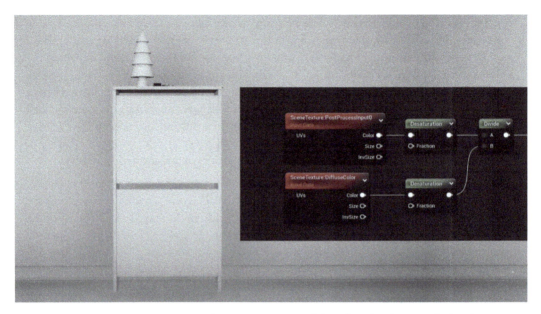

Figure 9.27 – A look at the nodes we've created and the effect they have on the level

Performing the previous operation will allow us to effectively analyze the lighting of the scene, simply because we are now taking the final look at the level and dividing it by its unlit version. We'll use that information to drive the appearance of our toon shader effect, as that rendering style usually relies on flat-shaded materials with sharp changes in tonality that illustrate how lighting affects a model. Let's take care of performing those calculations next.

10. Continue by creating an **If** node and connecting the result of the previous **Divide** function to its **A** input pin.

11. Next, create a **Scalar Parameter** and name it something along the lines of **Cutoff Point**. Connect it to input pin **B** of the previous **If** node.

12. In terms of both the **A > B** and **A == B** input pins of that same **If** node, connect them to the **Color** output of the **Scene Texture** node being driven by **Diffuse Color**.

13. After that, create a **Multiply** node and connect it to the **Color** output of the **Scene Texture** being driven by **Diffuse Color** as well.

14. Without leaving the previous **Multiply** node, set its **B** value to **0.5**.

15. Finally, connect the output of the **If** node to the **Emissive Color** property of our material.

The material graph should now look something like this:

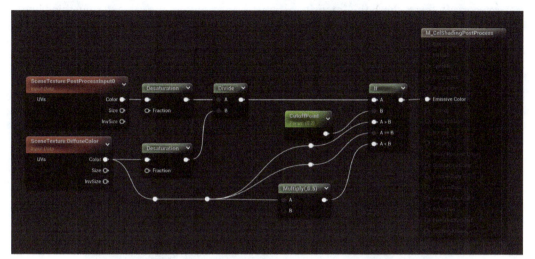

Figure 9.28 – A final look at the material graph

As you see, this new node contains five different inputs: **A**, **B**, **A > B**, **A == B**, and **A < B**. The first two (**A** and **B**) are called the conditions, which the node takes care of evaluating for us, whereas the other three inputs (**A > B**, **A == B**, and **A < B**) are there for us to specify what should happen if one of those conditions is met.

We went ahead and fed the **A** condition with the values from the lighting calculations we performed in *steps 4* through *9*, and we compared those against a static **Scalar Parameter** we named **Cutoff Point**, which represents a specific brightness value. Three branching paths will get triggered depending on the result of that comparison: in both the **A > B** and **A == B** routes, activated when the incoming lighting values are lighter or equal to the specified cutoff point, we'll let the scene render using the simple **Diffuse Color** information we obtained from the **Scene Texture** node; in the **A < B** path, we chose to use a darker version instead, as we multiplied that same **Diffuse Color** information by **0.5** – making the scene darker. The result of this operation will basically create a two-tone version of the level, making it similar to the cel-shading techniques this recipe takes inspiration from.

Lastly, all we need to do is apply the material to the scene to see the effects in action.

16. Click on the **Apply** and **Save** buttons and exit the Material Editor.

17. After that, head back to the main viewport and select the **Post Process Volume** actor present in the scene.

18. Focusing on the **Details** panel, scroll down to the **Global Illumination** section and set **Method** to **None**.

19. Next, move down to the **Rendering Features** section and expand the **Ambient Occlusion** category. In there, check the **Intensity** parameter and decrease the value to something like **0.25** – or disable it altogether by setting it to **0**.

20. After that, continue in the **Rendering Features** section but move up to the **Post Process Materials** category. Next, click on the + button next to the **Array** property to add an entry to it.

21. A new drop-down menu should have appeared after clicking on the previous + button: select the **Asset Reference** option in it and assign the material we've created in this recipe.

With all of that done, we should now be looking at a much-changed scene, something similar to the one shown in the next figure:

Figure 9.29 – A final look at the flat-shaded version of the level

As you can see, the scene now looks much flatter than before, where each model only displays a maximum of two shades depending on how brightly lit its surface is. Effects such as this one will allow you to apply specific looks to your levels, letting you move away from the standard photo-realistic approach. Be sure to play around with post process materials to see what other interesting effects you can come up with!

How it works...

Now that we've created a simple cartoon post process material, let's take a bit of time to review what we've created, the logic behind the nodes we've placed, and see how we can expand things even further by applying the knowledge we've gathered in some of the previous recipes.

To start with, let's take a quick look at the idea behind the technique that we've implemented. As we know, Unreal's default approach to rendering a scene is a photo-realistic one: the different systems and calculations that the engine performs when it wants to render a level try to ground themselves in the way light works in the real world, however many optimizations they need to implement to actually enable that. To move away from this state of things, we need to come up with our own systems that can modify the look of an entire level. Luckily for us, the engine puts several render passes at our disposal that we can use to come up with innovative ways of rendering a scene.

In the example that we've just tackled, the information that we took advantage of was the look of the scene and the diffuse information contained in our materials, both accessible through the two **Scene Texture** nodes we've placed. Combining both allowed us to mask different areas of the models according to how brightly they were being lit, something that we then used to shade those sections in a very simple manner: one tone for the bright areas and another one for the darker ones.

Of course, the approach we took when creating a toon shader effect was a simple one, as we only focused on the post process material that changed the way the scene was being lit. The effect could be expanded upon if we tackled other areas beyond the rendering pipeline. For instance, something important to consider when working with this rendering style is the setup of the materials themselves. To clarify that, think for a moment about the texture we used for the wooden floor: even though we are lighting it in a cartoonish way, the image used by that model is still grounded in reality, and it is not the look that a wooden floor would have if it were drawn by hand. These types of considerations have to be taken into account when working with cel-shaded scenes.

Finally, something else to consider is how we could have expanded upon our work to maybe include some of the other effects we've tackled in past recipes. One particular technique that would be very interesting to implement is the outline material we tackled in the *Highlighting interactive elements* recipe back in *Chapter 8*. Even though we used that material to highlight objects, we could rely on the same techniques we learned then to outline the models that we are interested in, much like what we see happen in cartoons, manga, or anime. Be sure to test yourself and try to combine both techniques if you feel up to the challenge!

See also

Even though we've created a toon-shading post process material, the approach we've taken in this recipe isn't the only one that would allow us to create such an effect. Unlike photo-realistic rendering, which is a physically-based operation, cartoon rendering is a technique that tries to mimic what many artists have drawn over the years, and as such, there are many ways in which we can tackle it – as many as there are drawing styles.

With that in mind, I'd like to leave you with another material that puts a spin on the work that we've done in this recipe. Its name is **M_CelShadingPostProcess_Alt**, and you can find it in the assets folder for this recipe – that is, **Content | Assets | Chapter09 | 09_06**. The difference between this and the material we've worked on in this recipe is that this other version has the ability to display more than two shades of the diffuse color, thanks to some custom math that takes the output of the luminance that we've calculated in *steps 1* through *9* and creates a stepped gradient with it. The gradient has been normalized into a 0 to 1 scale, and we then use those values to multiply against the **Diffuse Color** node of the scene, just like we've done in *step 13*. Make sure to check its material graph to learn about a different way of tackling this effect – and see the results here:

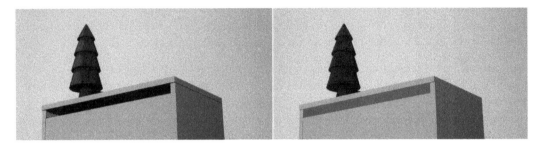

Figure 9.30 – A look at the alternate and the original post process material (left and right)

Summary

And that's the end of the book!

I hope you've had a great time reading it and that you've learned a lot from the different recipes that we've tackled. The intention has always been to provide you with as much detail and knowledge as possible, in a practical way, while tackling real materials that you may encounter in a real-time project.

Throughout the book, we've mixed a bit of practice with theory, which will have hopefully given you not just the knowledge that you need to tackle the examples that you've seen so far but also empower you to create any materials that you can think of and come up with variations of those. Having said that, something that I would also like to encourage you to do is to keep on using those same materials and techniques that you've created and try to produce new shaders by mixing and matching the different things that you've learned.

As a farewell note, let me say thanks and extend my gratitude for reading this book and getting this far. It's been a pleasure and an honor to author it, and I really hope that you've found it useful.

All the best,

Brais.

Index

Symbols

T

texture

packing 106

texture atlases 232

materials, optimizing through 232-235

reference link 235

texture coordinates

matching, across different models 261-267

textures

blending, based on distance between
camera and objects 75-79

textures, for mobile platforms

reference link 225

tiling texture 224

timeline graph 295

parts 295, 296

translucency

types 87

translucent

glass, creating 82-87

values, using for refraction 88, 89

Translucent Blend Mode 88

translucent materials

material instances, benefits 88

transparency

reference link 89

U

UE4_Logo texture 222

UI editor

features 211

Unlit Shading Model 96

references 98

V

Vector 54

vector parameter values 55

vertex colors

used, for adjusting appearance
of material 174-179

Vertex Painting 174

technique 180

video

playing, on in-game TV 284-288

Virtual Shadow Maps 29

reference link 29

VR

forward shading renderer, using 229-231

Packtpub.com

Subscribe to our online digital library for full access to over 7,000 books and videos, as well as industry leading tools to help you plan your personal development and advance your career. For more information, please visit our website.

Why subscribe?

- Spend less time learning and more time coding with practical eBooks and Videos from over 4,000 industry professionals
- Improve your learning with Skill Plans built especially for you
- Get a free eBook or video every month
- Fully searchable for easy access to vital information
- Copy and paste, print, and bookmark content

Did you know that Packt offers eBook versions of every book published, with PDF and ePub files available? You can upgrade to the eBook version at packtpub.com and as a print book customer, you are entitled to a discount on the eBook copy. Get in touch with us at customercare@packtpub.com for more details.

At www.packtpub.com, you can also read a collection of free technical articles, sign up for a range of free newsletters, and receive exclusive discounts and offers on Packt books and eBooks.

Other Books You May Enjoy

If you enjoyed this book, you may be interested in these other books by Packt:

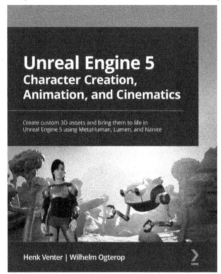

Unreal Engine 5 Character Creation, Animation, and Cinematics

Henk Venter | Wilhelm Ogterop

ISBN: 9781801812443

- Create, customize, and use a MetaHuman in a cinematic scene in UE5.

- Model and texture custom 3D assets for your movie using Blender and Quixel Mixer.

- Use Nanite with Quixel Megascans assets to build 3D movie sets.

- Rig and animate characters and 3D assets inside UE5 using Control Rig tools.

- Combine your 3D assets in Sequencer, include the final effects, and render out a high-quality movie scene.

- Light your 3D movie set using Lumen lighting in UE5.

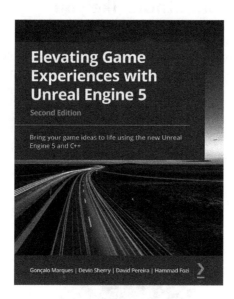

Elevating Game Experiences with Unreal Engine 5

Gonçalo Marques | David Pereira | Devin Sherry | Hammad Fozi

ISBN: 9781803239866

- Create a fully functional third-person character and enemies.
- Implement navigation with keyboard, mouse, and gamepad.
- Program logic and game mechanics with collision and particle effects
- Explore AI for games with Blackboards and behavior trees.
- Build character animations with animation blueprints and montages.
- Polish your game with stunning visual and sound effects.

Packt is searching for authors like you

If you're interested in becoming an author for Packt, please visit `authors.packtpub.com` and apply today. We have worked with thousands of developers and tech professionals, just like you, to help them share their insight with the global tech community. You can make a general application, apply for a specific hot topic that we are recruiting an author for, or submit your own idea.

Share Your Thoughts

Now you've finished *Unreal Engine 5 Shaders and Effects Cookbook*, we'd love to hear your thoughts! Scan the QR code below to go straight to the Amazon review page for this book and share your feedback or leave a review on the site that you purchased it from.

`https://packt.link/r/1-837-63308-8`

Your review is important to us and the tech community and will help us make sure we're delivering excellent quality content.

Download a free PDF copy of this book

Thanks for purchasing this book!

Do you like to read on the go but are unable to carry your print books everywhere?

Is your eBook purchase not compatible with the device of your choice?

Don't worry, now with every Packt book you get a DRM-free PDF version of that book at no cost.

Read anywhere, any place, on any device. Search, copy, and paste code from your favorite technical books directly into your application.

The perks don't stop there, you can get exclusive access to discounts, newsletters, and great free content in your inbox daily

Follow these simple steps to get the benefits:

1. Scan the QR code or visit the link below

https://packt.link/free-ebook/9781837633081

2. Submit your proof of purchase
3. That's it! We'll send your free PDF and other benefits to your email directly